Biotechnology in Agriculture and Forestry

Edited by
T. Nagata (Managing Editor)
H. Lörz
J. M. Widholm

Biotechnology in Agriculture and Forestry

Volumes already published and in preparation are listed at the end of this book.

Biotechnology in Agriculture and Forestry 57

Plant Metabolomics

Edited by
K. Saito, R.A. Dixon, and L. Willmitzer

With 96 Figures, 29 in Color, and 10 Tables

Springer

Series Editors

Professor Dr. Toshiyuki Nagata
University of Tokyo
Graduate School of Science
Department of Biological Sciences
7-3-1 Hongo, Bunkyo-ku
Tokyo 113-0033, Japan

Professor Dr. Horst Lörz
Universität Hamburg
Biozentrum Klein Flottbek
Zentrum für Angewandte Molekularbiologie
der Pflanzen (AMP II)
Ohnhorststraße 18
22609 Hamburg, Germany

Professor Dr. Jack M. Widholm
University of Illinois
285A E.R. Madigan Laboratory
Department of Crop Sciences
1201 W. Gregory
Urbana, IL 61801, USA

Volume Editors

Professor Dr. Kazuki Saito
Chiba University
Graduate School of Pharmaceutical Sciences
Yayoi-cho 1-33, Inage-ku
Chiba 263-8522, Japan;
RIKEN Plant Science Center
Yokohama 230-0045, Japan

Professor Dr. Richard A. Dixon
Plant Biology Division
Samuel Roberts Noble Foundation
2510 Sam Noble Parkway
Ardmore, OK 73401, USA

Professor Dr. Lothar Willmitzer
Max Planck Institute
of Molecular Plant Physiology
Am Mühlenberg 1
14476 Golm, Germany

Library of Congress Control Number: 2005936763

ISSN 0934-943X
ISBN-10 3-540-29781-2 Springer Berlin Heidelberg New York
ISBN-13 978-3-540-29781-9 Springer Berlin Heidelberg New York

Springer is a part of Springer Science + Business Media
springer.com

Printed in Germany

Editor: Dr. Dieter Czeschlik, Heidelberg, Germany
Desk Editor: Dr. Andrea Schlitzberger, Heidelberg, Germany
Cover design: *design&production* GmbH, Heidelberg, Germany
Typesetting and production: LE-TEX Jelonek, Schmidt & Vöckler GbR, Leipzig, Germany
Printed on acid-free paper 31/3152 5 4 3 2 1 0

Preface

Metabolomics is a rapidly-emerging sector of post-genome research. The metabolome (a set of all metabolites of an organism) represents not only the ultimate phenotype of cells by the perturbation of gene expression and the modulation of protein functions caused by the environment or mutations, but the metabolome can also feed back on gene expression and protein function. Therefore, metabolomics plays a key role for understanding cellular systems. Metabolomics is applied to a variety of biological fields from medical science to agriculture. Nevertheless, metabolomics research is particularly important in the plant field because plants collectively produce a huge variety of chemical compounds, far more than animals and even microorganisms. The number of all metabolites in the plant kingdom is estimated at 200,000 or more. In addition, most of the human-beneficial properties of plants, be they foods, medicinal resources, or industrial raw materials, are ascribed to plant metabolites.

This book aims to review the current status of plant metabolomics research. Since metabolomics itself is a new field, no such comprehensive book has yet been published. The chapters are divided into three sections: analytical technology, bioinformatics, and applications. These represent three major elements of metabolomics research. Each chapter provides cutting-edge information contributed by leading researchers from throughout the world.

We hope that this book will be a landmark for plant metabolomics research into the future and will give beneficial guidance to graduate students and researchers in academia, industry, and technology transfer organizations. Since metabolomics is still a growing discipline, further technology development in chemical analysis and bioinformatics will be required. We look forward to breakthrough technology innovations in metabolomics, and yet unforeseen findings and applications in plant science.

Finally, we would like to acknowledge our contributors who have enthusiastically put their efforts to ensure the high scientific quality of this volume. We also would like to thank our colleagues at Springer.

January 2006

Kazuki Saito,
Richard A. Dixon,
and Lothar Willmitzer

Contents

Section I Analytical Technology

Section II Bioinformatics

Section III Applications

List of Contributors

L. Achnine
Plant Biology Division, Samuel Roberts Noble Foundation, 2510 Sam Noble Parkway, Ardmore, OK 73401, USA

P. Ahiahonu
Phenomenome Discoveries Inc., 204–407 Downey Road, Saskatoon, Saskatchewan, Canada S7N 4L8

M. Altaf-Ul-Amin
Department of Bioinformatics and Genomics, Graduate School of Information Science, Nara Institute of Science and Technology (NAIST), Takayama-cho 8916–5, Ikoma, Nara 630–0101, Japan

M. Arita
Department of Computational Biology, Graduate School of Frontier Sciences, The University of Tokyo, 5–1–5 Kashiwanoha, Kashiwa, 277–8561 Japan, e-mail: arita@k.u-tokyo.ac.jp

H. Asahi
Department of Bioinformatics and Genomics, Graduate School of Information Science, Nara Institute of Science and Technology (NAIST), Takayama-cho 8916–5, Ikoma, Nara 630–0101, Japan

M.H. Beale
The National Centre for Plant and Microbial Metabolomics, Rothamsted Research, West Common, Harpenden, Herts. AL5 2JQ, UK, e-mail: mike.beale@bbsrc.ac.uk

R.J. Bino
Plant Research International, P.O. Box 16, 6700 AA Wageningen, The Netherlands

C. Böttcher
Leibniz Institute of Plant Biochemistry, Weinberg 3, 06120 Halle/Saale, Germany

P.M. Bramley
School of Biological Sciences, Royal Holloway, University of London, Egham, Surrey, TW20 0EX, UK

F. Carrari
Max Planck Institute of Molecular Plant Physiology, Am Mühlenberg 1, 14476, Golm, Germany

Y.H. Choi
Division of Pharmacognosy, Section Metabolomics, Institute of Biology, Leiden University, P.O. Box 9502, 2300RA, Leiden, The Netherlands

S. Clemens
Leibniz Institute of Plant Biochemistry, Weinberg 3, 06120 Halle/Saale, Germany, e-mail: sclemens@ipb-halle.de

R.B. Croteau
Institute of Biological Chemistry, Washington State University, Pullman, WA 99164, USA

T. Daskalchuk
Phenomenome Discoveries Inc., 204–407 Downey Road, Saskatoon, Saskatchewan, Canada S7N 4L8, e-mail: info@phenomenome.com

B.E. Deavours
Plant Biology Division, Samuel Roberts Noble Foundation, 2510 Sam Noble Parkway, Ardmore, OK 73401, USA

R.A. Dixon
Plant Biology Division, Samuel Roberts Noble Foundation, 2510 Sam Noble Parkway, Ardmore, OK 73401, USA, e-mail: radixon@noble.org

A.R. Fernie
Max Planck Institute of Molecular Plant Physiology, Am Mühlenberg 1, 14476, Golm, Germany, e-mail: fernie@mpimp-golm.mpg.de

H. Foerster
Carnegie Institution, Department of Plant Biology, 260 Panama Street, Stanford, CA 94305, USA

M. Franz
Leibniz Institute of Plant Biochemistry, Weinberg 3, 06120 Halle/Saale, Germany

P.D. Fraser
School of Biological Sciences, Royal Holloway, University of London, Egham, Surrey, TW20 0EX, UK, e-mail: p.bramley@rhul.ac.uk

Y. Fujiwara
Department of Computational Biology, Graduate School of Frontier Sciences, The University of Tokyo, 5–1–5 Kashiwanoha, Kashiwa, 277–8561 Japan

E. Fukusaki
Department of Biotechnology, Graduate School of Engineering, Osaka Univ, 2–1 Yamadaoka, Suita, 565–0871, Japan,
e-mail: fukusaki@bio.eng.osaka-u.ac.jp

I.A. Graham
CNAP, Department of Biology (Area 7), University of York, PO Box 373, York YO10 5YW, UK

J. Gullberg
Umeå Plant Science Centre, Department of Forest Genetics and Plant Physiology, Swedish University of Agricultural Sciences, SE-901 87 Umeå, Sweden

R.D. Hall
Plant Research International, P.O. Box 16, 6700 AA Wageningen, The Netherlands, e-mail: robert.hall@wur.nl

D. Heath
Phenomenome Discoveries Inc., 204–407 Downey Road, Saskatoon, Saskatchewan, Canada S7N 4L8

M.Y. Hirai
RIKEN Plant Science Center, Suehiro-cho 1–7–22, Tsurumi-ku, Yokohama, Kanagawa 230–0045, Japan

T. Hirayama
International Graduate School of Arts and Sciences, Yokohama City University, 1-7-29 Suehiro, Tsurumi-ku, Yokohama, 230-0045 Japan

T. Ikegami
Department of Polymer Science and Engineering, Kyoto Institute of Technology, Matsugasaki, Sakyo-ku, Kyoto, 606–8585, Japan

A.I. Johansson
Umeå Plant Science Centre, Department of Forest Genetics and Plant Physiology, Swedish University of Agricultural Sciences, SE-901 87 Umeå, Sweden

P. Jonsson
Research Group for Chemometrics; Organic Chemistry, Department of Chemistry, Umeå University, SE-901 87 Umeå, Sweden

K.-M. Oksman-Caldentey
VTT Biotechnology, P.O. Box 1500, 02044 VTT, Finland,
e-mail: Kirsi-Marja.Oksman@vtt.fi

S. Kanaya
Department of Bioinformatics and Genomics, Graduate School of Information Science, Nara Institute of Science and Technology (NAIST), Takayama-cho 8916–5, Ikoma, Nara 630–0101, Japan,
e-mail: skanaya@gtc.naist.jp

R.E.B. Ketchum
Institute of Biological Chemistry, Washington State University, Pullman, WA 99164, USA, e-mail: rketchum@wsu.edu

J. Kikuchi
RIKEN Plant Science Center, 1-7-22 Suehiro-cho, Tsurumi-ku, Yokohama, 230-0045 Japan, e-mail: kikuchi@psc.riken.jp

H.K. Kim
Division of Pharmacognosy, Section Metabolomics, Institute of Biology, Leiden University, P.O. Box 9502, 2300RA, Leiden, The Netherlands,
e-mail: verpoort@chem.leidenuniv.nl

J. Kopka
Max Planck Institute of Molecular Plant Physiology, Am Mühlenberg 1, 14476 Potsdam-Golm, Germany, e-mail: Kopka@mpimp-golm.mpg.de

K. Kurokawa
Department of Bioinformatics and Genomics, Graduate School of Information Science, Nara Institute of Science and Technology (NAIST), Takayama-cho 8916–5, Ikoma, Nara 630–0101, Japan

T.R. Larson
CNAP, Department of Biology (Area 7), University of York, PO Box 373, York YO10 5YW, UK, e-mail: trl1@york.ac.uk

B. Mehrotra
Virginia Bioinformatics Institute, Virginia Polytechnic Institute and State University, Washington St., MC 0477, Blacksburg, Virginia 24061, USA

P. Mendes
Virginia Bioinformatics Institute, Virginia Polytechnic Institute and State University, Washington St., MC 0477, Blacksburg, Virginia 24061, USA, e-mail: mendes@vt.edu

T. Moritz
Umeå Plant Science Centre, Department of Forest Genetics and Plant Physiology, Swedish University of Agricultural Sciences, SE-901 87 Umeå, Sweden, e-mail: thomas.moritz@genfys.slu.se

Y. Nakamura
Department of Bioinformatics and Genomics, Graduate School of Information Science, Nara Institute of Science and Technology (NAIST), Takayama-cho 8916–5, Ikoma, Nara 630–0101, Japan

Y. Nakanishi
Intec Web and Genome Informatics Corporation

M. Naoumkina
Plant Biology Division, Samuel Roberts Noble Foundation, 2510 Sam Noble Parkway, Ardmore, OK 73401, USA

D. Ohta
Department of Plant Genes and Physiology, Graduate School of Agriculture and Biological Sciences, Osaka Prefecture University, Gakuen-cho 1–1, Sakai, Osaka 599–8531, Japan

M. Orešič
VTT Biotechnology, P.O. Box 1500, 02044 VTT, Finland

S.Y. Rhee
Carnegie Institution, Department of Plant Biology, 260 Panama Street, Stanford, CA 94305, USA, e-mail: rhee@acoma.stanford.edu

C.H. Ric de Vos
Plant Research International, P.O. Box 16, 6700 AA Wageningen, The Netherlands

H. Rischer
VTT Biotechnology, P.O. Box 1500, 02044 VTT, Finland

K. Saito
Chiba University, Graduate School of Pharmaceutical Sciences, Yayoi-cho 1-33, Chiba 263-8522, Japan
RIKEN Plant Science Center, Suehiro-cho 1–7–22, Tsurumi-ku, Yokohama, Kanagawa 230–0045, Japan, e-mail: ksaito@faculty.chiba-u.jp

N. Sakurai
Kazusa DNA Research Institute, 2–6–7 Kazusa-kamatari, Kisarazu, Chiba 292–0818, Japan

N. Schauer
Max Planck Institute of Molecular Plant Physiology, Am Mühlenberg 1, 14476, Golm, Germany

D. Scheel
Leibniz Institute of Plant Biochemistry, Weinberg 3, 06120 Halle/Saale, Germany

D. Shibata
Kazusa DNA Research Institute, 2–6–7 Kazusa-kamatari, Kisarazu, Chiba 292–0818, Japan, e-mail: shibata@kazusa.or.jp

Y. Shinbo
New Energy and Industrial Technology Development Organization, Toshima, Tokyo 170–6028, Japan

L.W. Sumner
The Samuel Roberts Noble Foundation, 2510 Sam Noble Parkway, Ardmore, OK 73401, USA, e-mail: lwsumner@noble.org

H. Suzuki
Kazusa DNA Research Institute, 2–6–7 Kazusa-kamatari, Kisarazu, Chiba 292–0818, Japan

N. Tanaka
Department of Polymer Science and Engineering, Kyoto Institute of Technology, Matsugasaki, Sakyo-ku, Kyoto, 606–8585, Japan

C. Tissier
Carnegie Institution, Department of Plant Biology, 260 Panama Street, Stanford, CA 94305, USA

T. Tohge
RIKEN Plant Science Center, Suehiro-cho 1–7–22, Tsurumi-ku, Yokohama, Kanagawa 230–0045, Japan

T. Tokimatsu
Kazusa DNA Research Institute, 2–6–7 Kazusa-kamatari, Kisarazu, Chiba 292–0818, Japan

R.N. Trethewey
metanomics GmbH and metanomics Health GmbH, Tegeler Weg 33, 10589 Berlin, Germany, e-mail: richard.trethewey@metanomics.de

J. Trygg
Research Group for Chemometrics; Organic Chemistry, Department of Chemistry, Umeå University, SE-901 87 Umeå, Sweden

E. v. Roepenack-Lahaye
Leibniz Institute of Plant Biochemistry, Weinberg 3, 06120 Halle/Saale, Germany

H.A. Verhoeven
Plant Research International, P.O. Box 16, 6700 AA Wageningen, The Netherlands

R. Verpoorte
Division of Pharmacognosy, Section Metabolomics, Institute of Biology, Leiden University, P.O. Box 9502, 2300RA, Leiden, The Netherlands

J.L. Ward
The National Centre for Plant and Microbial Metabolomics, Rothamsted Research, West Common, Harpenden, Herts. AL5 2JQ, UK,
e-mail: Jane.ward@bbsrc.ac.uk

L. Willmitzer
Max Planck Institute of Molecular Plant Physiology, Am Mühlenberg 1, 14476, Golm, Germany

E. Willscher
Leibniz Institute of Plant Biochemistry, Weinberg 3, 06120 Halle/Saale, Germany

Y. Yamazaki
Phenomenome Discoveries Inc., 204–407 Downey Road, Saskatoon, Saskatchewan, Canada S7N 4L8

P. Zhang
Carnegie Institution, Department of Plant Biology, 260 Panama Street, Stanford, CA 94305, USA

Section I Analytical Technology

I.1 Gas Chromatography Mass Spectrometry

J. Kopka[1]

1 Introduction

GC-MS technology has been used for decades in studies which aim at the exact quantification of metabolite pool size and metabolite flux. Exact quantification has traditionally been focused on a single or small set of predefined target metabolites. Today GC-MS is one of the most widely applied technology platforms in modern metabolomic studies. Since early applications in unravelling the mode of action of herbicides (Sauter et al. 1988) it has experienced a renaissance (Fig. 1) in post-genomic, high-throughput fingerprinting and metabolite profiling of genetically modified (e. g. Roessner et al. 2001a,b, 2002; Fernie et al. 2004) or experimentally challenged plant samples (e. g. Cook et al. 2004; Kaplan et al. 2004; Urbanczyk-Wochniak and Fernie 2005). Metabolic phenotyping and analysis of respective phenocopies by metabolite profiling has become an integral part of plant functional genomics (Fiehn et al. 2000b; Roessner et al. 2002; Fernie et al. 2004). The essence of metabolite profiling, namely the non-biased screening of biological samples for changes of metabolite levels relative to control samples, has been thoroughly discussed earlier and is clearly distinguished from fingerprinting approaches and the concept of exact quantification (Fiehn et al. 2000b; Sumner et al. 2003; Birkemeyer et al. 2005).

GC-MS-based metabolome profiling analysis is on the verge of becoming a routine technology. This fact substantially contributes to the development of metabolomics as a fourth integral part of the Rosetta stone for functional genomics and molecular physiology (Trethewey et al. 1999; Fiehn et al. 2000b; Trethewey 2004). Nevertheless, GC-MS technology is already challenged again by new bottlenecks and demands for improved data sets which are optimised for the mathematical modelling tools currently developed in the fields of bioinformatics and biological systems analysis.

The challenges of modern, multi-parallel, GC-MS based metabolite analysis are manifold: (i) automation of sample preparation, wet chemistry and data processing after acquisition for increased throughput and reproducibility, (ii) extension of the analytical scope of metabolomics studies, for example by combined analysis of single samples using multiple analytical technology platforms, and combined analysis with the proteome and transcriptome

[1] Max Planck Institute of Molecular Plant Physiology, Am Mühlenberg 1, 14476 Potsdam-Golm, Germany, e-mail: Kopka@mpimp-golm.mpg.de

Biotechnology in Agriculture and Forestry, Vol. 57
Plant Metabolomics (ed. by K. Saito, R.A. Dixon, and L. Willmitzer)

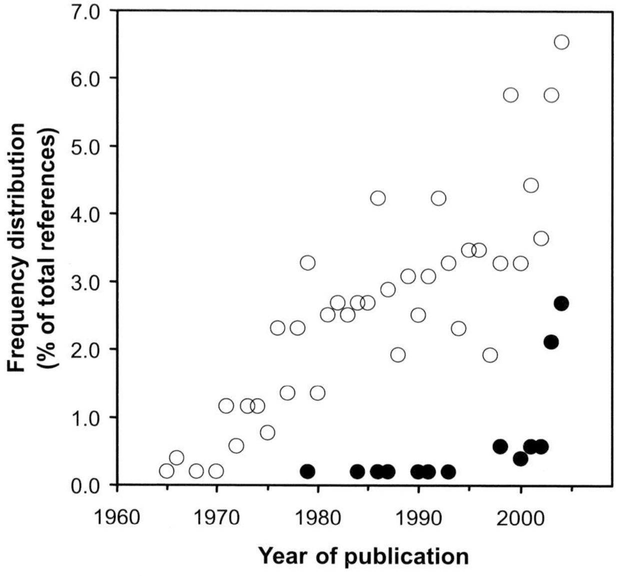

Fig. 1. Literature survey of publications which associate the concepts, "metabolite", "profiling", and "gas chromatography" performed on 1/2005. A total of ~500 citations without conference proceedings, abstracts and book chapters were found. The frequency of publications in all biological sciences (*open circles*) is compared to the contribution by plant metabolomics community (*closed circle*)

(Weckwerth et al. 2004b), (iii) profiling of trace compounds, or signalling molecules in the presence of bulk metabolites (Mueller et al. 2002; Birkemeyer et al. 2003; Schmelz et al. 2003, 2004), (iv) increasing accuracy in multi-parallel metabolite quantification (Birkemeyer et al. 2005), (v) combining profiling and flux analyses (Roessner-Tunali et al. 2004), (vi) establishment of quantitative repeatability, unambiguous nomenclature and comparability between analyses performed in different laboratories or using different analytical technology platforms (Schauer et al. 2005), and (vii) finally – perhaps the most important challenge of all metabolomic investigations – the identification of the unidentified majority of metabolic components from metabolite profiling experiments (Fiehn et al. 2000a; Schauer et al. 2005).

In agreement with the focus of this chapter the above challenges have predominantly analytic or technical motivation. The breakthrough of metabolomic investigations, however, will depend on the access to hitherto unavailable fundamental insights into metabolic and systems interactions. Increasingly integrative studies which consider the metabolome, proteome, transcriptome, and genome evolution of an organism have been initiated and are to be expected. Promising steps have been made – using GC-MS technology – towards network analysis (Fiehn 2003; Weckwerth et al. 2004a) and correlation studies between or within metabolome and transcriptome constituents (Urbanczyk-Wochniak et al. 2003; Steinhauser et al. 2004; Kopka et al. 2005). A detailed discussion of these general aspects including GC-MS studies and beyond can be found in the applications section of this book.

2 GC-MS Profiling Technology in a Nutshell

Metabolite profiling with GC-MS involves six general steps:

1. *Extraction* of metabolites from the biological sample, which should be as comprehensive as possible, and at the same time avoid degradation or modification of metabolites (e. g. Kopka et al. 2004).
2. *Derivatisation* of metabolites making them amenable to gas chromatography. Metabolites which are not volatile per se require chemical modification prior to GC analysis.
3. *Separation* by GC. High resolution GC can also be highly reproducible as it involves automated sample injection robotics, highly standardised conditions of gas-flow, temperature programming, and standardised capillary column material.
4. *Ionisation* of compounds as they are eluted from the GC. Electron impact (EI) ionisation is most widely used, as it is the technology which is least susceptible to suppression effects and produces reproducible fragmentation patterns.
5. Time resolved *detection* of molecular and fragment ions. Mass separation and detection can be achieved with different mass-detection devices, including sector field detectors, quadrupole detectors (QUAD), ion trap technology, and time-of-flight detectors (TOF). The choice of detectors depends on the targeted analytical niche. GC-MS systems with QUAD detection are most widely spread for routine analysis. Ion trap technology allows MS×MS (two-dimensional MS) analysis for structural elucidation and targeted quantification of trace compounds (e. g. Mueller et al. 2002). TOF detection can either be tuned to fast scanning rates (van Deursen et al. 2000) or to high mass precision comparable to sector field systems. Fast scanning GC-TOF-MS enables the, today, most advanced technology in the GC-MS field, namely two dimensional GC×GC-TOF-MS (two-dimensional GC-TOF-MS) (Ryan et al. 2004; Sinha et al. 2004a–c).
6. *Acquisition and evaluation* of GC-MS data files. All GC-MS system manufacturers provide software which is tuned for targeted, quantitative metabolite analysis. The targeted approach involves unequivocal identification of predefined metabolites by expected chromatographic retention times and mass-spectral fragmentation patterns and quantitative calibration by authentic standard concentrations. Recent software developments support the non-targeted analysis of GC-MS patterns, and the full evaluation of all resolved compounds. This feature of GC-MS allows discovery of novel hitherto

unknown metabolites. As we are far from knowing all possible metabolites of a given organism, non-biased, truly comprehensive data evaluation is the most essential requirement of metabolite profiling.

2.1 Chemical Derivatisation and Chromatography

The principles of fast metabolic sample inactivation and nondestructive extraction are common to all metabolome analyses. In contrast to all other technologies GC-MS is inherently restricted to volatile and temperature-stable compounds. The scope of GC-MS for metabolite analysis is limited by the typical temperature range of commercial capillary columns, for example up to 320–350 °C. The lower temperature range is determined by ambient temperature, but cold trapping devices and isothermal GC allow analysis of low molecular weight gases and highly volatile metabolites. GC received a considerable extension of applications through the development of a highly versatile tool box of derivatisation reagents, which chemically transform non-volatile metabolites into volatile analytes for GC-MS analysis (e. g. Knapp 1979; Blau and Halket 1993; Toyo'oka 1999). To date, GC-MS profiling of metabolites in plants has largely been confined to compounds, recovered in the methanol-water phase after methanol-water/chloroform extraction of tissues (Fiehn et al. 2000a; Roessner et al. 2000; Duran et al. 2003; Barsch et al. 2004; Gullberg et al. 2004; Strelkov et al. 2004; Broeckling et al. 2005). Although not all hydrophilic compounds can be volatilised by derivatisation, the following classes of compounds are detected routinely: amino-, organic-, and aromatic-acids, amines, sugars up to trisaccharides, alcohols and polyols, and some monophosphorylated metabolites.

The current limitations of metabolite preparation and derivatisation strategy, namely methoxyamination with subsequent direct trimethylsilylation of predominantly polar metabolites, call for extension. Application of other technology platforms is an obvious route and will be discussed in the following chapters. Here a short appraisal of the potential of chemical derivatisation is attempted. Four main types of reaction schemes will be discussed.

1. *Alkoxyamination* by reagents, such as methoxyamine $CH_3{-}O{-}NH_2$, stabilises carbonyl moieties in native metabolite structures, but forms E- and Z-isomers of the $-N{=}C<$ double-bond substituents. Keto-enol tautomerism is suppressed, as is the decarboxylation of unstable β-carbonyl-carboxylic acids. In addition, the formation of acetal- or ketal-structures in aqueous solution is inhibited. These equilibrium reactions generate multiple intramolecular and water adducts, for example the typical α- and β-conformers of reducing sugars. Ether- and ester-conjugates are mostly stable when exposed to methoxyamine reagent and maintain conformation. So far other alkoxy-reagents – for example hydroxylamine, ethyloxyamine, or benzyloxyamine – have not been exploited for systematic discovery of metabolites with carbonyl moieties:

2. *Silylation* reagents classify into those which introduce either a trimethylsilyl (TMS) moiety, $-Si(CH_3)_3$, or a dimethyl-(*tert*-butyl)-silyl (TBS) moiety, $-Si(CH_3)_2-C(CH_3)_3$. TMS reagents have been well investigated and are known to have the widest derivatisation spectrum (Little 1999; Halket et al. 2005). TMS has the potential to substitute all exchangeable, "acidic" protons of a metabolite. Steric hindrance of TMS substitution is rare but common with the bulkier TBS reagent. The benefit of the TBS reagent is higher tolerance for the presence of water and clear mass spectral fragmentation. However, vicinal diols, which typically occur in sugars, are only partially derivatised.
3. *Alkylation* reactions, mostly methylation, are widely used to derivatise carboxylic acids and alcohols. The enormous reactivity of available reagents – some allow for flash derivatisation during hot GC injection – leads to transalkylation of ester-bonds und consequently breaks down complex metabolites, such as glycero- and phospholipids. Alkylation of sugars leads to derivatives which are more volatile than the TMS derivatives and therefore allow analysis of higher sugar oligomers.
4. *Acylation* reactions, mostly acetylation or trifluoro-acetylation, are less reactive than transalkylation. Reagents usually form stable ester and amide bonds and break down only activated metabolic intermediates, e. g. thioesters.

In conclusion further developments of alternate GC-MS profiling techniques need to employ more selective combinations of metabolite fractionation and derivatisation schemes. Solid phase extraction can be explored to partition and concentrate metabolites amenable to alternate subsequent derivatisation. On the other hand, vapour phase extraction (VPE) for the separation and concentration of volatile derivatisation products prior to GC injection may prove promising (Schmelz et al. 2003, 2004). VPE has the potential to be a robust technique and was shown to operate with a range of commonly used reagents.

2.2 Mass Detection and Quantitative Calibration Techniques

One of the major criticisms and pitfalls of metabolome analyses is best explained by so-called matrix effects. This well-known effect describes unexpected losses or increased recovery of metabolites in complex extracts compared to pure authentic preparations. Matrix effects on one hand are caused by the presence of compounds which either specifically inhibit extraction or chemical analysis of metabolites. Positive matrix effects can stabilise otherwise labile compounds in the presence of suitable chemicals. Typical examples are suppression effects of soft ionization techniques, for example electrospray ionization (ESI) or matrix assisted laser desorption ionization (MALDI). Electron-impact ionization (EI) typically used in GC-MS profiling is not susceptible to suppression. Instead GC injection is the crucial step which may cause discriminations, especially in view of the complex and rather crude extracts which are typically injected.

So far, only exemplary – albeit time demanding – thorough tests for unexpected matrix effects have been performed with selections of chemically

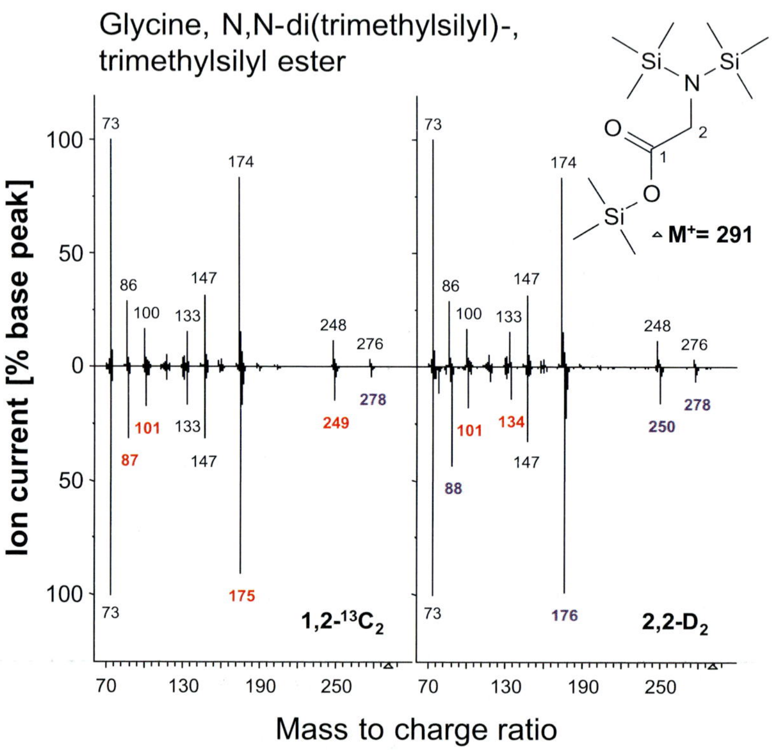

Fig. 2. Mass spectra of deuterated and ^{13}C labeled MSTs help structural elucidation and recovery analysis of metabolites. Labeled and non-labeled MSTs of Glycine *N,N*-di(trimethylsilyl)-, trimethylsilyl ester are shown. *Oryza sativa* L. cv. Nipponbare was labelled in vivo using deuterated water or $^{13}CO_2$. MSTs representing the fully labeled mass isotopomers demonstrate presence of two carbon atoms (*left panel*) and two non-exchangeable hydrogen atoms (*right panel*). Mass fragments which exhibited a mass shift of 1 amu (*red*) or 2 amu (*blue*) are indicated

▶ **Fig. 3a–c.** Mass spectral deconvolution of deuterated mass isotopomers. Succinic acid di(trimethylsilyl) ester was partially labelled in vivo by exposing *Oryza sativa* L. cv. Nipponbare to deuterated water. Metabolite profiles were performed on a Pegasus II GC-TOF-MS system (LECO, St. Joseph, MI, USA) with 20 scans s^{-1}. Mass spectra were deconvoluted using ChromaTOF software version 1.00, with baseline offset just above noise, smoothing and peak width set to 10 and 2 scans, respectively: **a** selected ion traces of non-deuterated (D_0, $m/z = 247$) and deuterated (D_{1-4}, $m/z = 248 - 251$) $M^+ - 15$ mass fragments. Mass fragments at 252 and 253 amu are carbon mass isotopomers of D_4; **b** peak area compared to deconvoluted peak height. Peak area integration does not allow differentiation of contributions by carbon mass isotopomers; **c** deconvoluted mass spectra of D_{0-4}. *Inset* shows partial deconvolution of D_{0-4} carbon mass isotopomers and missing carbon mass isotopomers of D_{1-3}

diverse, representative metabolites (e.g. Roessner-Tunali et al. 2003; Gullberg et al. 2004). Therefore technologies are required to improve quantitative standardisation for the comparison of increasingly diverse biological samples and experimental conditions.

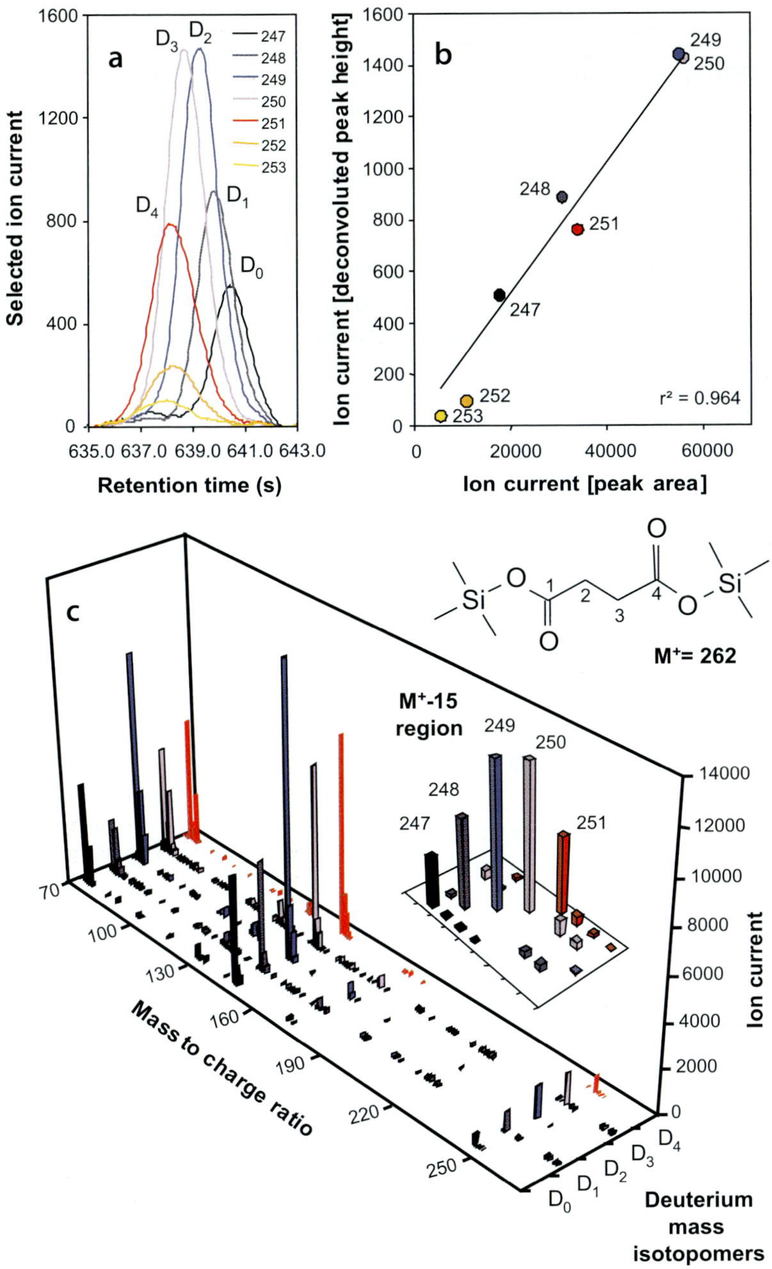

For this purpose, full saturating ^{13}C in vivo labelling was developed using yeast which is one of the most important organisms in systems biology (e. g. Stephanopoulos et al. 2004). Metabolites of yeast were demonstrated to be fully labelled when provided with an exclusive carbon source, such as U-^{13}C-glucose (Mashego et al. 2004; Birkemeyer et al. 2005). Refer to Birkemeyer et al. (2005) for detailed discussion of potential applications for ^{13}C-labelled metabolomes. Similar approaches are possible in plants (Figs. 2 and 3).

In short, standardised in vivo labelled extracts of yeast or other microorganisms can substitute the rather small number of chemically synthesised mass isotopomers used in earlier studies (Fiehn et al. 2000a; Gullberg et al. 2004). Typically a standardised labelled reference sample is combined in equal amounts with non-labelled experimentally challenged samples. The advantages of this approach are (i) the presence of a mass isotopomer for all identified but also all hitherto non-identified metabolites, (ii) the concentration of each mass isotopomer is inherently adjusted to the endogenous metabolite concentration, (iii) metabolic components can easily be distinguished from laboratory contaminations, and (iv) recovery of all metabolic components can be determined with the appropriate mass isotopomer.

Thus metabolite profiling will achieve the same level of transcriptome and proteome experiments, which utilize differential fluorescent probes or differential isotope coded tagging, respectively. In conclusion, comprehensive in vivo isotope labelling will help to establish quantitative between laboratory comparability of GC-MS based metabolome experiments. More importantly, we expect metabolome experiments with full mass isotopomer standardisation to be also independent of the mass spectrometric platform, e. g. CE-MS, LC-MS, or possibly even MALDI-TOF-MS.

3 Short Excursion into Nomenclature and Definitions

Concise and unambiguous description of GC-MS metabolite profiling results requires clear definitions. The definitions suggested within this section are biased towards the specifics of GC-MS technology but may also be applied to other technology platforms. This section is intended as a contribution to the ongoing process of unifying data formats and concepts within the field of plant metabolomics (e. g. Fiehn 2002; Bino et al. 2004; Jenkins et al. 2004).

3.1 Metabolite and Analyte

Routine GC-MS profiling analysis (Fiehn et al. 2000b; Roessner et al. 2000) has an upper size exclusion limit which is roughly equivalent to a persilylated trisaccharide derivative (MW:1296), hexatriacontane (MW:506), or hentriacontanoic acid trimethylsilylester (MW:523). Even though it may appear tempting, metabolite and analyte are best not defined by molecular weight.

A *metabolite* may be described as a compound which is internalised, chemically converted or secreted by an organism, but is not synthesised by DNA replication, transcription, or translation. Post-processing events of DNA, RNA and proteins, such as DNA methylation, RNA splicing, sequence specific protease cleavage or post-translational modification are not attributed to the metabolome. The origin of a metabolite is not exclusively dependent on the biosynthetic capacity of an organism or delimited by the genomic inventory. Metabolites may readily be exchanged between organisms, for example in plant microbe interactions, and – like drugs or pesticides – can today be of anthropogenic/xenobiotic origin.

In contrast to LC- or CE-MS, GC-MS analysis requires clear distinction between metabolite and analyte, because – depending on choice of chemical derivatisation – metabolites may be chemically transformed before quantification. The term *analyte* may be used to address the chemical structure and compound which is submitted to GC-MS and finally detected and quantified. An analyte can be identical with the metabolite, if the metabolite is not chemically derivatised. Single metabolites may have more than one analyte, if the chosen derivatisation reaction generates more than one derivative, for example methoxyamination (see above). In these cases *preferred* and *alternate* analytes exist for quantification. Analytes of one metabolite may differ in abundance, i. e. a *major* and one, even multiple, *minor* analytes may exist. Standardisation by stable mass isotopomers corrects the quantification errors which may arise from unforeseen matrix effects on analyte ratios during chemical derivatisation of GC injection.

Different metabolites may be chemically transformed into the same analyte structure. In addition a single analyte may arise from inadequate chromatographic separation of isomers. For example, the biochemically distinct stereoisomeric structures of DL-amino acids are only separated by specialised chiroselective chromatography. These analytes have *composite* properties in contrast to absolutely *specific* analytes.

These concepts are not unique to GC-MS technology. Analyte sensitivity, accuracy, and potentially composite analyte properties need to be thoroughly considered in MS-MS applications, non-chiroselective capillary electrophoresis or liquid chromatography, and in cases of adduct-formation or multiply charged ions.

3.2 Mass Spectral Tag (MST) and Mass Fragment

GC-MS metabolite profiles resolve hundreds of analytes, which represent metabolites, but also internal standard substances and laboratory contaminations. Typical GC-MS profiles may contain approximately 100 identified analytes of metabolites. The chemical structure of the majority of GC-MS analytes, however, is still unknown. Each new biological object or experimental condition still gives rise to new, hitherto unidentified, chemical components.

Because in non-biased analysis of GC-MS profiles identified and unidentified components are equally important, we created the term *mass spectral tag* (MST), i. e. a mass spectrum which is characterised by a specific chromatographic retention and by repeated occurrence in a single or multiple types of biological samples (Colebatch et al. 2004; Desbrosses et al. 2005). MSTs represent analytes. MSTs can be identified, in other words, unequivocally linked to a chemical structure. The use of MSTs allows uncoupling of metabolite profiling experiments from the time consuming process of chemical identification. MSTs can be used to track analytes in different experiments or laboratories (Schauer et al. 2005). Thus MST identification can be performed even years after the first discovery.

MSTs of GC-EI-MS profiles are composed of multiple characteristic *mass fragments* in constant relative abundances. In most cases residual, non-fragmented molecular ions are rare or even absent. In consequence GC-MS allows selection of multiple mass fragments which all represent the same MST and exhibit the same quantitative changes. Typically one quantifying mass fragment (QM) and a set of specific, supporting qualifying mass fragments are selected in GC-MS analysis (Halket et al. 2005). The criteria for the proper choice of QMs are equal to the choice of a preferred analyte. QMs need to be selective, i. e. not composite, in the context of the complexity of co-eluting MSTs. Therefore, the best QM is the most abundant among the available selective mass fragments.

3.3 Response and Relative Quantification of Metabolite Pools

GC-MS metabolite profiling studies monitor relative changes in metabolite pool sizes and but also allow insight into flux, i. e. the dynamic turnover of metabolite pools or metabolite substructures (e. g. Fischer and Sauer 2003; Sauer 2004; Roessner-Tunali et al. 2004). Flux experiments are easily distinguished from above mentioned saturating in vivo labelling experiments. Flux experiments monitor the initial kinetics of labelling and thus stable isotopes are only partially incorporated into metabolite pools. In contrast, saturating in vivo labelling reaches the endpoint of a completely stable isotope labelled metabolome.

MSTs are quantified by ion currents of QMs which are recorded after analyte ionization, fragmentation and mass separation. Ion currents in GC-MS are monitored either by peak area or peak height. Both measurements need to be baseline corrected for electronic and chemical noise. The resulting corrected values are defined to be what we call *responses*, i. e. X_{QM} of fragment QM (Colebatch et al. 2004; Desbrosses et al. 2005). The fragment response is routinely normalised to the amount of the sample, for example fresh or dry weight. In addition each response is corrected for recovery effects, which may occur at any step of the analytical process between metabolic inactivation of the sample and final recording of ion currents. Different levels of recovery correction exist: (i) correction by extract and sample volume, (ii) correction by addition of a constant amount of a representative internal standard compound (IS), and (iii) normalization by chemically identical, but stable-isotope labelled mass isotopomers of each metabolite. The *normalised response* (N_{QM}) is, consequently, $N_{QM} = X_{QM} \times X_{IS}^{-1} \times \text{sample weight}^{-1}$, where X_{IS} ideally represents a mass isotopomer response of QM. In a further step, the normalised response of a fragment, N_{QM}, is divided by the average relative response of QM as determined in a set of reference samples, $\text{avg}N_{QM(ref)}$. The resulting quotient, $R_i = N_{QM} \times \text{avg}N_{QM(ref)}^{-1}$, is called *response ratio* R_i. R_i describes the x-fold changes in metabolite pools sizes relative to the reference samples. Typical reference samples are taken at the start of a time series experiment or are mock-treated biological controls.

In GC-MS profiling analyses the standard deviation of normalised responses is dependent on the chemical nature of metabolite and analyte. Average relative standard deviations (RSD) of 10% (Weckwerth et al. 2004b) or 13.8% (5.5–33.4%; Gullberg et al. 2004) were reported for replicate GC-MS analyses. These analyses included extraction as well as derivatisation and were performed using representative analytes. Use of isotope labelled standardisation was reported to reduce RSD further to approximately 6.9–9.7% residual experimental variance (Gullberg et al. 2004).

4 Present Challenges of GC-MS Profiling

4.1 Standardisation of GC-MS Systems

GC-MS profiles, with the exception of GC×GC-TOF-MS data, are in essence three-dimensional and comprise a chromatographic time-resolved axis, a second coordinate axis which represents the mass to charge ratio (m/z, $z = 1$ in GC-MS with only rare exceptions), and an intensity axis which monitors the ion current (IC) and thus the abundance of molecules or mass fragments. A substantial breakthrough for GC-MS analyses was the early establishment of generally accepted calibration substances and procedures, so-called tuning routines, which allowed comparison of mass spectra from GC-MS systems of virtually all manufacturers and from different hyphenated mass detection technologies. In addition the widely used electron-impact ionisation technique (EI) ensured stable fragmentation ratios, which are in first approximation independent of analyte concentration. However, comparability was only achieved by restriction to 1 amu precision.

The chromatography axis is less standardised, not least because of multiple types of available capillary GC-columns which have different chromatographic properties and thus serve different separation problems. In addition slight changes in temperature program, pressure and flow settings of both carrier gas and injection technique, as well as slight production differences of column manufacturers cause minor but perceptible changes in retention time. Retention time indices (RI), based on homologous series of internal reference substances, such as *n*-alkanes, have been introduced early to aid GC analyses (Kovàts 1958). Use of an *n*-alkane RI system in GC-MS metabolite profiling substantially improves the reproducibility of the chromatography axis. The currently achievable accuracy of RI prediction was recently investigated in three different profiling laboratories which use the same type of capillary column but different GC-MS systems (Schauer et al. 2005). In this investigation the possibility of predicting RIs of more than 100 identified analytes was tested. Mathematical regression resulted in an average accuracy of ±5.4 RI units.

The IC intensity axis in GC-MS is standardised for high vs low mass discrimination. The GC-MS tuning includes processes which ensure constant ratios of high vs low mass intensities. However, mass spectra which are recorded by either QUAD-MS or fast scanning GC-TOF-MS detection may differ in this respect. Fast scanning GC-TOF-MS systems (e. g. Pegasus II MS system, LECO, St. Joseph, MI, USA) have increased sensitivity of small mass fragments and reduced sensitivity in the high mass range.

4.2 Deconvolution and Alignment of Mass Spectral Tags

The principal challenge in GC-MS profiling analysis is the automated unravelling of the multiple partially co-eluting MSTs which comprises a GC-MS chromatogram. One of the fundamental advances in GC-MS technology has been the development of algorithms and software for the so-called deconvolution of mass spectra from GC-MS chromatograms (Halket et al. 1999; Stein 1999; Shao et al. 2004; AnalyzerPro, http://www.spectralworks.com), specific software for the deconvolution of fast scanning GC-TOF-MS data files, e. g. ChromaTOF software used by Vreuls et al. (1999), Veriotti and Sacks (2001) or Jonsson et al. (2005), and ongoing developments for automated processing of GC×GC-TOF-MS chromatograms (Ryan et al. 2004; Sinha et al. 2004a,b). The inherent steps of deconvolution are (i) mass resolved baseline subtraction of electronic and chemical noise, (ii) assignment of retention time and/or retention time indices (RI) to chromatographic peak apices and respective MSTs, and (iii) accurate separation of MSTs from closely co-eluting analytes, the most challenging and advanced but still error-prone part (Fig. 3).

Even though automated mass spectral deconvolution has fundamentally facilitated GC-MS analyses of complex mixtures, accuracy and limitations of respective software have so far not been systematically compared and assessed. Typical errors of mass spectral deconvolution are (i) accidental generation of MSTs due to noise fluctuations, (ii) deconvolution of multiple MSTs from a single component, (iii) incomplete MSTs which lack one or multiple mass fragments (Fig. 3c), and (iv) chimeric MSTs, i. e. composite mass spectra of co-eluting compounds. The co-elution problem of complex mixtures has been fundamentally improved by introduction of fast scanning GC-TOF-MS and is today technically best solved by GC×GC-TOF-MS, using a set of two capillary GC columns with alternate phase-polarity (Sinha et al. 2004c).

Reliable alignment of identical MSTs in sets of consecutive GC-MS chromatograms is required for rapid, repeatable and automated comparative high-throughput analysis of large samples sets. So far, software solutions and novel algorithm developments for the alignment of complex mixtures depend on close to constant chromatographic retention within series of consecutive GC-MS chromatograms (Duran et al. 2003; Jonsson et al. 2004; metAlign, http://www.metalign.nl). Indeed, consistent run-to-run retention times are considered to be crucial to the application of chemometrics on complex mixtures, especially in the field of two-dimensional separations (Sinha et al. 2004c).

In conclusion, automated mass spectral deconvolution of GC-MS profiles appears to be in principal solved by both GC×GC technology and deconvolution algorithm, but the optimum solution still has to be found (Halket et al. 2005). In contrast prediction of chromatographic shifts in complex mixtures with highly dynamic range of concentrations has not been satisfyingly solved. As there is currently no solution – other than recalibration with pure standard substances – addressing the problem of RI shifts will be crucial for future GC-MS based metabolite profiling and identification of MSTs.

4.3 Identification of Mass Spectral Tags

Identification of MSTs requires chromatographic separation as well as mass spectrometric information (Wagner et al. 2003), mainly because plants like microorganisms contain a multitude of isomeric metabolites (e. g. Barsch et al. 2004; Stephanopoulos et al. 2004; Strelkov et al. 2004). These isomers give rise to MSTs, which can be chromatographically resolved but have almost identical mass spectra. Today, GC-MS appears to have found a generally accepted standard for mass spectral comparison. The NIST mass spectral search and comparison software (Ausloos et al. 1999; Stein 1999) has been integrated into the customised operating software of most GC-MS manufacturers. The GC-MS technology is in this respect more advanced than LC-MS (Halket et al. 2005). However, mass spectral search and comparison software, which harbours information on chromatographic retention in what we suggested to call MSRI libraries (Wagner et al. 2003; Kopka et al. 2005) and the automated utilization of this information for probability based matching, would be highly desirable. The new version NIST05 (National Institute of Standards and Technology, Gaithersburg, MD, USA) of a mass spectral search and comparison software now makes RI information available but currently does not utilize RI for automated matching. The result of a hitherto manual inventory of *Oryza sativa* L. cv. Nipponbare leave profiles is shown in Fig. 4.

Two different approaches exist for the identification of unknown MSTs from GC-MS profiles: (i) the "bottom up" approach in which metabolites of interest to a particular researcher are analysed by the purchase of authentic standard

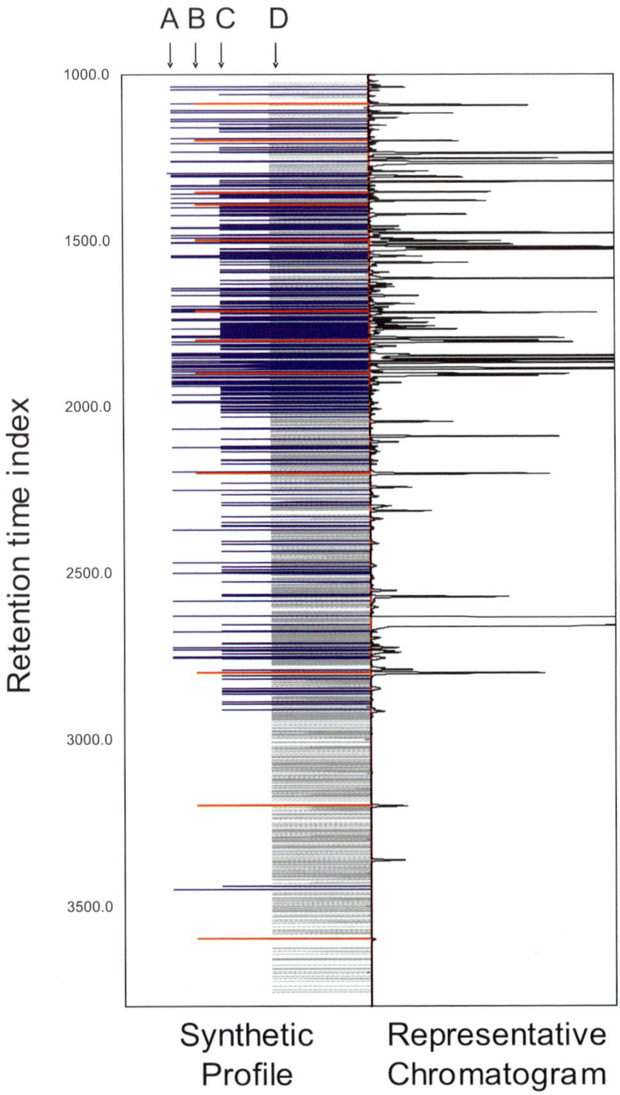

Fig. 4. Synthetic and representative GC-MS profiles of *Oryza sativa* L. cv. Nipponbare leaves: A – 132 identified MSTs representing 109 known metabolites; B – 12 added internal standard substances; C – 148 unidentified MSTs which match previous MSRI library entries; D – all previously observed MSTs present in the MSRI library at GMD (http://csbdb.mpimp-golm.mpg.de/gmd.html)

substances, which are subsequently mapped by standard addition experiments onto established standardised GC-MS systems, and (ii) the "top down" approach whereby structural elucidation is performed on a hitherto unknown, but important target MST. The work of "top down" structural identification is highly time-demanding and involves preparative enrichment, purification, spectroscopic, mass spectral and NMR analyses of the preparation and finally chemical synthesis of the predicted structure. As a consequence the "bottom up" approach prevails in most laboratories and "top down" identification is currently restricted to potentially novel signalling compounds or marker substances of specific biological samples and experimental conditions.

In order to avoid unnecessary "top down" investigations reliable identification by prior standard addition experiments is essential. MSRI library collections of mass spectra (Kopka et al. 2005), which comprise frequently observed identified and non-identified MSTs, appear to represent the most effective means to pool the identification efforts currently performed in many laboratories around the world (Schauer et al. 2005). Identified and yet unidentified, MSTs can efficiently be shared by public resources such as GMD@CSBDB (http://csbdb.mpimp-golm.mpg.de/gmd.html). In addition mass spectral identifications and chromatographic sequence of analyte elution can be transferred between laboratories. "Bottom up" identifications performed in parallel may be used for inter-laboratory confirmation of identifications and reduce the risk of unnecessary structural elucidation projects.

Currently the MSRI libraries available from GMD@CSBDB include in total more than 2000 evaluated mass spectral data sets obtained using GC-QUAD- and GC-TOF-MS technology platforms with 1089 non-redundant and 360 identified MSTs. Future efforts at GMD aim to refine mass spectral quality and annotation, and will add stable isotope labelled variants of MSTs (e.g. Fig. 2) for improved mass spectral interpretation of unidentified MSTs. The number of identified analytes and metabolites will continuously be extended and annotations updated.

Acknowledgements. I would like to thank A.R. Fernie, A. Erban, Max Planck Institute of Molecular Plant Physiology, Potsdam-Golm, Germany for critically reading and discussing my manuscript. My thanks extend to Prof. Dr. Le Tran Binh, Institute of Biotechnology (IBT), Hanoi, Vietnam, for sharing his expertise in rice cultivation. This work was supported by the Max-Planck society, and the Bundesministerium für Bildung und Forschung (BMBF), grant PTJ-BIO/0312854.

References

Ausloos P, Clifton CL, Lias SG, Mikaya AI, Stein SE, Tchekhovskoi DV, Sparkman OD, Zaikin V, Zhu D (1999) The critical evaluation of a comprehensive mass spectral library. J Am Soc Mass Spectrom 10:287–299

Barsch A, Patschkowski T, Niehaus K (2004) Comprehensive metabolite profiling of *Sinorhizobium meliloti* using gas chromatography-mass spectrometry. Funct Integrat Genomics 4:219–230

Bino RJ, Hall RD, Fiehn O, Kopka J, Saito K, Draper J, Nikolau BJ, Mendes P, Roessner-Tunali U, Beale MH, Trethewey RN, Lange BM, Wurtele ES, Sumner LW (2004) Potential of metabolomics as a functional genomics tool. Trends Plant Sci 9:418–425

Birkemeyer C, Kolasa A, Kopka J (2003) Comprehensive chemical derivatization for gas chromatography-mass spectrometry-based multi-targeted profiling of the major phytohormones. J Chromatogr A 993:89–102

Birkemeyer C, Luedemann A, Wagner C, Erban A, Kopka J (2005) Metabolome analysis: the potential of in vivo labeling with stable isotopes for metabolite profiling. Trends Biotechnol 23:28–33

Blau K, Halket JM (1993) Handbook of derivatives for chromatography, 2nd edn. Wiley, New York

Broeckling CD, Huhman DV, Farag MA, Smith JT,May GD, Mendes P, Dixon RA, Sumner LW (2005) Metabolic profiling of *Medicago truncatula* cell cultures reveals the effects of biotic and abiotic elicitors on metabolism. J Exp Bot 56:323–336

Colebatch G, Desbrosses G, Ott T, Krusell L, Kloska S, Kopka J, Udvardi MK (2004) Global changes in transcription orchestrate metabolic differentiation during symbiotic nitrogen fixation in *Lotus japonicus*. Plant J 39:487–512

Cook D, Fowler S, Fiehn O, Thomashow MF (2004) A prominent role for the CBF cold response pathway in configuring the low-temperature metabolome of *Arabidopsis*. Proc Natl Acad Sci USA 101:15243–15248

Desbrosses G, Kopka J, Udvardi MK (2005) Legume metabolomics: development of GC-MS resources for functional genomics of plant-microbe interactions. Plant Physiol 137:1302–1318

Duran AL, Yang J, Wang L, Sumner LW (2003) Metabolomics spectral formatting, alignment and conversion tools (MSFACTs). Bioinformatics 19:2283–2293

Fernie AR, Trethewey RN, Krotzky AJ, Willmitzer L (2004) Metabolite profiling: from diagnostics to systems biology. Nat Rev Mol Cell Biol 5:763–769

Fiehn O (2002) Metabolomics – the link between genotypes and phenotypes. Plant Mol Biol 48:155–171

Fiehn O (2003) Metabolic networks of *Cucurbita maxima* phloem. Phytochem 62:875–86

Fiehn O, Kopka J, Trethewey RN, Willmitzer L (2000a) Identification of uncommon plant metabolites based on calculation of elemental compositions using gas chromatography and quadrupole mass spectrometry. Anal Chem 72:3573–3580

Fiehn O, Kopka J, Dörmann P, Altmann T, Trethewey RN, Willmitzer L (2000b) Metabolite profiling for plant functional genomics. Nat Biotechnol 18:1157–1161

Fischer E, Sauer U (2003) Metabolic flux profiling of *Escherichia coli* mutants in central carbon metabolism using GC-MS. Eur J Biochem 270:880–891

Gullberg J, Jonsson P, Nordström A, Sjöström M, Moritz T (2004) Design of experiments: an efficient strategy to identify factors influencing extraction and derivatization of *Arabidopsis thaliana* samples in metabolomic studies with gas chromatography/mass spectrometry. Anal Biochem 331:283–295

Halket JM, Przyborowska A, Stein SE, Mallard WG, Down S, Chalmers RA (1999) Deconvolution gas chromatography mass spectrometry of urinary organic acids – potential for pattern recognition and automated identification of metabolic disorders. Rapid Commun Mass Spectrom 13:279–284

Halket JM, Waterman D, Przyborowska AM, Patel RKP, Fraser PD, Bramley PM (2005) Chemical derivatization and mass spectral libraries in metabolic profiling by GC/MS and LC/MS/MS. J Exp Bot 56:219–243

Jenkins H, Hardy N, Beckmann M, Draper J, Smith AR, Taylor J, Fiehn O, Goodacre R, Bino RJ, Hall RD, Kopka J, Lane GA, Lange BM, Liu JR, Mendes P, Nikolau BJ, Oliver SG, Paton NW, Rhee S, Roessner-Tunali U, Saito K, Smedsgaard J, Sumner LW, Wang T, Walsh S, Wurtele ES, Kell DB (2004) A proposed framework for the description of plant metabolomics experiments and their results. Nature Biotechnol 22:1601–1606

Jonsson P, Gullberg J, Nordström A, Kusano M, Kowalczyk M, Sjöström M, Moritz T (2004) A strategy for identifying differences in large series of metabolomic samples analyzed by GC/MS. Anal Chem 76:1738–1745

Jonsson P, Johansson AI, Gullberg J, Trygg J, A J, Grung B, Marklund S, Sjostrom M, Antti H, Moritz T (2005) High-throughput data analysis for detecting and identifying differences between samples in GC/MS-based metabolomic analyses. Anal Chem 77: 5635–5642

Kaplan F, Kopka J, Haskell DW, Zhao W, Schiller KC, Gatzke N, Sung DY, Guy CL (2004) Exploring the temperature-stress metabolome of *Arabidopsis*. Plant Physiol 136:4159–4168

Knapp DR (1979) Handbook of analytical derivatization reactions. Wiley, New York
Kopka J, Fernie AF, Weckwerth W, Gibon Y, Stitt M (2004) Metabolite profiling in Plant Biology: Platforms and Destinations. Genome Biol 5(6):109–117
Kopka J, Schauer N, Krueger S, Birkemeyer C, Usadel B, Bergmüller E, Dörmann P, Gibon Y, Stitt M, Willmitzer L, Fernie AR, Steinhauser D (2005) GMD@CSBDB: The Golm Metabolome Database. Bioinformatics 21:1635–1638
Kovàts ES (1958) Gas-chromatographische Charakterisierung organischer Verbindungen: Teil 1. Retentionsindices aliphatischer Halogenide, Alkohole, Aldehyde und Ketone. Helv Chim Acta 41:1915–1932
Little JL (1999) Artifacts in trimethylsilyl derivatization reactions and ways to avoid them. J Chromatorgr A 844:1–22
Mashego MR, Wu L, van Dam JC, Ras C, Vinke JL, van Winden WA, van Gulik WM, Heijnen JJ (2004) MIRACLE: mass isotopomer ratio analysis of U-^{13}C-labeled extracts. A new method for accurate quantification of changes in concentrations of intracellular metabolites. Biotech Bioeng 85:620–628
Mueller A, Duechting P, Weiler EW (2002) A multiplex GC-MS/MS technique for the sensitive and quantitative single-run analysis of acidic phytohormones and related compounds, and its application to *Arabidopsis thaliana*. Planta 216:44–56
Roessner U, Wagner C, Kopka J, Trethewey RN, Willmitzer L (2000) Simultaneous analysis of metabolites in potato tuber by gas chromatography-mass spectrometry. Plant J 23:131–142
Roessner U, Luedemann A, Brust D, Fiehn O, Linke T, Willmitzer L, Fernie AR (2001a) Metabolic profiling allows comprehensive phenotyping of genetically or environmentally modified plant systems. Plant Cell 13:11–29
Roessner U, Willmitzer L, Fernie AR (2001b) High-resolution metabolic phenotyping of genetically and environmentally diverse plant systems – identification of phenocopies. Plant Physiol 127:749–764
Roessner U, Willmitzer L, Fernie AR (2002) Metabolic profiling and biochemical phenotyping of plant systems. Plant Cell Rep 21:189–196
Roessner-Tunali U, Hegemann B, Lytovchenko A, Carrari F, Bruedigam C, Granot D, Fernie AR (2003) Metabolic profiling of transgenic tomato plants overexpressing hexokinase reveals that the influence of hexose phosphorylation diminishes during fruit development. Plant Physiol 133:84–99
Roessner-Tunali U, Lui J, Leisse A, Balbo I, Perez-Melis A, Willmitzer L, Fernie AR (2004) Flux analysis of organic and amino acid metabolism in potato tubers by gas chromatography-mass spectrometry following incubation in ^{13}C labelled isotopes. Plant J 39:668–679
Ryan D, Shellie R, Tranchida P, Casilli A, Mondello L, Marriott P (2004) Analysis of roasted coffee bean volatiles by using comprehensive two-dimensional gas chromatography-time-of-flight mass spectrometry. J Chromatogr A 1054:57–65
Sauer U (2004) High-throughput phenomics: experimental methods for mapping fluxomes. Curr Opin Biotechnol 15:58–63
Sauter H, Lauer M, Fritsch H (1988) Metabolite profiling of plants – a new diagnostic technique. Abstr Pap Am Chem Soc 195:129
Schauer N, Steinhauser D, Strelkov S, Schomburg D, Allison G, Moritz T, Lundgren K, Roessner-Tunali U, Forbes MG, Willmitzer L, Fernie AR, Kopka J (2005) GC-MS libraries for the rapid identification of metabolites in complex biological samples. FEBS Lett 579:1332–1337
Schmelz EA, Engelberth J, Alborn HT, O'Donnell P, Sammons M, Toshima H, Tumlinson JH (2003) Simultaneous analysis of phytohormones, phytotoxins, and volatile organic compounds in plants. Proc Natl Acad Sci USA 100:10552–10557
Schmelz EA, Engelberth J, Tumlinson JH, Block A, Alborn HT (2004) The use of vapor phase extraction in metabolic profiling of phytohormones and other metabolites. Plant J 39:790–808
Shao XG, Wang GQ, Wang SF, Su QD (2004) Extraction of mass spectra and chromatographic profiles from overlapping GC/MS signal with background. Anal Chem 76:5143–5148

Sinha AE, Fraga CG, Prazen BJ, Synovec RE (2004a) Trilinear chemometric analysis of twodimensional comprehensive gas chromatography-time-of-flight mass spectrometry data. J Chromatogr A 1027:269–277
Sinha AE, Hope JL, Prazen BJ, Nilsson EJ, Jack RM, Synovec RE (2004b) Algorithm for locating analytes of interest based on mass spectral similarity in GC×GC-TOF-MS data: analysis of metabolites in human infant urine. J Chromatogr. A 1058:209–215
Sinha AE, Prazen BJ, Synovec RE (2004c) Trends in chemometric analysis of comprehensive two-dimensional separations. Anal Bioanal Chem 378:1948–1951
Stein SE (1999) An integrated method for spectrum extraction and compound identification from gas chromatography/mass spectrometry data. J Am Soc Mass Spectrom 10:770–781
Steinhauser D, Usadel B, Luedemann A, Thimm O, Kopka J (2004) CSB.DB: a comprehensive systems-biology database. Bioinformatics 20:3647–3651
Stephanopoulos G, Alper H, Moxley J (2004) Exploiting biological complexity for strain improvement through systems biology. Nat Biotechnol 22:1261–1267
Strelkov S, von Elstermann M, Schomburg D (2004) Comprehensive analysis of metabolites in *Corynebacterium glutamicum* by gas chromatography/mass spectrometry. Biol Chem 385:853–861
Sumner LW, Mendes P, Dixon RA (2003) Plant metabolomics: large-scale phytochemistry in the functional genomics era. Phytochemistry 62:817–836
Toyo'oka T (1999) Modern derivatization methods for separation science. Wiley, New York
Trethewey RN (2004) Metabolite profiling as an aid to metabolic engineering in plants. Curr Opin Plant Biol 7:196–201
Trethewey RN, Krotzky AJ, Willmitzer L (1999) Metabolic profiling: a Rosetta stone for genomics? Curr Opin Plant Biol 2:83–85
Urbanczyk-Wochniak E, Fernie AR (2005) Metabolic profiling reveals altered nitrogen nutrient regimes have diverse effects on the metabolism of hydroponically-grown tomato (*Solanum lycopersicum*) plants. J Exp Bot 56:309–321
Urbanczyk-Wochniak E, Luedemann A, Kopka J, Selbig J, Roessner-Tunali U, Willmitzer L, Fernie AR (2003) Parallel analysis of transcript and metabolic profiles: a new approach in systems biology. EMBO Reports 4:989–993
Van Deursen MM, Beens J, Janssen HG, Leclercq PA, Cramers CA (2000) Evaluation of time-of-flight mass spectrometric detection for fast gas chromatography. J Chromatogr A 878:205–213
Veriotti T, Sacks R (2001) High-speed GC and GC/time-of-flight MS of lemon and lime oil samples. Anal Chem 73:4395–4402
Vreuls RJJ, Dallüge J, Brinkman UAT (1999) Gas chromatography – time-of-flight mass spectrometry for sensitive determination of organic microcontaminants. J Microcolumn Sep 11:663–675
Wagner C, Sefkow M, Kopka J (2003) Construction and application of a mass spectral and retention time index database generated from plant GC/EI-TOF-MS metabolite profiles. Phytochem 62:887–900
Weckwerth W, Loureiro ME, Wenzel K, Fiehn O (2004a) Differential metabolic networks unravel the effects of silent plant phenotypes. Proc Natl Acad Sci USA 18:7809–7814
Weckwerth W, Wenzel K, Fiehn O (2004b) Process for the integrated extraction, identification and quantification of metabolites, proteins and RNA to reveal their co-regulation in biochemical networks. Proteomics 4:78–83

I.2 Current Status and Forward Looking Thoughts on LC/MS Metabolomics

L.W. Sumner[1]

1 Introduction

The metabolome can be viewed as the consequential end products of gene expression and the goal of metabolomics includes the comprehensive evaluation of the metabolome (Trethewey et al. 1999; Fiehn et al. 2000; Trethewey 2001; Oliver et al. 2002; Sumner et al. 2003). Quantitative and qualitative measurements of large numbers of cellular metabolites provide a high-resolution biochemical phenotype of an organism which can be used to monitor and assess gene function (Fiehn et al. 2000) or a system's response (Weckwerth 2003). Although mRNA/transcripts represent a mechanism for information transmission from the genome to the cellular machinery for protein synthesis, mRNA levels do not always correlate well with protein levels (Gygi et al. 1999). Furthermore, once translated a protein may or may not be enzymatically active as post translational modifications, protein sorting, protein–protein interactions, and controlled proteolysis all contribute to the regulation of active enzyme levels. Due to these factors, changes in the transcriptome or the proteome may not always lead to alterations in the metabolic phenotype. In addition, the majority of transcript and protein annotations are currently inferred based on sequence or structural similarity. It is estimated that less than 10% of annotated genes have experimental evidence supporting assigned function and thus, the accuracy of these annotations are of some uncertainty (Somerville and Somerville 1999; Somerville and Dangl 2000). In the absence of functionally annotated database information, transcript or protein profiling often yields limited information. For example, transcriptomics or proteomics often reveal the differential accumulation of a hypothetical or unannotated protein; however, without annotation it is very difficult to infer biological context. Microarray or proteomics experiments may also yield putative or generic protein identifications such as a putative peroxidase or peroxidase-like protein. These generic annotations have limited information as many of these enzymes are promiscuous and/or involved in a large number of different reactions. However, metabolomics has the ability to reveal that the accumulated peroxidase/enzyme is more specifically related to lignification or to another specific biochemistry. Thus, profiling the metabolome may actually provide the most direct and "functional" information of the "omics" technologies.

[1] The Samuel Roberts Noble Foundation, 2510 Sam Noble Parkway, Ardmore, OK 73401, USA, e-mail: lwsumner@noble.org

Biotechnology in Agriculture and Forestry, Vol. 57
Plant Metabolomics (ed. by K. Saito, R.A. Dixon, and L. Willmitzer)

The plant metabolome is quite complex with current estimates on the order of 15,000 metabolites within a given species and over 200,000 different metabolites within the plant kingdom (Dixon 2001; Hartman et al. 2005). Due to the chemical complexity of the plant metabolome, it is generally accepted that a single analytical technique will not provide comprehensive visualization of the metabolome, and therefore, multiple technologies are generally employed. The selection of the most suitable technology is generally a compromise between speed, chemical selectivity and instrumental sensitivity. Tools such as nuclear magnetic resonance spectroscopy (NMR) are rapid, highly selective, and non-destructive, but have relatively lower sensitivity. Other methods such as capillary electrophoresis (CE) coupled to laser induced fluorescence (LIF) detection are highly sensitive, but have limited chemical selectivity. Chromatographically coupled mass spectrometry methods such as gas chromatography (GC)/mass spectrometry (MS) and liquid chromatography (LC)/MS offer the best combination of sensitivity and selectivity, and therefore are central to most metabolomics approaches. Mass selective detection provides highly specific chemical information including molecular mass and/or characteristic fragment ion(s) information that are directly related to chemical structure. This information can be utilized for compound identification through spectral matching with data compiled in libraries for authentic compounds or used for de novo structural elucidation. Further, chemically selective MS information can be obtained from extremely small metabolite quantities with limits of detection in the pmole and fmole level for many primary and secondary plant metabolites.

GC/MS has proven capability for profiling large numbers of metabolites with reports covering several hundred to slightly more than a thousand various components (Fiehn et al. 2000; Roessner et al. 2000, 2001; Birkemeyer et al. 2003; Wagner et al. 2003; Broeckling et al. 2005; Schauer et al. 2005; Welthagen et al. 2005). The term component is used because a large number of metabolites often yield more than one derivatized component which are observed in the GC/MS analysis. The achievable range and number of metabolites profiled by GC/MS can be attributed to the high separation efficiencies of long (30–60 m) capillary GC columns (i. e. $N \geq 250{,}000$ for 60 m). These high efficiencies enable the separation of very complex mixtures, and with mass selective detection, qualitative identification of a significant proportion of these compounds is achievable. This makes GC/MS a very efficient and cost effective metabolomics tool. A major prerequisite for GC/MS is sample volatility which is necessary to enable separation in the gas phase. Analytes may be innately volatile or chemically derivatized to yield volatile compounds. Unfortunately, there exist a large number of metabolites which are not amenable to GC/MS even following derivatization. These include compounds such as phenylpropanoid and other natural product glycosides whose labile glycosidic bonds degrade during heating and vaporization. Thus, alternative techniques are necessary and especially so for the study of secondary metabolism.

Liquid introduction techniques for mass spectrometry such as electrospray ionization, atmospheric chemical ionization, and photo ionization remove the necessity for chemical derivatization. Thus, aqueous samples can be analyzed with minimal sample processing or even directly from the tissue source (Takats et al. 2004). Further, these techniques allow for the analyses of more labile and larger metabolites, and for the coupling of liquid separation technologies to mass spectrometry. Therefore high performance liquid chromatography (HPLC) and CE are readily coupled to mass spectrometry to yield powerful tools for targeted metabolic profiling and non-targeted metabolomics.

The utility of LC/MS emanates from the coupling of a 'universal' separation technology to a selective and sensitive mass analyzer detector. HPLC is commonly considered a universal separation technique because of its applicability to a broad range of chemical classes with a diversity of physical and chemical properties. For example, HPLC has been utilized for the analysis of ionic compounds, inorganics, volatile organics, polar organics, non-polar organics, lipids, amino acids, carbohydrates, nucleotides, carotenoids, phenylpropanoids, hormones, peptides, proteins, and the list goes on. The major point is that HPLC can be used for many of those compounds commonly analyzed by GC and many more. LC/MS also removes the need for derivatization and thus, complex samples can be analyzed directly or with minimal sample processing. As a result of these favorable properties, it is not surprising that LC/MS and LC coupled to tandem mass spectrometry (LC/MS/MS) have become popular tools for metabolism investigations.

HPLC is performed on various scales utilizing different column sizes. General values are provided in Table 1 for preparative, analytical, micro, capillary and nano-scale modes of HPLC. Generally, preparative scale HPLC is used for compound(s) purification and analytical scale is traditionally used for the quantitative analyses of plant extracts. However, smaller scale technologies (micro, capillary, nano) are now commercially available for quantitative analyses. These smaller scale separations offer significant sensitivity enhancements, and thus reduce the amount of material necessary for analysis. Further, capillary and nano HPLC often offer increased chromatographic resolution. Unfortunately as the separation scale gets smaller it becomes more difficult to reproducibly generate mobile phase gradients and the retention time variance increases. However, this problem is continually decreasing as novel instrumentation and approaches become available.

Table 1. General liquid chromatographic scales

Scale	Column internal diameter	Flow rate
Preparative	2.1–>200 mm	10 mL/min,
Analytical (*conventional*)	2.1–4.6 mm	1.0 mL/min
Micro	1.0 mm	200 µL/min
Capillary	300 µm–1 mm id	4 µL/min
Nano	25–300 µm id	200 nL/min

2 Chromatography Theory

Currently, the chromatographic performance of HPLC, relative to GC and CE, is lower, and there is a significant need for improvement. However, to discuss this issue and possible improvements in detail, several terms must be defined. A number of quantifiers are used to assess chromatographic performance. These include resolution (R_s), selectivity (α), efficiency (N), and peak capacity (n) which are defined below:

1. *Resolution* (R_s) is a quantifier of the degree of separation between mixture components, i. e. two peaks t_a and t_b with peak widths at the base w_a and w_b. A resolution of 1 indicates that two adjacent peaks are baseline resolved. Resolution can also be expressed as a function of the theoretical plate number (N) and selectivity (α) as defined below in Eq. (1):

$$R_s = \frac{2(t_b - t_a)}{w_a + w_b} = \frac{2\Delta t_R}{w_a + w_b} \qquad R_s = \frac{\sqrt{N}}{4}\left(\frac{\alpha - 1}{\alpha}\right)\left(\frac{k_2}{1 + k_2}\right) \tag{1}$$

2. *Selectivity* (α), which is also referred to as the separation factor, is a ratio of the retention or capacity factor (k') of two peaks. The capacity factor is a relative retention parameter that has been normalized using the void elution time (t_v) or volume (V_v) and is therefore independent of column geometry – see Eq. (2). The void value is the volume or time of an unretained component. The selectivity parameter provides a quantifier of the relative separation of two components. Selectivity can be altered based on the chemical composition of the stationary phase, stationary phase manufacturer, mobile phase, and pH:

$$\alpha = \frac{k'_2}{k'_1} \qquad k'_1 = \frac{t_1 - t_v}{t_v} = \frac{V_1 - V_2}{V_v} \tag{2}$$

3. *Column efficiency* is usually quantified based upon a column's theoretical plate number (N) which is unitless and a measure of band broadening per unit time – see Eq. (3). This can be practically quantified using retention time (t_R) and peak width. Peak width can be defined at the base (W_b) or at half height ($w_{1/2}$) as they are directly related if one assumes a Gaussian peak shape, i. e. $W_b = 1.698\ w_{1/2} = 4\sigma$ where σ equals the standard deviation of the peak. Alternatively, plate number can be calculated using the column resolution (R) and selectivity (α).
4. *Separation efficiency* is also quantified using a normalized theoretical plate number based on column length, i. e. (N/L) with units of plates/m. The theoretical plate number can be dramatically increased by decreasing the peak width. Plate number and efficiency are also related to particle size (d_p) and column length (L) as described below:

$$N = \left(\frac{t'_R}{\sigma}\right)^2 = 16\left(\frac{t'_R}{W_b}\right)^2 = 5.54\left(\frac{t'_R}{w_{1/2}}\right) = \frac{16R^2}{(1-\alpha)^2} = \frac{L}{d_p} \tag{3}$$

5. *Peak capacity* (n) is a measure of the maximum number of theoretical peaks resolvable by the chromatographic system based on optimum performance and equal variation in the partitioning of all components in the mixture – see Eq. (4). The peak capacity is a good parameter for estimating the maximum number of compounds resolvable by a given chromatographic system. Ideally this value should approach or exceed the number of compounds that need to be separated, i. e. the number of metabolites:

$$n = \frac{\sqrt{N}}{4R} \ln\left(\frac{t_2}{t_1}\right) + 1 \qquad (4)$$

3 Limitations of Current Metabolic Profiling Approaches and Proposed Solutions to Advance Metabolomics

Currently, the major limitation of metabolomics is its inability to comprehensively profile all of the metabolome. This inability is directly related to the chemical complexity of the metabolome, the biological variance inherent in most living organisms, and the dynamic range limitations of most instrumental approaches (Sumner et al. 2003). Many biological responses to altered gene expression or to environmental stimuli result in both quantitative and qualitative changes in metabolite pools. Understanding these responses is most dependent upon the qualitative identification of the altered metabolite. Quantitative measurements are also important, as both temporal and spatial changes in metabolite concentrations are expected; however this information is of little use if it cannot be assigned to a specific metabolite or biological process. Thus, comprehensive qualitative and quantitative analysis of all metabolites within a cell, tissue or organ is the ultimate goal of metabolomics; however, this is still a very ambitious goal and far from a reality for any system. Bino and colleagues (Bino et al. 2004) proposed two major objectives to increase the comprehensive nature of metabolomics. They were:

1. Increase the current capacity for metabolite separation and differentiation (i. e. the number of resolvable components within the complex metabolome mixture) using multi-dimensional separations.
2. Increase the number of identifiable metabolites through the generation of spectral libraries, high resolution accurate mass measurements, and tandem mass spectrometry.

Unfortunately, the separation of complex metabolome mixtures is still quite challenging. Currently, analytical scale HPLC (4.6 × 250 mm) is most commonly used for natural product analyses; however, the upper peak capacities (i. e. theoretical number of maximum peaks resolvable based on optimum performance) of these systems is approximately 300 (Tanaka et al. 2004). Based on this estimate, a maximum of 300 components could be resolved in a best case scenario; however in practice this value is seldom achieved and more realistic peak capacities are between 100 and 200. Thus, current HPLC technologies are limiting the comprehensive scope of metabolomics. Separation efficiencies can be improved by altering selectivity, increasing column lengths, reducing particle sizes, increasing temperature, and/or alternative column materials. Alternatively, the utilization of multidimensional chromatography offers increased HPLC peak capacities of greater than 1000 to provide more comprehensive coverage of plant natural products (Tanaka et al. 2004). Each of methods to increase HPLC efficiency is discussed below.

Typically, improving selectivity is the best approach to improving chromatographic resolution. Selectivity is based upon the chemical or physical interaction properties that are fundamental to the separation process. More precisely, the separation selectivity of specific components can be optimized by the appropriate choice of column materials, mobile phases, and/or manufacturer. Various generic and proprietary materials are available for various chromatographic modes for HPLC. Example modes include ion-exchange, normal-phase, reverse-phase, hydrophilic interaction, and size exclusion chromatography. All HPLC columns are not equal, and different particles, particle sizes, surface modification chemistries, surface coverage, and packing processes vary significantly from manufacturer to manufacturer. These parameters dramatically influence chromatographic performance.

Often selectivity is optimized for a targeted set of analytes as a means of increasing resolution. However, in more complex mixtures associated with global metabolomics-based approaches, improved selectivity for one class of compounds often results in decreased selectivity for others. Thus, techniques (e. g. reverse-phase chromatography) with a broad range of selectivity are most likely to be the best choices for metabolomics.

One of the simplest means of increasing resolution is to increase the number of theoretical plates. Since the plate number is directly proportional to the column length (Eq. (3)), one needs only to increase the column length to increase resolution. However, Eq. (1) tells us that R is proportional to the square root of N. Thus, to achieve a 2× increase in resolution, we would have to square the column length. For example a 250 mm long column would need to be extended to 625 cm (i. e. 25 × 25 cm) for a twofold increase in resolution. Unfortunately, this is not a practical solution as the operating pressure is directly proportional to the column length. Equation (5) defines the relationship between pressure (ΔP), column length (L), analyte diffusion coefficient (D_m), particle size (d_p), mobile-phase viscosity (η), and column permeability (K^o):

$$\Delta P = \left(\frac{L \nu D_m}{d_p} \right) \frac{\eta}{K^o} \tag{5}$$

If a typical column of 25 cm has an operational pressure of 3000 pounds per square inch (p.s.i.), then a twofold resolution increase obtained by squaring the column length $(25\,\text{cm})^2$ would require an operational pressure of 75,000 p.s.i. (i. e. 3,000 p.s.i. × 25). Although this illustrates the advantage of very high pressure liquid chromatography which has been achieved by select groups using custom apparatuses (MacNair et al. 1997, 1999; Tolley et al. 2001; Patel et al. 2004; Shen et al. 2005), commercial pumps do not operate at these pressures (most commercial HPLC pumps have a 5,000-p.s.i. limit). Therefore, significant resolution enhancements achieved through longer columns is limited for most researchers. With that said, several companies (i. e. Waters and JASCO) have recently introduced 15,000-p.s.i. HPLC pumps.

Equation (5) reveals that the pressure differential is proportional to the mobile phase viscosity (η). Thus, lowering of the mobile phase viscosity (η) by increasing the temperature can lower the operational pressure and allow the use of longer columns for resolution enhancement (Djordjevic et al. 1998, 1999, 2000). Selectivity is also affected by temperature and additional efficiency can be achieved by heating alone. However, one must ensure analyte thermal stability if elevated temperature separations are to be employed.

Equation (5) also shows that the pressure is a function of the column permeability (K^o). New monolithic columns offer greater permeability and lower pressures, thus allowing for the use of longer columns. The continuous bed stationary phases of these columns consist of porous polymeric materials generated from silica or organic materials such as acrylamide, styrene, acrylate, or methacrylate monomers which result in lower back-pressure than packed particles. The lower back-pressure allows for the use of longer columns and hence greater efficiencies. Several groups have reported on the use of up to 1 m capillary columns (Que and Novotny 2002; Legido-Quigley et al. 2003; Tolstikov et al. 2003; Tanaka et al. 2004) and this technology looks promising.

Plate number and efficiency are also related to particle size (d_p) and column length (L) as shown in Eq. (3). This equation shows that decreasing the particle size increases the theoretical plate number/efficiency (MacNair et al. 1997, 1999; Tolley et al. 2001; Shen et al. 2005). However, Eq. (5) shows again that pressure increases with smaller particle size. Fortunately, new commercial ultra-high pressure liquid chromatography pumps (UPLC) are now available from multiple manufacturers that allow the use of smaller particles in the range of 1–2 μm. These instruments offer substantial resolution enhancements with plate numbers on the order of several hundred thousand and peak capacities in excess of 400 (Wilson et al. 2005). In addition to increased resolution, UPLC also offers higher speed separations as the optimum flow velocity has a significantly broader range which allows for increased flow rates without significant loss of resolution (Wilson et al. 2005). Estimates of up to ninefold increases in flow rates without significant loss of resolution have been suggested (Wilson et al. 2005). It is important to note that ultra-high pressure separations result in increased frictional heating; however this can be reduced by down-scaling the chromatography dimensions with the heating being negligible in columns of less than 1 mm (MacNair et al. 1997).

4 Future Directions and Forward-Looking Thoughts

Although several of the above principles can be used to achieve enhanced chromatographic resolution, the resolution enhancements are still far from that which is needed for very complex metabolomics mixtures. To separate these mixtures, peak capacities of thousands to tens of thousands are necessary. Currently, only multidimensional chromatographic methods offer peak capacities of this magnitude (Mondello et al. 2002; Evans and Jorgenson 2004). Multidimensional chromatography utilizes combinations of two or more separation mechanisms with different selectivity, e. g. ion-exchange and reverse-phase or capillary electrophoresis and reverse-phase LC. These systems offer enhanced resolution due to the utilization of multiple columns with independent chemistries which expands the selectivity of the method. Recall that selectivity improvements can dramatically improve resolution. The maximum peak capacity of a multidimensional system is the product of the two or more individual separation dimensions. For example, a realistic system that has a peak capacity in the first dimension (n_y) of 150 and the peak capacity in the second dimension (n_z) of 50, then the total maximum peak capacity of the multidimensional system is $n_y \times n_z = 150 \times 50 = 7500$. If one considers that an individual metabolome consists of 15,000 metabolites, then one recognizes that this is a considerable increase in comprehensive coverage.

Multidimensional LC-LC separations have been capitalized upon in the area of proteomics and are often referred to as multidimensional protein identification technology (i. e. MUDPIT; Washburn et al. 2001; Wolters et al. 2001); however multidimensional separations have only recently been pursued for metabolomics using GC×GC/time-of-flight (TOF)-MS (Welthagen et al. 2005). Unfortunately, these complex separations will come with increased analysis times, but I believe they will be worth the additional temporal costs.

The above discussion focuses on homogenous multidimensional separations (i. e. LC×LC/MS or GC×GC/MS, but multidimensional LC×GC separations are possible. In fact, the combination of these technologies is commonly referred to as unified chromatography (Chester and Parcher 2001; Chester and Pinkston 2002; Wells et al. 2002, 2003; Luo et al. 2003) and often associated with supercritical fluid chromatography (Chester and Parcher 2001; Chester and Pinkston 2002; Mondello et al. 2002; Wells et al. 2002, 2003; Luo et al. 2003). Although this technology is conceptually exciting, it is still somewhat empirically limited. Another possible LC×GC approach would be to couple HPLC with ion mobility mass (IMS) TOF-MS spectrometry (Verbeck et al. 2002; Guevremont 2004; Liu et al. 2004; Shvartsburg et al. 2005). In this configuration, analytes are ionized as they elute from the HPLC and an electrostatic field propels the analyte ions through a gas field maintained at elevated, atmospheric, or sub ambient pressures. Ions of different size and geometric structure traverse the gas field at different rates dependent upon their charge and collisional cross section therefore allowing separation. The LC-IMS method has been demonstrated for proteomics (Lee et al. 2002; Matz et al. 2002; Liu et al.

2004) and more recently applied to metabolite analyses (Kapron et al. 2005). Extension of this concept to metabolomics will surely occur.

The above text discusses multidimensional chromatographic approaches in an on-line context. However, multidimensional approaches can also be pursued in an off-line, multiplexed, or parallel approach. For example, fractions can be collected off-line using a separate HPLC. The fractions can then be concentrated and reinjected onto an on-line LC/MS system. Alternately, fractions of the same samples could be injected onto a series of parallel systems using different methods (i. e. GC/MS, LC/MS, or various selective modes of each performed with different column selectivities). This is our current approach. For example, samples are fractionated and/or enriched and then the polar and lipophilic fractions are analyzed by GC/MS. In addition methanolic extracts are analyzed for phenolic/saponin content. An interesting concept would be to design a multiplexed system, with multiple chromatographic-mass spectrometry systems operating in an integrated manner. For example, a multiplexed chip system with each chip having a slightly different selectivity and independent mass analyzer could be designed to increase the comprehensive coverage. Such a system with on-line enrichment could also be used to address dynamic range limitations that currently exist for specific compound classes such as phytohormones.

If higher resolution chromatography is obtained, mass analyzers must also be employed with compatible scans speeds to record data for compounds eluting in very short temporal periods. It is expected that LC peak widths of 1–5 s will be routine in the very near future. For accurate quantification, it is commonly accepted that the sampling rate should be sufficient to capture 10 data points across the eluting peak with higher sampling rates being beneficial. Thus, sampling rates should be less than 0.1 s or greater than 10 Hz. This is achievable with current TOF-MS analyzers. It is worth mentioning that quadrupole based mass analyzers, including traps, can approach these speeds; however, TOF mass spectrometers equipped with delayed extraction and ion-reflectrons also offer improved mass accuracy over quadrupoles.

Improvements in the accuracy of the mass analyzer can further enhance metabolite differentiation, elemental composition determination, identification, and allow for the profiling of greater numbers of metabolites. Mass accuracy is directly related to the mass resolution or the ability of the mass analyzer to resolve compounds of different m/z values. Mass resolution is defined in Eq. (6) and is a function of mass (M) divided by the peak width (ΔM) which is most commonly defined at half-height:

$$R_m = \frac{M}{\Delta M} \tag{6}$$

Often, LC/MS is performed with ion-traps or quadrupole mass analyzers that yield mass accuracies in the range of 1.0–0.1 Da. Unfortunately, many metabolites have similar nominal masses which can not be differentiated at this level of mass accuracy. For example, the important natural products genistein and medicarpin have similar nominal masses of 270, but have different accurate masses of 270.2390 ($C_{15}H_{10}O_5$) and 270.2830 ($C_{16}H_{14}O_4$) respectively due to different chemical compositions. If one could measure their mass with sufficient accuracy, then one could differentiate these compounds in the mass domain even if they could not be physically separated in the chromatographic domain. This mass differentiation can be achieved at a mass resolution ($M/\Delta M$) greater than 6136. Compounds with closer accurate masses such as rutin ($C_{27}H_{30}O_{16}$ = 610.5180) and hesperidin ($C_{28}H_{34}O_{15}$ = 610.5620) would require a higher mass resolution of 13,864 for their differentiation. Mass resolutions on the order of 10,000 can be achieved with modern TOF-MS analyzers, and resolutions in excess of 100,000 with sub-part-per-million mass accuracies (i. e. less than 0.001 at m/z of 1,000 Da) are achievable with Fourier transform ion cyclotron mass spectrometry (FTMS). Newer technologies, such as Thermo Electron Corporation's Orbitraps are currently surfacing that also offer high-resolution solutions. Although high resolution accurate mass measurements have great advantages, this technology is still rather costly.

Interestingly, a significant argument can be made that accurate mass measurements significantly reduce the need for ultra-high resolution separations due to the enhanced separation in the mass domain. However if the chromatography step is omitted or compressed significantly, then ion suppression, competitive ionization, and other matrix affects become increasingly more problematic. I personally believe that both improved chromatographic resolution and accurate mass measurements offer the best solution and that the combination of these techniques will provide greater comprehension and confidence in our ability to profile the metabolome. Further, I also believe that the needed magnitude of enhancements in chromatographic resolution can only be achieved with multidimensional approaches.

References

Bino RJ, Hall RD, Fiehn O, Kopka J, Saito K, Draper J, Nikolau BJ, Mendes P, Roessner-Tunali U, Beale MH, Trethewey RN, Lange BM, Wurtele ES, Sumner LW (2004) Potential of metabolomics as a functional genomics tool. Trends Plant Sci 9:418–425

Birkemeyer C, Kolasa A, Kopka J (2003) Comprehensive chemical derivatization for gas chromatography-mass spectrometry-based multi-targeted profiling of the major phytohormones. J Chromatogr A 993:89–102

Broeckling CD, Huhman DV, Farag MA, Smith JT, May GD, Mendes P, Dixon RA, Sumner LW (2005) Metabolic profiling of Medicago truncatula cell cultures reveals the effects of biotic and abiotic elicitors on metabolism. J Exp Bot 56:323–336

Chester T, Parcher JF (2001) Blurring the Boundaries. Science 291:502–503

Chester T, Pinkston J (2002) Supercritical fluid and unified chromatography. Anal Chem 74:2801–2811

Dixon RA (2001) Phytochemistry in the genomics and post-genomics eras. Phytochemistry 57:145–148

Djordjevic N, Houdiere F, Fowler P (1998) High temperature and temperature programming in capillary HPLC. Biomed Chromatogr 12:153–154

Djordjevic N, Fitzpatrick F, Houdiere F, Lerch G, Rozing G (2000) High temperature and temperature programming in capillary electrochromatography. J Chromatogr A 887:245–252
Djordjevic NM, Fowler PWJ, Houdiere F (1999) High temperature and temperature programming in high-performance liquid chromatography: Instrumental considerations. J Microcolumn Separ 11:403–413
Evans C, Jorgenson J (2004) Multidimensional LC-LC and LC-CE for high-resolution separations of biological molecules. Anal Bioanal Chem 378:1952–1961
Fiehn O, Kopka J, Dormann P, Altmann T, Trethewey RN, Willmitzer L (2000) Metabolite profiling for plant fuctional genomics. Nat Biotechnol 18:1142–1161
Guevremont R (2004) High-field asymmetric waveform ion mobility spectrometry: a new tool for mass spectrometry. J Chromatogr A 1058:3–19
Gygi SP, Rochon Y, Franza BR, Aebersold R (1999) Correlation between protein and mRNA abundance in yeast. Mol Cell Biol 19:1720–1730
Hartman T, Kutchan TM, Strack D (2005) Evolution of metabolic diversity. Phytochemistry 66:1198–1199
Kapron J, Jemal M, Duncan G, Kolakowski B, Purves R (2005) Removal of metabolite interference during liquid chromatography/tandem mass spectrometry using high-field asymmetric waveform ion mobility spectrometry. Rapid Commun Mass Spectrom 19:1979–1983
Lee Y, Hoaglund-Hyzera C, Srebalus Barnes C, Hilderbrand A, Valentine S, Clemmer D (2002) Development of high-throughput liquid chromatography injected ion mobility quadrupole time-of-flight techniques for analysis of complex peptide mixtures. J Chromatogr B Anal Technol Biomed Life Sci 782:343–351
Legido-Quigley C, Marlin N, Melin V, Manz A, Smith N (2003) Advances in capillary electrochromatography and micro-high performance liquid chromatography monolithic columns for separation science. Electrophoresis 24:917–944
Liu X, Plasencia M, Ragg S, Valentine S, Clemmer D (2004) Development of high throughput dispersive LC-ion mobility-TOFMS techniques for analysing the human plasma proteome. Brief Funct Genomic Proteomic 3:177–186
Luo Z, Xiong Y, Parcher J (2003) Chromatography with dynamically created liquid "stationary" phases: methanol and carbon dioxide. Anal Chem 75:3557–3562
MacNair J, Lewis K, Jorgenson J (1997) Ultrahigh-pressure reversed-phase liquid chromatography in packed capillary columns. Anal Chem 69:983–989
MacNair J, Patel K, Jorgenson J (1999) Ultrahigh-pressure reversed-phase capillary liquid chromatography: isocratic and gradient elution using columns packed with 1.0-micron particles. Anal Chem 71:700–708
Matz L, Dion H, Hill H (2002) Evaluation of capillary liquid chromatography-electrospray ionization ion mobility spectrometry with mass spectrometry detection. J Chromatogr A 946:59–68
Mondello L, Lewis AC, Bartle KD (2002) Multidimensional Chromatography. Wiley, Chichester, UK
Oliver DJ, Nikolau B, Wurtele ES (2002) Functional genomics: high-throughput mRNA, protein, and metabolite analyses. Metab Eng 4:98–106
Patel K, Jerkovich A, Link J, Jorgenson J (2004) In-depth characterization of slurry packed capillary columns with 1.0-microm nonporous particles using reversed-phase isocratic ultrahigh-pressure liquid chromatography. Anal Chem 76:5777–5786
Que A, Novotny M (2002) Separation of neutral saccharide mixtures with capillary electrochromatography using hydrophilic monolithic columns. Anal Chem 74:5184–5191
Roessner U, Wagner C, Kopka J, Trethewey RN, Willmitzer L (2000) Simultaneous analysis of metabolites in potato tuber by gas chromatography-mass spectrometry. Plant J 23:131–142
Roessner U, Luedemann A, Brust D, Fiehn O, Linke T, Willmitzer L, Fernie AR (2001) Metabolic profiling allows comprehensive phenotyping of genetically or environmentally modified plant systems. Plant Cell 13:11–29
Schauer N, Steinhauser D, Strelkov S, Schomburg D, Allison G, Moritz T, Lundgren K, Roessner-Tunali U, Forbes M, Willmitzer L, Fernie A, Kopka J (2005) GC-MS libraries for the rapid identification of metabolites in complex biological samples. FEBS Lett 579:1332–1337

Shen Y, Zhang R, Moore R, Kim J, Metz T, Hixson K, Zhao R, Livesay E, Udseth H, Smith R (2005) Automated 20 kpsi RPLC-MS and MS/MS with chromatographic peak capacities of 1000–1500 and capabilities in proteomics and metabolomics. Anal Chem 77:3090–3100

Shvartsburg A, Tang K, Smith R (2005) Optimization of the design and operation of FAIMS analyzers. J Am Soc Mass Spectrom 16:2–12

Somerville C, Dangl J (2000) Plant biology in 2010. Science 290:2077–2078

Somerville C, Somerville S (1999) Plant functional genomics. Science 285:380–383

Sumner LW, Mendes P, Dixon RA (2003) Plant metabolomics: large-scale phytochemistry in the functional genomics era. Phytochemistry 62:817–836

Takats Z, Wiseman JM, Gologan B, Cooks RG (2004) Mass spectrometry sampling under ambient conditions with desorption electrospray ionization. Science 306:471–473

Tanaka N, Kimura H, Tokuda D, Hosoya K, Ikegami T, Ishizuka N, Minakuchi H, Nakanishi K, Shintani Y, Furuno M, Cabrera K (2004) Simple and comprehensive two-dimensional reversed-phase HPLC using monolithic silica columns. Anal Chem 76:1273–1281

Tolley L, Jorgenson J, Moseley M (2001) Very high pressure gradient LC/MS/MS. Anal Chem 73:2985–2991

Tolstikov VV, Lommen A, Nakanishi K, Tanaka N, Fiehn O (2003) Monolitichic silica-based capillary reversed-phase liquid chromatography/electrospray mass spectrometry for plant metabolomics. Anal Chem 75:6737–6740

Trethewey RN (2001) Gene discovery via metabolic profiling. Curr Opin Biotechnol 12:135–138

Trethewey RN, Krotzky AJ, Willmitzer L (1999) Metabolic profiling: a rossetta stone for genomics? Curr Opin Plant Biol 2:83–85

Verbeck G, Ruotolo B, Sawyer H, Gillig K, Russell D (2002) A fundamental introduction to ion mobility mass spectrometry applied to the analysis of biomolecules. J Biomol Tech 13:56–61

Wagner C, Sefkow M, Kopka J (2003) Construction and application of a mass spectral and retention time index database generated from plant GC/EI-TOF-MS metabolite profiles. Phytochemistry 62:887–900

Washburn M, Wolters D, Yates J (2001) Large-scale analysis of the yeast proteome by multidimensional protein identification technology. Nat Biotechnol 19:242–247

Weckwerth W (2003) Metabolomics in systems biology. Annu Rev Plant Biol 54:669–689

Wells P, Zhou S, Parcher J (2002) Gas-liquid chromatography with a volatile "stationary" liquid phase. Anal Chem 74:2103–2111

Wells P, Zhou S, Parcher J (2003) Unified chromatography with CO2-based binary mobile phases. Anal Chem 75:18A–24A

Welthagen W, Shellie RA, Spranger J, Ristow M, Zimmermannn R, Fiehn O (2005) Comprehensive two-dimensional gas chromatography-time-of-flight mass spectrometry (GC×GC-TOF) for high resolution metabolomics: biomarker discovery on spleen tissue extracts of obese NZO compared to lean C57BL/6 mice. Metabolomics 1:65–73

Wilson I, Nicholson J, Castro-Perez J, Granger J, Johnson K, Smith B, Plumb R (2005) High resolution "ultra performance" liquid chromatography coupled to oa-TOF mass spectrometry as a tool for differential metabolic pathway profiling in functional genomic studies. J Proteome Res 4:591–598

Wolters D, Washburn M, Yates J (2001) An automated multidimensional protein identification technology for shotgun proteomics. Anal Chem 73:5683–5690

I.3 Plant Metabolomics Strategies Based upon Quadrupole Time of Flight Mass Spectrometry (QTOF-MS)

H.A. Verhoeven[1,2], C.H. Ric de Vos[1,2], R.J. Bino[1,2], and R.D. Hall[1,2]

1 Introduction

The growing interest in the use of metabolomics technologies in plant research has come about both due to the broad value of such approaches in almost every field of plant science and also through the improvements in instrumentation and bioinformatics tools which has been realised in recent years. A comprehensive overview of all the various technologies available is beyond the scope of this chapter but the reader is referred to other chapters in this volume or to a number of recent reviews for information on the different approaches (Fiehn 2001, 2002; Sumner et al. 2003; Goodacre et al. 2004). However, MS-based strategies, and in particular in combination with GC or LC separation technology, are proving most popular as these combine very high analytical precision with an equally high detection sensitivity. This enables reliable measurements to be made down to the femtomolar range (Fernie 2003). Furthermore, recent advances in electronics and computing have given rise to the development of yet a new generation of mass spectrometers to supplement the traditional magnetic sector and scanning quadrupole instruments that have been around for several decades now. In this new generation, instruments based on ion traps and time-of-flight (generally referred as TOF) are the most prominent. In particular, the TOF instruments have become popular due to their relatively simple construction and their capacity to be combined with a number of other technologies to enable multi-dimensional analysis. This has resulted in an unprecedented expansion of our metabolomic capabilities. For example, the fast spectral acquisition capacity of TOF instruments has resulted in approximately 1000 components being detected in leaf extracts and an analytical capacity of 1000 samples per month has been achieved (Weckwerth et al. 2001). Such sample numbers and breadth of metabolite detection represent the arrival of true metabolomics research in the true sense of the word. Since that time, our capacity for metabolomic analyses has continued to improve. The high mass accuracy, high resolution, good dynamic range and the large diversity of detectible masses possible with TOF instruments, in association with their intrinsic high sensitivity, are therefore the main reasons behind the many applications, first in the field of proteomics, and now also in metabolomics.

[1] Plant Research International, P.O. Box 16, 6700 AA Wageningen, The Netherlands
[2] Centre for BioSystems Genomics, P.O. Box 98, 6700 AB Wageningen, The Netherlands, e-mail: robert.hall@wur.nl

Biotechnology in Agriculture and Forestry, Vol. 57
Plant Metabolomics (ed. by K. Saito, R.A. Dixon, and L. Willmitzer)

2 The Technology

Time of flight was already introduced in the early 1960s but was quickly replaced by other approaches. This was due to the lack of sufficiently fast electronics needed to process data on a nanosecond scale. Thirty years later in the early 1990s, the development of high megahertz and even gigahertz digital circuits led to the dramatic increase in the application of TOF technology. This, combined with new developments in the area of sample introduction and ionisation of (macro)molecules, has subsequently led to many new applications of (TOF-based) mass spectrometry in the fields of biology and pharmacy.

A TOF instrument serves as the main mass analyser, and its principle is based on ions with different mass/charge ratios having different flight times in a field-free drift zone once they have been accelerated by a very short electric pulse from the electrodes of an accelerator: lighter ions travel faster through the measurement chamber than the heavier ones. A thorough discussion on the physical principles can be found in, for example, Guilhaus (1995). In most TOF instrument designs, ions are detected using Micro Channel Plates (MCP), which, on capturing the ions, generate a cascade of electrons to amplify the signal so that it can be detected by the associated electronics (see Fig. 1 for a schematic representation of the various parts). Several ion recorders have been used with the various designs of hybrid TOF mass spectrometer. The two most widely used are the time-to-digital converter (TDC) and the transient recorder or analogue to digital converter (ADC) (Chernushevich et al. 2001). The type of detector affects both the dynamic range of the signals that can be measured and also the mass accuracy. In a TDC, every individual ion generates a pulse. This pulse is shaped into a digital signal, of which the rising flank is used for timing. The time passed since the start of the ion accelerating pulse and its arrival at the MCP is stored in the memory. This system is very accurate over the entire mass range and is optimally suited for the accurate timing of low ion counts. It is, however, less suitable or even unsuitable for the detection of ions arriving simultaneously at the MCP since these will be recorded as being single events and this will thus lead to an underestimation of the signal. TDCs also suffer from an additional limitation concerning the detection of ions. During the time required to process one pulse, the detector is 'blind' to new incoming pulses. This so-called dead time, not only leads to a further underestimation of the signal, but also it causes a shift in the observed m/z value towards lower values. This can lead to serious deviations from the true accurate mass at high to very high signal intensities. These problems occur to a lesser extent in instruments equipped with an ADC, since these machines can sample the analog output of the MCP at very high frequencies, thus providing multiple data points per observed m/z value. In this way, multiple ions arriving at the same time will lead to a linear increase in peak area. In some designs, TDC and ADC are both used to combine the high mass accuracy of the TDC at low ion rates, with the high dynamic range and accurate m/z value measurements of an ADC at high ion rates.

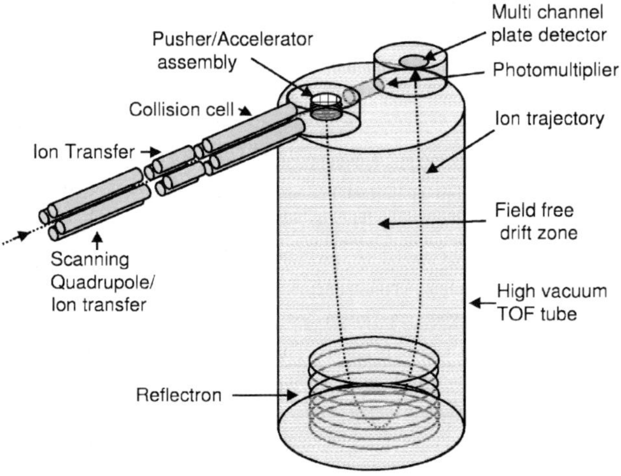

Fig. 1. Schematic diagram showing the main components of a typical quadrupole/time of flight configuration. Ions enter the instrument on the left, and pass through the first quadrupole. This can be operated either in ion transfer mode, which allows all ions to pass, or in selective mode, which is used for precursor scanning and alignment of the different quadrupoles. The ion transfer is a special quadrupole, intended to separate the operating pressures in the different compartments of the quadruple section. In the collision cell, a collision gas can be present in order to induce fragmentation of the incoming ions. When no gas is present molecular ions will be detected. Subsequently, molecular ions and/or charged fragments enter the TOF tube, where they are collected and pushed into the drift zone. During this transition, ions are accelerated in an electric field in the accelerator assembly, consisting of several ion lenses, which determines their kinetic energy. The ions now all follow a trajectory towards the reflectron, consisting of a pile of cylindrical ion lenses at different potentials, which causes them to be repelled towards the detector, the multi channel plate. Here the ions strike the surface of the detector, which finally converts the arrival of every single ion into a measurable electric current. Additional electronics is required to process the electrical signals and the timing between pushing the ions into the drift zone and their arrival at the detector

Regardless of the high mass accuracy, resolution and sensitivity, the application of TOF instruments in structure elucidation is quite limited due to the absence of filtering and scanning capabilities. Consequently, hybrid instruments have since been designed to cope with these shortcomings. These machines include the addition of ion trap(s), quadrupoles or combinations thereof, to the basic TOF analyser. One key example is the now increasingly well-known and widely used QTOF system. These instruments rely on the combination of two or more quadrupoles with a TOF analyser. The first quadrupole (Q1) serves as a mass filter or ion tunnel, depending on the operational mode, with the second quadrupole (Q2) serving as the collision cell for the fragmentation of the ions which have passed through Q1. This fragmentation is achieved using an electric field to accelerate the ions, in combination with a collision gas such as nitrogen or argon. Fragmentation can be controlled by varying the (very low) pressure of the collision gas and/or by varying the collision energy through altering the acceleration voltage of the cell. Collisions with the gas molecules also result in a cooling of the ions, which incurs that their kinetic energy is transferred. This results in a more homogeneous energy distribution of the individual ions, which in turn improves the mass accuracy capacity of the instrument. The ions and/or ion fragments are subsequently collected in the accelerator part of the TOF instrument where a very short pulse is applied to the electrodes of the chamber to eject the ions. In the case of orthogonal ejection, the differences in kinetic energy in the z-axis will be less than in case of forward ejection. The differences in kinetic energy are also further reduced in the reflectron lens, which repels the ions towards the detector. Here, ions with higher energy will travel further than lower energy ions, thus reducing the difference. A number of variations on this basic design have been created. These include, for example, the modification of the second quadrupole into a linear ion trap with axial ejection through the addition of a number of extra ion lenses (Hager 2002). As a result, new possibilities are created, such as the ability to store specific ions, which can be selectively ejected for complex MS/MS or MSn analyses (Hager 2002). The high mass accuracy can be further improved by using an internal (reference) standard that is sampled at regular intervals throughout the entire analysis period. This reference is then used to correct the instrument calibration on-the-fly (lock mass correction). Such a capacity for continuous (re)calibration is particularly useful, if not essential, in the case of long series of chromatographic runs where excellent, long-term stability of the mass accuracy can be continuously achieved down to ±5 ppm. This is significant as mass accuracy at or below this level allows us to predict the chemical composition of a given ion by using the small known differences in atomic masses of the various atomic elements. In this way, a first prediction can be made about the nature and identity of the molecular component. Combined with other data (retention time, N rule, stable isotope distribution of ^{13}C etc.) this can then enable the list of possible molecular identities to be reduced even further and thus come closer to translating MS output into named metabolites.

Combining the results obtained from several biological samples into a single comparative analysis is an arduous task that requires the precise alignment and matching of peaks representing the same compound over all chromatograms. Due to its relatively robust chromatography and compound separation efficiency, GC-(TOF)-MS of derivatized extracts is at present generally preferred over LC-MS in metabolomic studies (Fiehn et al. 2000; Roessner et al. 2001a,b; Fernie et al. 2004). Nevertheless, GC-MS is less suitable for semi-polar compounds among which are key classes of plant (secondary) metabolites including flavonoids, (glyco-)alkaloids, glucosinolates and saponins. Recent advances in techniques for improving resolution in LC by using capillary electrophoresis (Soga et al. 2002), hydrophilic interaction columns (Tolstikov and Fiehn 2002) and monolithic columns (Tolstikov et al. 2003) demonstrate the high potential which TOF technology has for LC-MS to complement GC-MS in unravelling metabolic profiles.

3 Data Analysis

Data analysis is perhaps the most crucial step in any metabolomics strategy and the importance of bioinformatics tools should not be underestimated. In a standard (ideal) approach, a whole range of standards would be used to assist in identifying through simple linkage, which peaks in an MS output represent which metabolites. However, as the vast majority of metabolites present in complex plant extracts are as yet unknown and are not commercially available, as is especially true for the secondary plant metabolites, this approach is unfeasible at present for a true untargeted metabolomics approach. Another strategy is therefore required which enables the automated and essentially blind direct comparison of large numbers of spectra. Since most datasets are very complicated, dedicated metabolomics software is needed for this purpose. Some of this software is already available but more still needs to be developed and this represents a major task for the next five years.

Data manipulation is essential for reliable metabolomics analyses and special attention has to be paid to aspects such as baseline correction and noise elimination. In addition, in the case of LC-MS, particular attention also needs to be given to reliable correction of local drifts in retention time and accurate mass. Different compositions of eluant can cause significant variation in baseline especially when using steep LC gradients. For the successful correction of such baseline fluctuations, the chromatogram has to contain a region without strong peaks. Digital filtering will enable the elimination of excess noise which would otherwise lead to the generation of erroneous (false) peaks. Some recent software packages are able to deal with a number of these problems in TOF data analysis. Another key element is the need to correct for retention time fluctuations. Unlike capillary gas chromatography which is generally very stable, liquid chromatography often suffers from relatively large, non-linear (localised) fluctuations in retention time. This can be due to small differences in pH, temperature, or the co-elution of components which interact differently with the stationary phase. Consequently, this problem prevents a simple direct comparison of different samples. A number of algorithms have been designed to correct for this phenomenon. One such approach, based on photodiode array type data, uses correlation optimised warping of the chromatograms to achieve alignment of shifted peaks in the chromatograms (Nielsen et al. 1998). For MS data, MetAlign™ software, in contrast, uses specific mass peaks with strong local maxima throughout the chromatogram as 'landmark peaks' with which to correct for chromatographic shifts over the entire series of analyses (Vorst et al. 2005). After correction, unbiased, direct spectral comparisons, based on mass peak intensities are possible and contrasting mass signals can be reliably identified and extracted. Differential chromatograms are produced from which all unchanging peaks have been removed to reveal the true extent of the differences between two (groups of) samples in one or both directions. This dedicated software can automatically handle hundreds of full scan MS datasets obtained by either LC or GC, and is independent of type of mass spectrometer. Another package, Markerlynx™, which is dedicated to Waters instruments, exploits the high mass accuracy and resolution of the (Q)TOF technology. Here, the distribution of specific ions in a predefined mass window, usually in a 20–50 ppm range, over the chromatogram is analysed and retention shifts are corrected within a predetermined retention time window. In this way, metabolites can be compared over many samples using high mass resolution. Both approaches have their own advantages and limitations. In our lab we more or less routinely use metAlign™ to process LC-QTOF and GC-TOF data. However, in cases where there are insufficient landmark peaks, metAlign™ is unable to perform a thorough alignment. When spectra are noisy or highly complex, the Markerlynx™ approach will likely give misalignments.

How to proceed further with the processing of corrected chromatograms is dependent upon the type of metabolomics analysis required. Within metAlign™ there is a tool to extract user-defined significant differences between two groups of samples based on the Student t-test. In many cases, where large sample numbers are being compared, multivariate analysis will be desirable. For this purpose, software originally developed for the analysis of microarray datasets can be advantageously applied, since metabolomic data share a number of problems similar to those which were encountered with microarray data. We use, for example, the GeneMaths software package (Vorst et al. 2005), which enables rapid statistical data analysis and provides clear graphic outputs of the results in the form of histograms and principle components plots. Other software packages should be equally valuable.

4 Application of QTOF MS-based Plant Metabolomics Analyses

4.1 Rapid Sample Profiling by Direct Flow Injection Analysis (DFI)

In plant metabolomics there is often an initial desire for a rapid pre-screening of the samples. This is especially the case when dealing with large sample numbers where only a limited number of individuals might be expected to be different. One can think here for example, of natural populations, potential mutants or the progeny from a breeding cross (Bino et al. 2005; Hall et al. 2005). Direct Flow Injection can effectively be used to get a rapid, overall impression of the composition of a biological extract. It is an unbiased analysis, and seeks to cover as many metabolites as possible in a single short run. The only selective property is the type of ionisation used, i. e. positive vs negative, ESI (ElectroSpray ionisation) vs APCI (Atmospheric Pressure Chemical Ionisation). The advantage of this approach is time-saving. When using chromatographic separation to prevent excessive component interaction, run times of 30–90 min are usually required (but see as an exception; Jander et al. 2004). In the case of DFI, run times of only 30 s to a few minutes (Goodacre et al. 2003) may be required. For DFI, a few microlitres of extract is introduced by the autosampler directly into the ion source and all ions with the corresponding charge are then analysed by the MS. In this case, TOF instruments have a clear advantage over scanning quadrupole instruments because their significantly higher resolution allows for the simultaneous detection of many ion species and, because no scanning is required, every individual ion can theoretically be captured. This inevitably results in a very rich mass spectrum which is further complicated by the many interactions which can occur between the different components of the sample during ionization. Furthermore, unstable ions can cause additional extensive ion vapour phase interactions. For these reasons, there was initially considerable scepticism of the potential value of DFI approaches for reliable metabolic analysis and these phenomena are extensively discussed elsewhere (Kebarle 2000; King et al. 2000). However, recent publications have shown that the mass spectrum data obtained in this way is actually highly reproducible and can effectively be used for a fast screening of complex extracts (Aharoni et al. 2002; Goodacre et al. 2002, 2003; Castrillo et al. 2003; Verhoeven et al., in preparation).

Data processing is in many cases the bottleneck for the successful deployment of this technology, and many applications rely on a dedicated approach to data processing. This was clearly demonstrated for example, by the MS analysis of unfractionated plant extracts of *Pharbitis* leaf sap (Goodacre et al. 2003). Correct data processing of the complex mass spectra was found crucial for reliable discrimination between the different physiological treatments used. Experiments performed in our laboratory resulted in similar conclusions. Five commercially available extracts from *Salix* were analysed using DFI in a QTOF MS in positive mode. A single total ion count (TIC) injection peak was observed, and all the masses obtained were combined into a single mass spectrum per sample. The aligned spectra were processed for noise elimination, baseline correction and then centroided to obtain the accurate masses of each m/z peak. These were then aligned in the m/z dimension using exact masses of known metabolites to correct for small fluctuations in exact mass due unavoidable minor (thermal) drift in the TOF tube. Intensities of the m/z peaks were log transformed, and exported to GeneMaths™ for multivariate analysis. Principle Components Analysis (PCA) revealed first (Fig. 2a) that the sample replicates (Samples 1 and 2) cluster close together reflecting the high reproducibility of the extraction and mass profiling techniques. Sample 3 is also clearly similar in overall composition to Sample 5, whereas Sample 4 is clearly distinct from all others. Sample 4 was found to have come from a different supplier. Differences in sample composition were readily detected by selecting the m/z values which were responsible for the separation of the samples in the PCA (Fig. 2b). This example indicates the usefulness of rapid screening for quality control of complex extracts without the need for more dedicated but time-consuming LC separation. In a similar manner Goodacre also used a rapid DFI approach to compare olive oil samples and to test for adulteration (Goodacre et al. 2002).

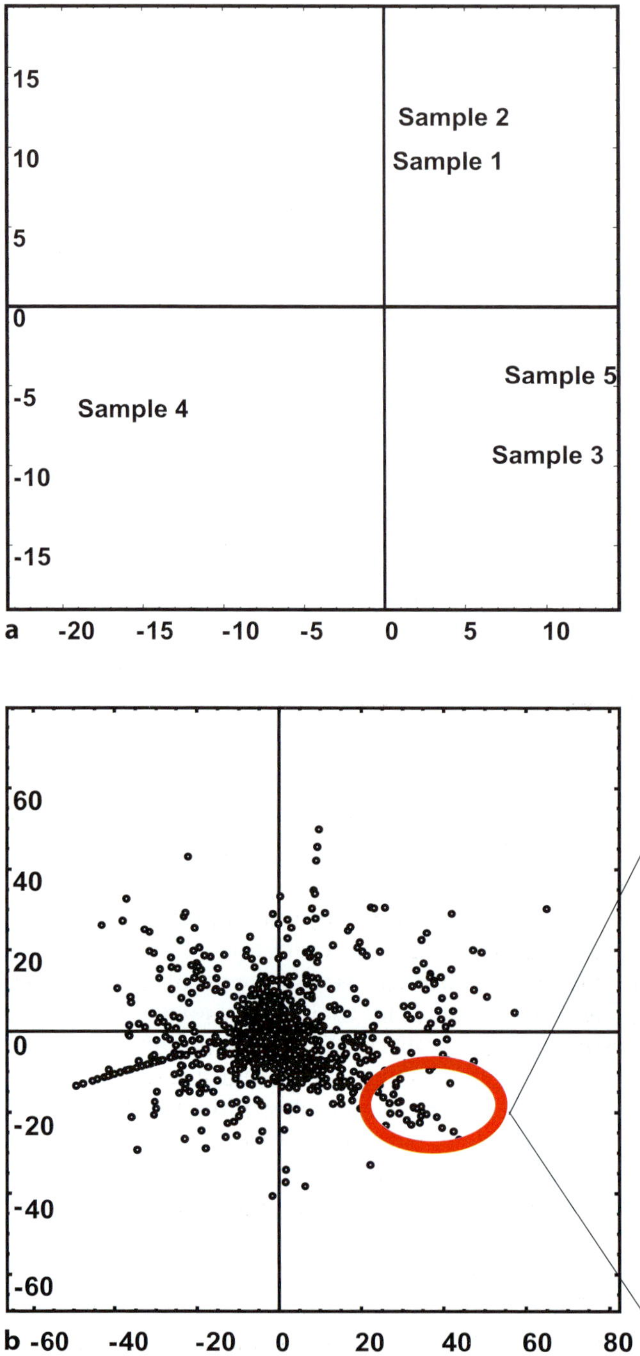

Sample 1	Sample 2	Sample 3	Sample 4	Sample 5	m/z value
0.00	0.00	0.92	0.00	1.43	137.15
0.00	0.00	2.03	0.00	1.84	147.06
0.00	0.00	2.20	0.00	2.15	162.13
0.00	0.00	1.73	0.00	1.58	177.06
0.00	0.00	1.63	0.00	2.20	183.02
0.00	0.00	1.71	0.00	0.00	195.13
0.00	1.20	1.78	0.00	1.65	210.04
1.21	0.00	2.03	0.00	2.02	225.05
0.00	0.00	2.52	0.00	0.00	233.11
0.00	0.00	1.50	0.00	0.00	239.16
0.00	0.00	2.11	0.00	0.00	261.12
1.72	1.97	3.08	2.45	3.61	267.1
0.00	0.00	1.72	0.00	1.32	267.19
0.00	0.00	1.67	0.00	1.51	277.23
0.00	0.00	1.56	0.00	0.00	281.18
0.00	0.00	1.57	0.00	0.00	282.14
0.00	0.00	0.69	0.18	1.71	284.57
0.00	0.00	1.72	0.00	1.75	295.24
0.00	0.00	1.73	0.00	0.87	373.24
0.00	0.00	1.09	0.00	1.60	419.28
2.77	2.70	3.96	2.55	3.80	463.11
0.00	0.00	1.57	0.00	0.87	463.26
2.11	2.05	3.30	2.15	3.14	464.11
0.36	0.68	1.34	0.76	2.39	488.64
0.00	0.44	1.65	0.00	0.49	496.27
0.00	0.00	1.57	0.00	0.00	518.33
0.00	0.00	2.32	0.00	0.70	520.35
0.00	0.00	1.83	0.00	0.00	521.36
1.80	1.73	3.21	1.93	4.12	567.12
1.37	1.42	2.70	1.81	3.58	568.13
0.00	0.00	1.21	0.21	2.04	700.71
0.00	1.00	1.59	0.00	1.86	731.07
0.91	0.88	2.20	0.00	1.59	850.14
1.31	1.33	3.10	1.51	2.57	887.25
0.82	0.88	2.72	1.31	2.24	888.26
1.10	1.11	2.47	0.00	2.04	889.14
0.95	0.82	2.01	0.00	1.69	890.12
0.00	0.00	1.96	0.00	0.00	953.33
0.00	0.00	2.11	0.00	0.00	957.3
0.00	0.00	1.81	0.00	0.00	958.31
0.00	0.00	0.94	0.00	1.80	965.32
0.00	0.00	1.93	0.00	2.34	975.31
0.00	0.00	2.63	0.00	0.00	991.29
0.49	0.80	2.38	0.00	0.00	992.3
0.00	0.00	1.48	0.00	1.90	995.31

◀ **Fig. 2. a** PCA plot of the entire set of detected mass peaks of 5 *Salix* samples. Samples 1 and 2 were experimental replicates taken from plant extracts of the same origin, but with different batch numbers. Samples 3, 4 and 5 were samples of unknown, but different origin. This figure shows that experimental variation (Samples 1 and 2) is low, Samples 3 and 5 are highly similar with respect to their overall composition while Sample 4 is distinctly different from the rest being placed on the other side of the PCA plot. **b** Detailed PCA of all mass peaks in the Samples 1 to 5. The area responsible for the grouping and positioning of Samples 3 and 5 in the *bottom right quadrant* is highlighted, and the corresponding mass peaks are shown as their logarithmic ratio on the right together with the m/z values of each. *Light grey*: low abundant mass peaks, *dark grey*: highly abundant mass peaks

4.2 QTOF MS Coupled to HPLC

As outlined above, direct infusion methods for a (Q)TOF-MS approach are relatively fast and simple approaches for obtaining metabolic composition fingerprints of multiple samples which can be used to get a preliminary estimation of the extent of similarities and differences between complex extracts. However, ion suppression phenomena may result in decreased detection sensitivity of some compounds, especially of those which ionize relatively poorly (discussed in Kebarle 2000). Moreover, the unavoidable consequences of direct infusion such as matrix-dependent ion suppression, adduct formation and unintended in-source fragmentation, may severely hamper the further detailed interpretation of the origins of the differential mass signals detected. This will thus limit possibilities for the subsequent metabolite identification involved. In addition, with DFI analyses it is, per definition, impossible to discriminate between molecular isomers or between a quasi-molecular ion and an ion having identical mass but which resulted from unintended in-source fragmentation. When such problems arise, or when a more detailed analysis of interesting samples (preselected, e. g. using a DFI approach) is required, LC separation can be used to reduce or solve some of these problems.

Separation of metabolites in complex extracts by liquid chromatography, prior to MS analysis, takes more time but nevertheless has a number of clear advantages. Sensitivity of detection for most compounds will be increased, the formation of adducts at the ionization source will be reduced and the detection of isomeric compounds will be improved. Isomer discrimination is especially important in plant metabolomics as plants are well-known to contain many (secondary) metabolites that may have identical accurate mass but different molecular structures. This is especially true for the large group of flavonoids, within which many compounds have the same elemental composition (and consequently the same accurate mass) whereas the chemical structures are quite distinct, e. g. kaempferol and scutellarein both of which have a neutral accurate mass of 286.04721. Furthermore, when using chromatographic separation, it is also possible to collect additional valuable structural information by applying, e. g. on-line tandem MS and/or by making use of other molecular characteristics such as UV-Vis absorbance and fluorescence which can be detected on-line prior to the molecules entering the MS. It is this combination of technologies which has made QTOF-based MS analysis a popular choice.

4.2.1 HPLC-PDA-ESI-QTOF-MS

The key to successful, full scale metabolomics analysis is the establishment of a technology platform which generates the maximum amount of reliable information in a single analytical run. For example, the LC-TOF MS based metabolomics system used in our laboratory incorporates a Waters Alliance 2795 HT autoinjector and HPLC pump system fitted with a column oven, a Waters 2996 photodiode array detector (PDA to give absorbance spectra in the 190–700 nm range) and a QTOF Ultima API mass spectrometer with MS/MS capability. In this system, four sets of data are therefore obtained simultaneously: UV/vis spectra, retention time, accurate mass and, when applied, MS/MS fragment information. A lock mass™ spray module is routinely connected to the ESI source in order to correct, on-the-fly, for any small measurement deviations from the exact mass (e. g. Wolff et al. 2001). We routinely use the synthetic peptide leucine enkephalin, which is continuously supplied by a separate low-flow HPLC pump as a reference lock mass in both positive and negative ESI measurement modes. By making combined use of accurate mass, MS/MS fragmentation information and absorbance characteristics, the number of possible elemental compositions and isomers can be narrowed down and essential structural information about a specific metabolite can be derived. For instance, most plant extracts contain multiple (poly)phenolic compounds, among which many isomers exist. Upon MS/MS fragmentation, many isomeric forms can already be distinguished, e. g. quercetin-rhamnoside (m/z 449.1079 in ESI positive mode) provides a fragment of 303.0505 while kaempferol-glucoside (also m/z 449.1079) provides a positively-charged fragment of 287.0556. However, with a QTOF, fragmentation experiments are not always conclusive. For instance, the glycosylated flavonoids kaempferol-3-*O*-glucoside and cyanidin-3-*O*-glucoside have identical mass and elemental composition of $C_{21}H_{20}O_{11}$, and show more or less similar MS/MS fragmentation patterns in ESI-positive mode with the loss of the glucosidic group leaving $C_{15}H_{10}O_6$ as the major fragment. However, in contrast, their UV-Vis absorbance characteristics are markedly different, with only the red-coloured cyanidin-glycoside having significant absorbance at wavelengths between 500 and 520 nm. This additional PDA information is therefore key to rapid isomer discrimination in this case.

A typical chromatogram obtained by reversed phase HPLC separation of a crude plant extract and subsequent on-line detection of eluting compounds by both PDA and QTOF MS is shown in Fig. 3. The observed mass of the metabolite eluting at retention time 23.48 min was 611.1596. Taking into account the uneven mass (indicating an even number of nitrogen atoms) and the isotopic distribution (indicating the absence of sulphur atoms), about 36 different elemental compositions are possible at 5 ppm accuracy (9 at 1 ppm accuracy)

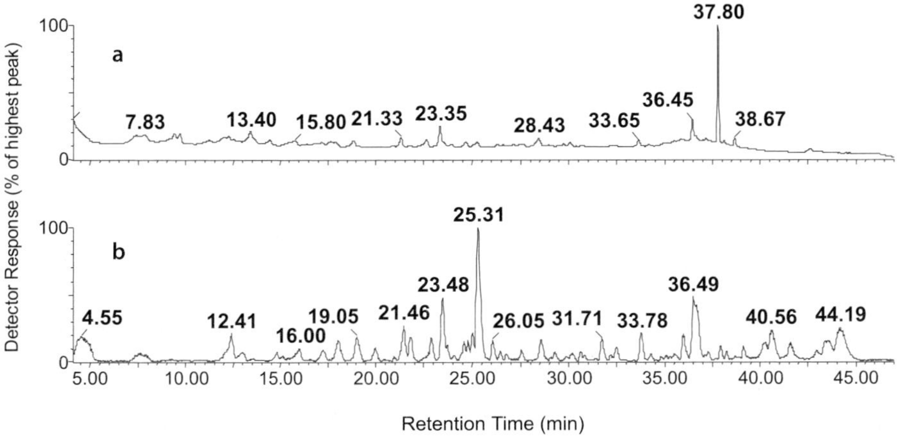

Fig. 3. Typical LC-PDA-QTOF MS chromatograms (base peak intensities) obtained by injection of 5 µl of an aqueous-methanol extract of tomato peel: **a** photodiode array signal (240–600 nm) **b** QTOF-ESI$^+$-MS signal (m/z 100–1500). Indicated are retention times of the most intense peaks

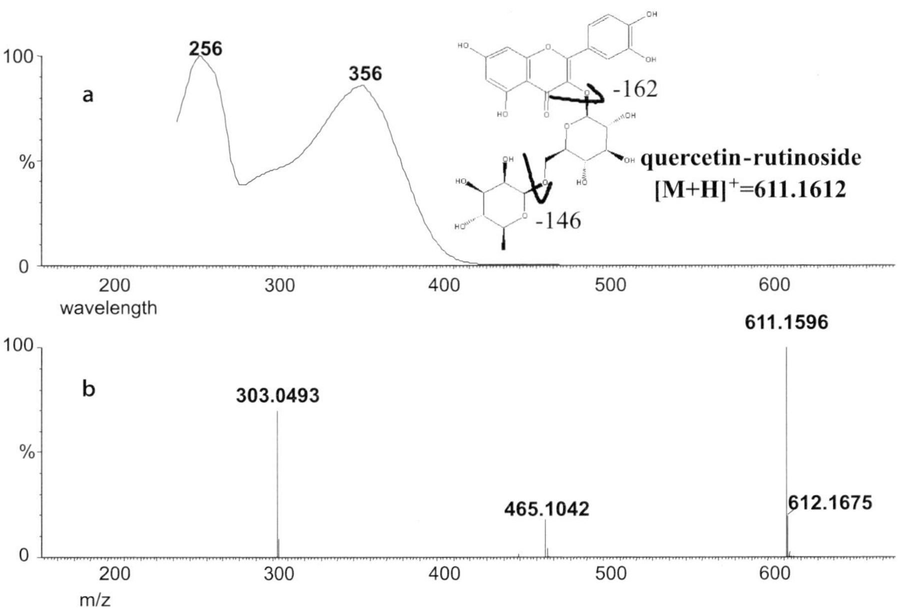

Fig. 4. **a** Absorbance spectrum of chromatographic peak at retention time 23.48 min. **b** QTOF-ESI-MS/MS spectrum of chromatographic peak at retention time 23.48 min. Observed accurate mass of parent ion $[M+H]^+$ = 611.1596 corresponds to an elemental composition of $C_{27}H_{31}O_{21}$ (−2.6 ppm) and its fragments obtained correspond to $C_{21}H_{21}O_{16}$ (+1.9 ppm) and $C_{15}H_{11}O_7$ (−3.9 ppm)

with parameter settings of $C \leq 75, H \leq 100, O \leq 75, N \leq 10$ and $P \leq 4$. Subsequent on-line LC-MS/MS fragmentation experiments (Fig. 4b) showed neutral losses of 146 and 162, and accurate mass fragments ($[M+H]^+$) of 465.1042 and 303.0493. The accurate mass and MS/MS fragmentation pattern correspond to a metabolite having an elemental composition of $C_{27}H_{30}O_{21}$, e. g. a diglycosylated anthraquinone, an (iso)flavonoid or a benzoyl-benzoic acid. The PDA spectrum of the same chromatographic peak (Fig. 4a) showed two absorbance maxima at around 255 and 360 nm, indicative of a flavonol-type flavonoid with at least two hydroxyl groups on the B ring (Markham 1989). The combination of accurate mass, MS/MS fragments and UV/vis absorbance spectrum indicates that the most likely candidate is quercetin-3-*O*-rutinoside. This is also supported by the knowledge that this flavonoid has been reported to be the major flavonol in tomato fruits (Muir et al. 2001; Le Gall et al. 2003). Subsequent comparison with an authentic standard indeed revealed identical chromatography, accurate mass, MS/MS fragmentation pattern as well as absorbance spectrum thus confirming peak identity.

4.2.2 Metabolomics to Characterize Tomato Mutants

The LC-PDA-QTOF-MS-based platform approach has been shown to be an effective, reproducible and sensitive method for non-targeted metabolomic profiling. Sample preparation, chromatographic system and accurate mass measurements have been optimized in order to screen hundreds of extracts in an unsupervised stable manner. After unbiased alignment of the chromatograms, the data are imported into multivariate analyses software to elucidate the biological variables underlying the data structure (Vorst et al. 2005). As an example of the power of non-targeted metabolomics approaches using LC-QTOF MS, we recently reported on the effect of a single mutation on the metabolic profile of ripe fruits of tomato (Bino et al. 2005). In tomato, several natural photomorphogenic mutants are known and these have been the subject of detailed physiological investigations. One mutant, carrying one of the *high pigment* (*hp-1*, $hp\text{-}1^w$, *hp-2*, $hp\text{-}2^j$, and $hp\text{-}2^{dg}$) mutations, is characterized by its exaggerated light responsiveness. Generally, these mutants have higher pigmentation levels in their hypocotyls, leaves and fruit in comparison to their semi-isogenic, wild-type counterparts (Levin et al. 2003; van Tuinen et al. 2005). The more intense colour of the fruits is a clear indication that these mutants accumulate more all-trans lycopene in their ripe fruits. However, by using a metabolomics approach it became clear that the metabolic perturbations in these fruit were much more extensive than just involving lycopene (Bino et al. 2005). The $hp\text{-}2^{dg}$ mutant and wild-type tomato plants (cv. Manapal) were grown simultaneously under controlled environmental conditions and fruit samples were pooled per plant. Different, complementary metabolic profiling techniques, including GC-MS and LC-MS, were applied to measure as many compounds as possible in ripe fruits. For non targeted profiling of non-volatile (semi-polar) compounds, aqueous methanol extracts were prepared and subjected to reversed phase HPLC using both PDA (240–600 nm) and QTOF-MS (m/z 100–1500; ESI positive and negative modes). A lock mass spray (sampled every 10 s) was used to enable accurate mass measurements. Raw data were processed by the metAlign™-software and mass traces were extracted (6168 in negative mode and 5401 in positive mode with a ratio of > 3) and aligned across all samples. Pair-wise comparisons (Student t-test) could then be used to determine statistically significant differences between *hp*-2^{dg} and wild-type fruit extracts. Differential chromatograms were produced from the original LC-MS software, and from these (Fig. 5) it was evident that the *hp*-2^{dg} mutation resulted in a significant increase in many compounds (246 mass signals in negative mode and 137 in positive mode were > twofold higher) and a decrease in only a small number of other compounds (57 mass signals being a factor 2 or more lower in negative mode and 5 mass signals twofold lower in positive mode). The metabolites corresponding to the differential masses were identified using accurate mass, MS/MS fragmentation experiments and absorbance spectra (PDA) information. In this way, it was possible to identify a number of phenolic compounds, flavonoids and alkaloids that were significantly increased in the *hp*-2^{dg} mutant (Bino et al. 2005) pointing clearly to a pleiotropic

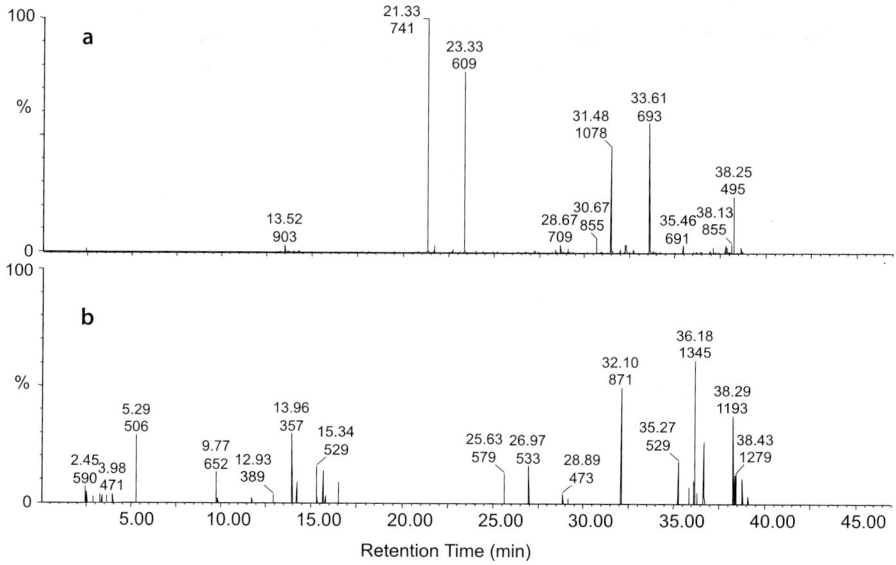

Fig. 5a,b. Metalign™-processed LC-QTOF MS chromatograms (recorded in ESI-negative mode) showing metabolites that are significantly different (Student t-test, $p < 0.01$; $n = 5$) between *hp*-2^{dg} and wild-type tomato fruits: **a** metabolites at least twofold higher in *hp*-2^{dg} than in wild-type; **b** metabolites at least twofold higher in wild-type than in *hp*-2^{dg}. Retention times and nominal masses of metabolites are indicated. 100% scale of y-axis (TIC) is 25,000 in **a** and 500 in **b**

effect of photomorphogenic mutations on tomato fruit metabolism which was much greater than was initially visible.

5 Conclusions and Future Prospects

The examples presented in this chapter clearly underline the versatility of hybrid TOF mass spectrometers, and their capabilities with regard to metabolic profiling, structure elucidation and compound identification, using accurate mass determinations and MS/MS fragmentation. The high sensitivity and mass resolution allows the rapid screening of complex plant extracts by DFI, suitable for semi quantitative high throughput (pre)screening. More detailed analysis is possible when MS detection is preceded by applying separation technologies such as LC. Data processing and efficient data handling are becoming more and more the bottleneck in the process, especially when high throughput screening is required and the currently available bioinformatics tools are inadequate. Another bottleneck is the low number of available reference compounds needed for definitive identification of differentially accumulating components. Key developments for the near future will therefore have to be made in these areas if true plant metabolomics strategies are to become routine. With better software and more easily mined databases we will be best equipped for the identification of the large numbers of the highly chemically diverse components typically present in complex plant extracts.

References

Aharoni A, de Vos CHR, Verhoeven HA, Maliepaard CA, Kruppa G, Bino RJ, Goodenowe DB (2002) Nontargeted metabolome analysis by use of Fourier Transform Ion Cyclon Mass Spectrometry. OMICS 6:217–234

Bino RJ, de Vos CHR, Lieberman M, Hall RD, Bovy A, Jonker HH, Tikunov Y, Lommen A, Moco S, Levin I (2005) The light-hyperresponsive high pigment-2^{dg} mutation of tomato: alterations in the fruit metabolome. New Phtyologist 166:427–438

Castrillo JI, Hayes A, Mohammed S, Gaskell SJ, Oliver SG (2003) An optimized protocol for metabolome analysis in yeast using direct infusion electrospray mass spectrometry. Phytochemistry 62:929–937

Chernushevich IV, Loboda AV, Thomson BA (2001) An introduction to quadrupole-time-of-flight mass spectrometry. J Mass Spectrom 36:849–865

Fernie AR (2003) Metabolome characterization in plant system analysis. Funct Plant Biol 30:111–120

Fernie AR, Trethewey RW, Krotzky AJ, Willmitzer L (2004) Metabolic profiling: from diagnostics to systems biology. Nature Rev Mol Cell Biol 5:763–769

Fiehn O (2001) Combining genomics, metabolome analysis and biochemical modelling to understand metabolic networks. Comp Funct Genom 2:155–168

Fiehn O (2002) Metabolomics-the link between genotypes and phenotypes. Plant Mol Biol 48:115–171

Fiehn O, Kopka J, Dörmann P, Altmann T, Trethewey RN, Willmitzer L (2000) Metabolic profiling for plant functional genomics. Nat Biotechnol 18:1157–1161

Goodacre R, Vaidyanathan S, Bianchi G, Kell DB (2002) Metabolic profiling using direct infusion electrospray ionisation mass spectrometry for the characterisation of olive oils. Analyst 127:1457–1462

Goodacre R, York EV, Heald JK, Scott IM (2003) Chemometric discrimination of unfractionated plant extracts analyzed by electrospray mass spectrometry. Phytochemistry 62:859–863

Goodacre R, Vaidyanathan S, Dunn WB, Harrigan GG, Kell DB (2004) Metabolomics by numbers: acquiring and understanding global metabolomics data. Trends Biotechnol 22:245–252

Guilhaus M (1995) Principles and instrumentation in Time-of-flight mass spectrometry. J Mass Spectrom 30:1519–1532

Hager JW (2002) A new linear ion trap mass spectrometer. Rapid Commun Mass Spectrom 16:512–526

Hall RD, de Vos CHR, Verhoeven HA, Bino RJ (2005) Metabolomics for the assessment of functional diversity and quality traits in plants. In: Harrigan G, Vaidyanathan S, Goodacre R (eds) Metabolic profiling. Kluwer Acad Publ, Dordrecht, Netherlands pp 31–44

Jander G, Norris SR, Joshi V, Fraga M, Rugg A, Yu S, Li L, Last RL (2004) Application of a high-throughput HPLC-MS/MS assay to Arabidopsis mutant screening; evidence that threonine aldolase plays a role in seed nutritional quality. Plant J 39:465–475

Kebarle P (2000) A brief overview of the present status of the mechanisms involved in electrospray mass spectrometry. J Mass Spectrom 35:804–817

King R, Bonfiglio R, Fernandez-Metzler C, Miller-Stein C, Olah T (2000) Mechanistic investigation of ionization supression in electrospray ionization. J A Soc Mass Spectrom 11:942–950

LeGall G, DuPont MS, Mellon FA, Davis AL, Collins GJ, Verhoeyen ME, Colquhoun IJ (2003) Characterization and content of flavonoid glycosides in genetically-modified tomato (*Lycopersicon esculentum*) fruits. J Agric Food Chem 51:2438–2446

Levin I, Frankel P, Gilboa N, Tanny S, Lalazar A (2003) The tomato dark green mutation is a novel allele of the tomato homolog of the *DEFOLIATED 1* gene. TAG 106:454–460

Markham KR (1989) Flavones, flavonols and their glycosides. In: Dey PM, Harborne JB (eds) Methods in plant biochemistry, vol 1. Academic Press, San Diego, USA, pp 197–235

Muir S, Collins GJ, Robinson S, Hughes S, Bovy A, de Vos CHR, van Tunen AJ, Verhoeyen ME (2001) Overexpression of petunia chalcone isomerase in tomato results in fruit containing increased levels of flavonoids. Nat Biotechnol 19:470–474

Nielsen N-PV, Carstensen JM, Smedsgaard J (1998) Aligning of single and multiple wavelength chromatographic profiles for chemometric data analysis using correlation optimised warping. J Chromatogr A 805:17–35

Roessner U, Willmitzer L, Fernie AR (2001a) High-resolution metabolic phenotyping of genetically and environmentally diverse potato tuber systems. Identification of phenocopies. Plant Physiol 127:746–764

Roessner U, Luedemann A, Brust D, Fiehn O, Linke T, Willmitzer L, Fernie A (2001b) Metabolic profiling allows comprehensive phenotyping of genetically or environmentally modified plant systems. Plant Cell 13:11–29

Soga T, Ueno Y, Naraoka H, Matsuda K, Tomita M, Nishioka T (2002) Pressure-assisted capillary electrophoresis electrospray ionization mass spectrometry for analysis of multivalent anions. Anal Chem 74:6224–6229

Sumner LW, Mendes P, Dixon RA (2003) Plant metabolomics: large-scale phytochemistry in the functional genomics era. Phytochemistry 62:817–836

Tolstikov VV, Fiehn O (2002) Analysis of highly polar compounds of plant origin: combining of hydrophilic interaction chromatography and electrospray ion trap spectrometry. Anal Biochem 301:298–307

Tolstikov VV, Lommen A, Nakanishi K, Tanaka N, Fiehn O (2003) Monolithic silica-based capillary reversed phase liquid chromatography / electrospray mass spectrometry for plant metabolomics. Anal Chem 75:6737–6740

Van Tuinen A, de Vos CHR, Hall RD, van der Plas LHW, Bowler C, Bino RJ (2005) Use of metabolomics for identification of tomato genotypes with enhanced nutritional value derived from natural light-hyperresponsive mutants. In: Jaiwal PK (ed) Improving the nutritional and therapeutic qualities of plants. (Plant Metabolic Engineering & Molecular pharming.) SciTech Publishers, Raleigh, USA (in press)

Vorst OF, de Vos CHR, Lommen A, Staps RV, Visser RGF, Bino RJ, Hall RD (2005) A non directed approach to the differential analysis of multiple LC/MS-derived metabolic profiles. Metabolomics 1:169–180

Weckwerth W, Tolstikov V, Fiehn O (2001) Metabolomic characterization of transgenic potato plants using GC/TOF and LC/MS analysis reveals silent metabolic phenotypes. Abstract: Proceedings of the 49th ASMS Conference on Mass spectrometry and Allied Topics (1–2)

Wolff JC, Eckers C, Sage AB, Giles K, Bateman R (2001) Accurate mass liquid chromatography / mass spectrometry on quadrupole orthogonal acceleration time-of flight mass analyzers using switching between separate sample and reference sprays. 2 Applications using the dual-electrospray ion source. Anal Chem 73:2605–2612

I.4 Capillary HPLC

T. Ikegami[1], E. Fukusaki[2], and N. Tanaka[1]

1 Introduction

Among many techniques employed for the separation and identification of metabolites, HPLC (high performance liquid chromatography)-MS (mass spectrometry) is most widely applicable to metabolomics, although the chromatographic efficiency is generally lower than that of the other separation techniques, GC (gas chromatography)-MS or CE (capillary electrophoresis)-MS. Recently, significant improvement was made to increase the separation capability of HPLC, which will help the analysis of complex metabolite samples. In the field of metabolomics, because of the importance of separation and detection of thousands of small molecules, micro HPLC techniques will become a common method of separation in the near future (Tomita and Nishioka 2003). In this article, the use of long capillary columns that give high separation efficiencies in micro HPLC system, and multidimensional HPLC that can provide even higher peak capacity will be described. Special attention will be paid to the examples of high efficiency HPLC separations made possible by monolithic silica columns composed of network type silica skeletons.

2 Monolithic Silica Columns for Micro HPLC

Micro HPLC systems with a monolithic silica capillary column possess the following advantages:

1. Small consumption of stationary and mobile phases
2. High detection sensitivity for a certain amount of samples
3. High speed separation with low pressure drop
4. The possible use of a long column with $1 \sim 2$ m that can provide around $100{,}000 \sim 200{,}000$ theoretical plates

along with some disadvantages:

[1] Department of Polymer Science and Engineering, Kyoto Institute of Technology, Matsugasaki, Sakyo-ku, Kyoto, 606–8585, Japan, e-mail: ikegami@kit.ac.jp, nobuo@kit.ac.jp
[2] Department of Biotechnology, Graduate School of Engineering, Osaka Univ, 2–1 Yamadaoka, Suita, 565–0871, Japan, e-mail: fukusaki@bio.eng.osaka-u.ac.jp

Biotechnology in Agriculture and Forestry, Vol. 57
Plant Metabolomics (ed. by K. Saito, R.A. Dixon, and L. Willmitzer)

Table 1. Column sizes, flow rates, linear velocities, and degrees of sample dilution

Column type	Inner diameter [mm (μm)]	Column volume[a] [μl]	Flow rate [μl/min]	t_0^a [s/10 cm]	Solvent linear velocity [mm/s]	Relative degree of dilution[b]
Conventional	4.6	1660	1000	70	1.4	2100
Semi-micro	2.0	314	200	66	1.5	400
Micro	1.0	78	50	66	1.5	100
	0.5 (500)	20	12.5	66	1.5	25
Micro-capillary	0.3 (300)	7.1	5	59	1.7	9
	0.2 (200)	3.1	2	66	1.5	4
	0.1 (100)	0.78	0.5	66	1.5	1
	0.05 (50)	0.20	0.12	69	1.5	0.25
	0.025 (25)	0.05	0.03	69	1.5	0.06

[a] Column lengths were 10 cm, total porosity was estimated as 0.70
[b] Column of id 100 μm is taken as a standard

1. Smaller sample capacities of a monolithic silica column than particle-packed columns
2. Necessity of skill and knowledge to operate a capillary HPLC system to obtain high separation efficiency, and insufficient supply of good columns and instruments for capillary HPLC

Particle-packed capillary columns have been employed for separations of analytes with or without the assistance of electroosmotic flow. It is possible to pack silica particles into a fused silica capillary equipped with a frit, but it is difficult to produce high efficiency and long-lasting columns using 1 ~ 2 μm particles (Novotony 1988; Knox and Grant 1991; Schmeer et al. 1995).

Recently, monolithic silica capillary columns have been reported to show higher separation efficiencies than particle packed columns (Ishizuka et al. 2000; Tanaka et al. 2000). They consist of network silica skeletons that can be prepared in capillaries by a sol-gel method. Monolithic silica columns of 4.6 mm ID (inner diameter), 0.2 mm ID, 0.1 mm ID and 0.05 mm ID are commercially available at present. Column sizes and flow rates to be employed are listed in Table 1.

2.1 Characteristics of Monolithic Silica Columns

Here, the features of monolithic silica capillary columns and the optimization of separation conditions will be described. The use of monolithic silica columns consisting of network silica skeletons and through-pores for micro HPLC was reported recently (Minakuchi et al. 1996; Tanaka et al. 2001). Monolithic silica capillary columns were reported to provide better separation efficiencies than particle-packed columns, and the use of these columns for proteomics and

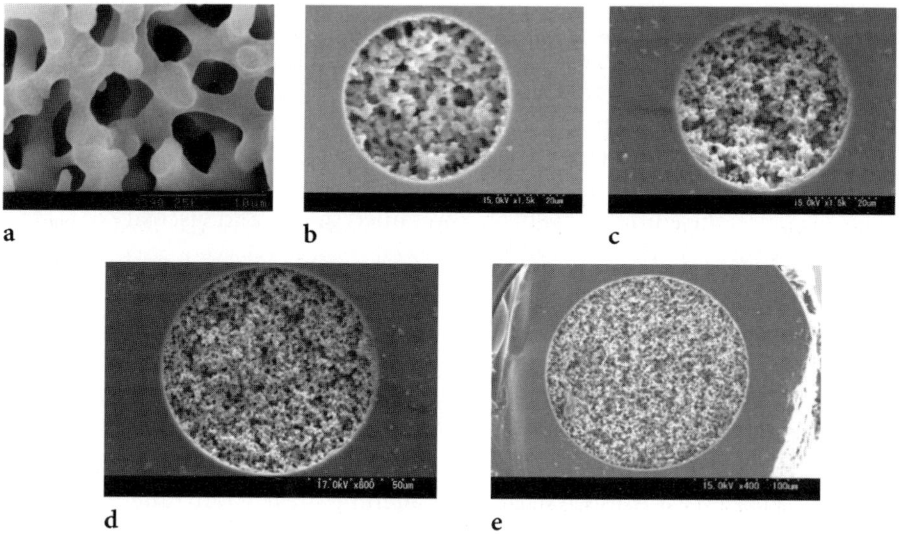

Fig. 1. Scanning electron microscope images of monolithic silica prepared from sol-gel methods: **a** monolithic silica prepared in a test tube; **b,c** monolithic silica prepared in 50 μm ID fused silica capillary; **d** monolithic silica prepared in 100 μm ID fused silica capillary; **e** monolithic silica prepared in 200 μm ID fused silica capillary tube

metabolomics seems to be attractive (Cabrera 2004). Monolithic silica columns are prepared by acid-catalyzed hydrolytic polymerization of alkoxysilanes in the presence of water-soluble polymers such as poly(ethylene glycol) (Tanaka et al. 2001; Cabrera 2004). Figure 1a shows a scanning electron microscope (SEM) image of monolithic silica prepared in a test tube, while Fig. 1b–e shows SEM images of monolithic silica columns prepared in fused silica capillaries with 50 ∼ 200 μm internal diameter (Motokawa et al. 2002).

Currently available monolithic capillary columns include organic polymer columns (Svec 2004) and chemically modified silica columns, and they have the following features. Monolithic polymer columns generally show higher permeability than particle-packed columns, and high efficiency for the separation of macromolecules (Svec et al. 2000). In the case of monolithic silica capillary columns, silica skeletons are covalently bonded to capillary walls. Thus, frits are not necessary to hold the skeletons in a column, and column length can be varied in the range of 5 ∼ 200 cm after preparation. Generally, the silica skeleton sizes are in the range of 1 ∼ 2 μm. As shown in Fig. 1a, monolithic columns have 3 ∼ 10 times bigger (through-pore size/skeleton size) ratio, 1 ∼ 3, than particle-packed columns with (through-pore size/particle size) ratio, 0.25 ∼ 0.4. Monolithic silica columns produce similar separation efficiencies to particle-packed columns at much lower pressure drop. At the same pressure drop, monolithic columns can provide higher separation efficiencies than particle-packed columns. Moreover, due to the small silica skeleton sizes, relatively high separation efficiencies can be expected at higher linear velocities (Minakuchi et al. 1997, 1998). In terms of separation impedance, total performance of columns (E), monolithic silica capillary columns can produce higher separation efficiencies, nearly 10 times greater than that of a particle-packed column (Motokawa et al. 2002). Separation impedance is given by Eq. (1) where N, ΔP, t_0 and η stand for number of theoretical plates, column back pressure, the elution time of an unretained solute, and viscosity of mobile phase, respectively (Bristow and Knox 1977):

$$E = t_0 \Delta P / N^2 \eta = (\Delta P/N)(t_0/N)(1/\eta) \tag{1}$$

Figure 2 shows chromatograms produced by monolithic silica capillary columns of 25 ~ 130 cm lengths modified with C18 stationary phase (Ishizuka et al. 2002).

The dilution factors for analytes are proportional to the internal diameters of columns squared assuming that band broadening (peak width) and resolutions are similar for various HPLC systems. Sample concentrations after the separation are higher in micro columns with smaller diameters, and the higher sample concentrations can lead to higher detection sensitivity (Table 1). Because lower flow rates can lead to higher ionization efficiencies and higher detection sensitivity in LC-ESI (electrospray ionization) MS system, development of high

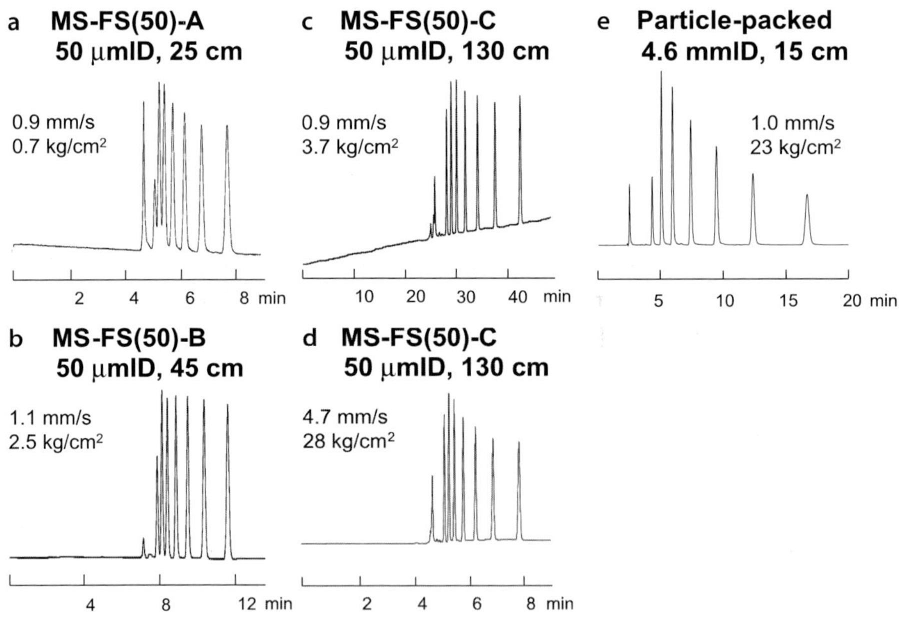

Fig. 2. Chromatograms obtained for alkylbenzenes ($C_6H_5(CH_2)_nH$, n = 0 – 6) by: **a–d** C18 monolithic silica capillary columns; **e** particle packed column (5 mm silica-C18 particles, Mightysil RP18)

efficiency micro HPLC system is an important issue for metabolomics studies (Schmidt et al. 2003).

2.2 Column Efficiencies and the Optimization of Separation Conditions

The number of theoretical plate N is a measure of the quality of a column and elution conditions, and is given by Eq. (2) from the retention time of a peak (t_R) and peak width at half height ($t_{w1/2} = 2.35\sigma$, σ being the standard deviation of a Gaussian peak). Resolution R_s is given by Eq. (4), that includes N, α (Eq. (5), selectivity, the ratio of retention factors of two adjacent peaks), and k (Eq. (3), a retention factor, distribution coefficient of a solute between stationary and mobile phases, i. e. the ratio of times ($t_R - t_0$) to t_0, the former stands for time the solute exists in mobile phase, and the latter stands for time the solute exists in stationary phase). For convenient separation and detection, the k values should be in a range of $2 \sim 5$:

$$N = (t_R/\sigma)^2 = 5.54(t_R/t_{w1/2})^2 = 16(t_R/t_w)^2 \quad (2)$$

$$k = (t_R - t_0)/t_0 \quad (3)$$

$$R_s = (N^{1/2}/4)[(\alpha - 1)/\alpha][k/(1 + k)] \quad (4)$$

$$\alpha = k_2/k_1 \quad (5)$$

$$\Delta P = \phi\eta uL/d_p^2(u = L/t_0) \quad (6)$$

ΔP is proportional to η, u (linear velocity of the mobile phase), and L (column length) while it is inversely proportional to d_p^2, where d_p stands for diameter of particle. Thus, a column packed with particles of small diameter leads to high separation efficiency, (greater N) at the expense of high column backpressure. Due to the drawback, an approach to get high efficiencies by reducing diameter of particles has a limit: since the pressure limit of a pump system is around $300 \sim 400$ bar with a normal operational pressure $100 \sim 200$ bar, the limit in particle sizes is in a range of $1 \sim 3$ µm. The flow resistance parameter ϕ in Eq. (6), is usually ca. 2000 for particle-packed columns, while ϕ values reach to $200 \sim 400$ in the case of monolithic silica columns (Giddings 1965; Bristow and Knox 1977).

A solute band is broadened when it travels outside a column due to parabolic flow profile in a tube as well as due to slow diffusion in the stagnant mobile phase existing in an injector, a detector, or connection tubing. Especially, for solutes of small retention factors which elutes in early part of a chromatogram, sample injection into a capillary column of $1 \sim 5\%$ of column volume has significant influence on band spreading, mainly caused by sample diffusion at orifice in an injector or by dead volume in all connection parts (Ikegami at al. 2004). The split-flow injection technique is practical and useful for micro HPLC with monolithic silica capillary columns in order to avoid the peak spreading during injection (Taniguchi and Murata 2002). Moreover, the use of weak eluents for sample injection is also effective to increase the separation efficiency: in the case of reversed-phase HPLC, sample solution can be prepared with water-rich solvent (Ikegami at al. 2004).

3 Applications of Monolithic Silica Columns to Metabolomics

Figure 3 shows chromatograms of leaf extracts of *Arabidopsis thaliana* by LC-ESI-MS using 30 $\sim$ 90 cm monolithic silica capillary columns modified with C18 stationary phase under gradient conditions, from aqueous ammonium acetate buffer (pH 5.5) to acetonitrile, MeCN (Tolstikov et al. 2003). A shallow gradient (large t_G, gradient time) with a long column has lead to better separation. The results indicate that improvement of separation by the use of the longer columns caused the reduction of ion suppression effect by introducing the solute bands separately into ES ionization interface. In the case of Fig. 3, the peak capacity provided by the long monolithic silica column is still not enough for complete separation, but it shows a feasible approach of using longer monolithic silica capillary columns to achieve higher separation efficiency avoiding ion suppression effect in the LC-ESI-MS system. This approach will result in longer separation time, but the amount and quality of information after the analysis of metabolites would be better than conventional LC-MS systems using particle packed columns. Connected monolithic columns (conventional size) in series showed good separation of polyprenol homologues (Bamba et al. 2004).

Mass spectrometry would often be used in metabolomics research due to its superiority in both quantification and qualification. However, mass spectrometry has a serious drawback named 'ionization suppression'. Ionization suppression is a phenomenon that presence of impurity at ionization might cause a serious impairment in qualitative accuracy (King et al. 2000; Müller et al. 2002). Coelution in chromatography might cause ionization suppression. Even the technology of capillary monolithic chromatography might not provide a perfect time separation that is one of the ideal solutions against ionization suppression. Recently, stable isotope dilution technology tends to be used as a practical tool to reduce an ionization suppression negative effect. A stable isotope dilution method employs isotopologues as an internal standard that would be separated not by chromatography but by mass spectrometry to provide an accurate comparative quantification. This principle is used in the proteomics research tool 'isotope coded affinity tags (ICAT)' (Han et al. 2001). A metabolic profiling of sulfur metabolite using ^{34}S was reported (Mougous et al. 2002). ^{13}C and ^{15}N stable isotope labeling techniques could be available in some case. In addition, post sampling stable isotope labeling would be also applicable, although D-labeling may face some difficulties (Zhang et al. 2001; Fukusaki et al. 2005). In future a combination of monolithic capillary chromatography and stable isotope diluted comparative quantification would be one of the de facto standard methods in metabolomics.

4 Two-Dimensional HPLC

Peak capacity (PC) given by Eq. (7) indicates the separation ability regarding how many solutes can be potentially separated by a chromatographic system. Retention times of the first solute and the last solute are given as t_1 and t_R respectively in Eq. (7). Separation methods such as ultrahigh-pressure liquid chromatography (UHPLC) and supercritical fluid chromatography (SFC) can produce a PC of ca. 300/h (Shen and Lee 1998; MacNair et al. 1999), while a conventional HPLC system gives a PC of 100 $\sim$ 200/h. In order to achieve far larger PC using conventional HPLC systems, multidimensional separation systems were shown to be effective. When two chromatographic systems with PC_x and PC_y are combined to form a two-dimensional (2D) chromatography system, PC for the total system can be theoretically estimated as a product of two PC values as Eq. (8) (Giddings 1991):

$$PC = 1 + (N^{1/2}/4)\ln(t_R/t_1) \quad (7)$$

$$PC_{2D} = PC_x \times PC_y \quad (8)$$

In comprehensive 2D-HPLC separations every fraction obtained from 1st-D separation is to be separated in 2nd-D HPLC, while the next fraction is eluted from 1st-D. Therefore the 2nd-D column should ideally be eluted at very high speed to meet the rate of fractionation at the 1st-D separation. The 2nd-D column should possess low-pressure drop and reasonable efficiency at high flow rate. In addition to high efficiency and high permeability, the 1st-D and 2nd-D columns must possess adequate difference in selectivity to effect 2D separations. Ideally the 1st-D and 2nd-D should have orthogonal selectivity or different separation mechanisms (Bushey and Jorgenson 1990; Köhne and Welsch 1999; Wagner et al. 2002; Venkataramani and Zelechonok 2003). Ion-exchange mode and reversed-phase mode, or size-exclusion mode and reversed-phase mode have often been combined to effect 2D separations for peptide mixtures in proteomics. Because a particle-packed column cannot be operated at adequately high flow rate, various approaches were taken in the past: (i) small columns were employed at 1st-D compared to 2nd-D, (ii) the first column was eluted slowly or intermittently, or (iii) two or more sets of chromatographs were used at the 2nd-D. Even with these methods, however, truly "two-dimensional" HPLC is hard to achieve due to the mixing of separation modes.

Figure 4 shows a scheme of 2D-HPLC and its working principle, in which the outlet tubing of the 1st-D column was connected to a loop of the 2nd-D injector to couple particle the packed 1st-D column and 2nd-D monolithic silica column run at higher linear velocity (Tanaka et al. 2004). In this case, the

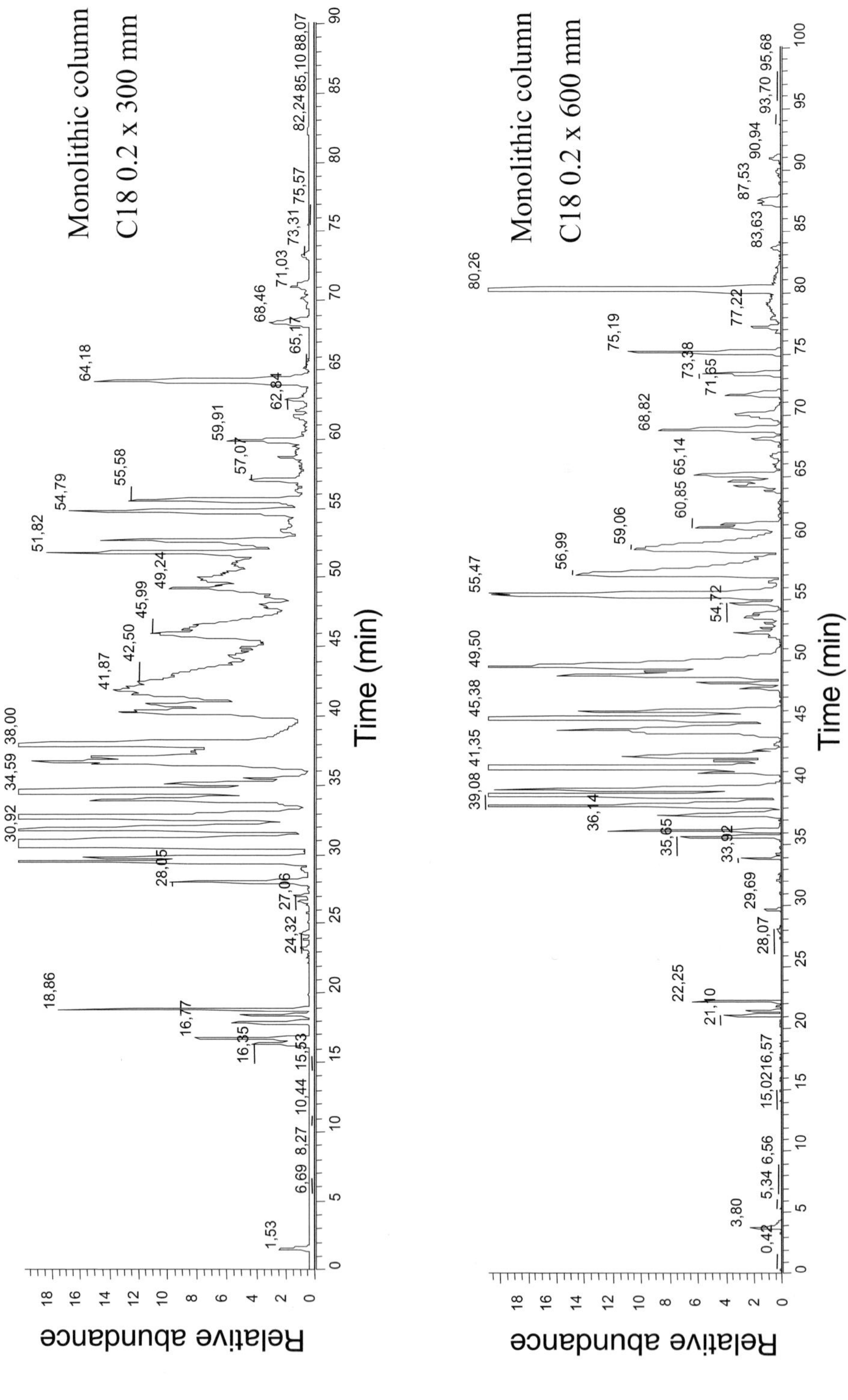
Monolithic column
C18 0.2 x 300 mm
Time (min)
Relative abundance
Monolithic column
C18 0.2 x 600 mm
Time (min)
Relative abundance

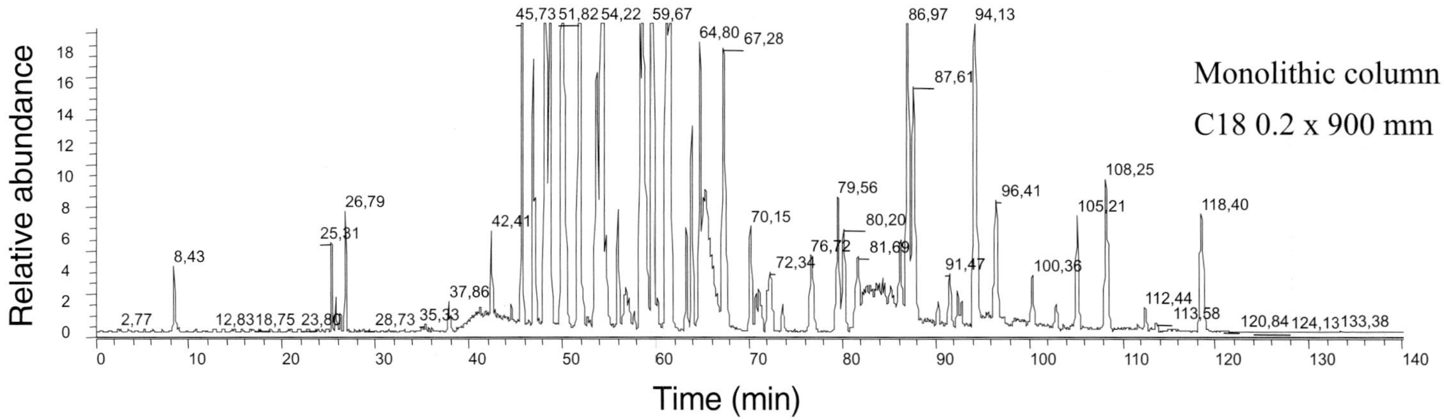

Fig. 3. Replicate injections of an Arabidopsis leaf methanol extract on capillary monolithic C18 columns in positive ionization fullscan MS, given as base peak chromatograms. *Upper panel* 0.2 mm ID, 300 mm long; *middle panel* 0.2 mm ID, 600 mm long; *lower panel* 0.2 mm ID, 900 mm long column; t_0, void volume

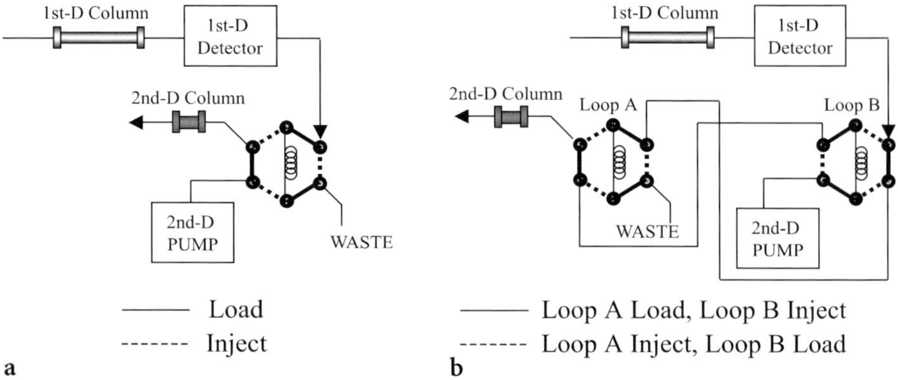

Fig. 4. **a** Tubing connection at 2nd-D injector of simple 2D-HPLC. **b** Tubing connection of two six-port valves used as 2nd-D injector

fraction from the 1st-D column is loaded and temporarily kept in a loop of the 2nd-D injector that results in mixing of separated peaks, but the flow rates of two HPLC systems can be controlled independently. The 2nd-D separation can be carried out at very high flow rate (for example, 10 ml/min for a 4.6 mm ID column) throughout the separation. The simplest 2D-HPLC in Fig. 4a produced PC = 1000 in reversed phase mode. When two six-port valves or a ten-port valve is used at the 2nd-D HPLC in Fig. 4b, all fractions can be subjected to the separation at the 2nd-D column to provide a comprehensive 2D-HPLC system resulting in so-called group separation, solutes of similar structural features appear as a group. Because of fast flow rate in the 2nd-D separation using a 4.6 mm ID column, the 2D-HPLC system consumed a lot of mobile phase solvent. In order to reduce the consumption of mobile phases, the sufficiently fast, simple 2D-HPLC using capillary columns has been examined (Kimura et al. 2004). The use of capillary column at 2nd-D leads to less solvent consumption and better MS detectability compared to a larger-sized column. Figure 5a shows a 2D chromatogram for the tryptic digest of BSA (Bovine serum albumin) obtained from total ion monitoring by ESI-TOF (Time of flight)-MS. From the 1st-D (2.1 mm ID, 5.0 cm long), 18 fractions were injected at 2-min intervals into the 2nd-D reversed-phase system (4.6 mm ID, 2.5 cm long), generating 18 chromatograms that were used to produce a 2D chromatogram. Figure 5b shows a 2D chromatogram obtained for the separation of tryptic digest of BSA using a capillary column (100 μm ID, 10 cm long) in the 2nd D separation. The number of spots distinguishable in vertical direction in Fig. 5b was greater than that in Fig. 5a. This is due to the higher column efficiency and longer gradient time in 2nd-D, along with greater MS detection sensitivity based on nearly optimum flow rate (3 μL/min) on the capillary column, the greater amount of sample introduced to the 2nd-D column because of the longer fractionation interval, and the smaller extent of dilution due to the use of small diameter column (Kimura et al. 2004).

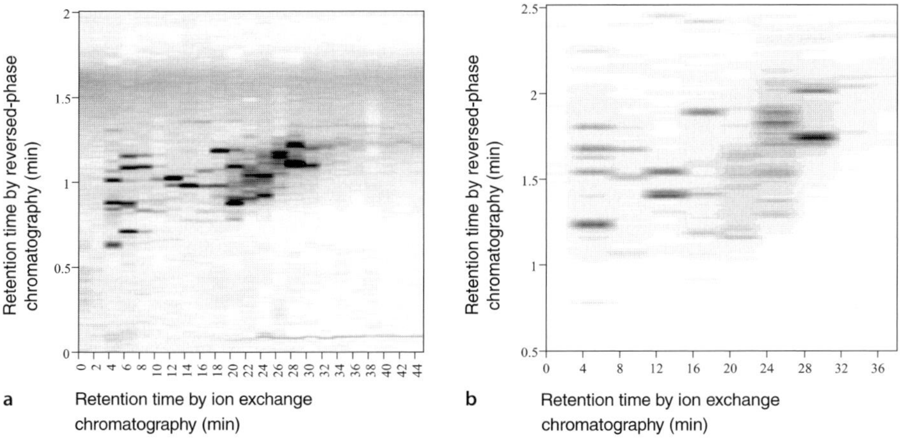

Fig. 5. Two-dimensional separation of tryptic digest of BSA in simple 2D-HPLC, 1st-D; MCI CQK-31S column (2.1 mm ID, 50 mm long), flow rate; 50 μl/min: **a** 2nd-D; monolithic silica-C18 column (4.6 mm ID, 25 mm long), flow rate; 5.0 ml/min; **b** 2nd D; C18 monolithic column (0.1 mm ID, 100 mm long), flow rate in a capillary column; 3.0 μl/min with a split flow/injection; linear velocity in the column; 7.7 mm/s. ESI-TOF-MS detection, total ion chromatogram for a mass range 400–2000

5 Combination of Reversed-Phase HPLC and Other Separation Modes

Since many compounds of similar properties are to be separated in proteomics, 2D-HPLC hyphenated to an MS system has been employed combining ion exchange mode and reversed phase mode, or size-exclusion mode and reversed-phase mode. In the case of metabolomics, combination of several different separation modes is preferable to separate a variety of substances. Reversed-phase mode is most often employed in HPLC, where chemically bonded stationary phases (C8, C18, and C30, etc.) have advantages in rapid equilibration with mobile phase, high separation efficiency, and high reproducibility in gradient.

Recently, hydrophilic LC (HILIC LC) (Alpert et al. 1994; Yoshida 1997) was shown to be effective for the separation of metabolomes utilizing the interaction between solutes and hydrophilic functional groups on the stationary phases. The selectivities of HILIC columns are similar to those of a conventional silica column, but HILIC columns have advantages over silica columns in the recovery of samples, and compatibility with mobile phases used in reversed-phase mode. Figure 6 shows a comparison of elution patterns of HILIC mode and reversed-phase mode in the separation of an extract from *Arabidopsis thaliana* (Tolstikov and Fiehn 2002). Since the solvent type for HILIC and reversed-phase mode are common, it is possible to combine the two separation modes to form multidimensional HPLC, although

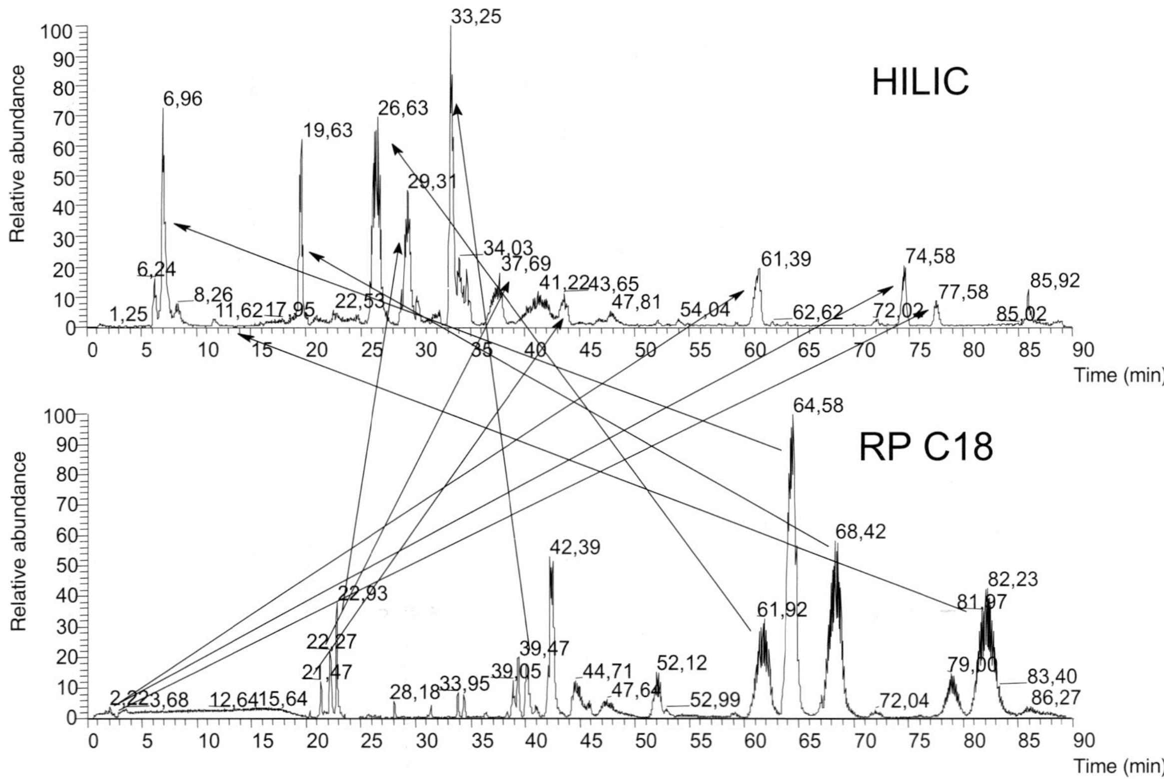

Fig. 6. Comparison of chromatograms of an *Arabidopsis thaliana* leaf methanol extract, obtained by HILIC-LC mode (*top panel*) and reversed-phase mode (*bottom panel*): Conditions (*top panel*) TSK Gel Amide 80, 4.6 mm ID, 150 mm long, gradient elution from MeCN to ammonium acetate buffer (6.5 mmol/l, pH 5.5), MeCN content (%) (time, min) 100 → 100(5) → 90(8) → 60(75) → 0(80), (*bottom panel*) C18 column, 4.6 mm ID, 150 mm long, gradient elution from ammonium acetate buffer (6.5 mmol/l, pH 5.5) to MeCN, MeCN content (%) (time, min) 0→ 0(15) → 95(40) → 100(60) → 100(80)

the compositions of mobile phases that controls the retention order are total opposites to each other. Capillary columns for HILIC LC are under development.

6 Outlook

Routine use of micro HPLC will need development of several important constituents; the reproducible preparation of high performance columns, small-volume pumps and gradient systems, and improvement of an injection system. Subjects to be studied are the development of high performance monolithic silica columns for variety of separation modes, multidimensional microLC systems, and optimization of an interface between LC and MS instruments. Large peak capacities realized by highly efficient microHPLC systems or multidimensional HPLC will greatly contribute to metabolomics studies when coupled with MS instruments and stable isotope dilution methodology.

References

Alpert AJ, Shukla M, Shukla AK, Zieske LR, Yuen SW, Ferguson MAJ, Mehlert A, Pauly M, Orlando R (1994) Hydrophilic-interaction chromatography of complex carbohydrates. J Chromatogr A 676:191–202

Bamba T, Fukusaki E, Nakazawa Y, Kobayashi A (2004) Rapid and high-resolution analysis of geometric polyprenol homologues by connected octadecylsilylated monolithic silica columns in high-performance liquid chromatography. J Sep Sci 27:293–296

Bristow PA, Knox JH (1977) Standardization of test conditions for high performance liquid chromatography columns. Chromatographia 10:279–289

Bushey MM, Jorgenson JW (1990) Automated instrumentation for comprehensive two-dimensional high-performance liquid chromatography of proteins. Anal Chem 62:161–167

Cabrera K (2004) Application of silica-based monolithic HPLC columns. J Sep Sci 27:843–852

Fukusaki E, Harada K, Bamba T, Kobayashi A (2005) An isotope effect on the comparative quantification of flavonoids by means of methylation-based stable isotope dilution coupled with capillary liquid chromatograph/mass spectrometry. J Biosci Bioeng 99:75–77

Giddings JC (1965) Dynamics of chromatography, part 1. Principles and theory. Dekker, New York

Giddings JC (1991) Unified separation science. Wiley-Interscience, New York, pp 126–128

Han DK, Eng J, Zhou H, Aebersold R (2001) Quantitative profiling of differentiation-induced microsomal proteins using isotope-coded affinity tags and mass spectrometry. Nat Biotechnol 19:946–951

Ikegami T, Dicks E, Kobayashi H, Morisaka H, Tokuda D, Cabrera K, Hosoya H, Tanaka N (2004) How to utilize the true performance of monolithic silica columns. J Sep Sci 27:1292–1302

Ishizuka N, Minakuchi H, Nakanishi K, Soga N, Nagayama H, Hosoya K, Tanaka N (2000) Performance of a monolithic silica column in a capillary under pressure-driven and electrodriven conditions. Anal Chem 72:1275–1280

Ishizuka N, Kobayashi H, Minakuchi H, Nakanishi K, Hirao K, Hosoya K, Ikegami T, Tanaka N (2002) Monolithic silica columns for high-efficiency separations by high-performance liquid chromatography. J Chromatogr A 960:85–96

Kimura H, Tanigawa T, Morisaka H, Ikegami T, Hosoya K, Ishizuka N, Minakuchi H, Nakanishi K, Ueda M, Cabrera K, Tanaka N (2004) Simple 2D-HPLC using a monolithic silica column for peptide separation. J Sep Sci 27:897–904

King R, Bonfiglio R, Fernandez-Metzler C, Miller-Stein C, Olah T (2000) Mechanistic investigation of ionization suppression in electrospray ionization. J Am Soc Mass Spectrom 11:942–950

Knox JH, Grant IH (1991) Electrochromatography in packed tubes using 1.5 to 50 μm silica gels and ODS bonded silica gels. Chromatographia 32:317–328

Köhne AP, Welsch T (1999) Coupling of a microbore column with a column packed with non-porous particles for fast comprehensive two-dimensional high-performance liquid chromatography. J Chromatogr A 845:463–469

MacNair JE, Patel KD, Jorgenson JW (1999) Ultrahigh-pressure reversed-phase capillary liquid chromatography: isocratic and gradient elution using columns packed with 1.0 mm particles. Anal Chem 71:700–708

Minakuchi H, Nakanishi K, Soga N, Ishizuka N, Tanaka N (1996) Octadecylsilylated porous silica rods as separation media for reversed-phase liquid chromatography. Anal Chem 68:3498–3501

Minakuchi H, Nakanishi K, Soga N, Ishizuka N, Tanaka N (1997) Effect of skeleton size on the performance of octadecylsilylated continuous porous silica columns in reversed-phase liquid chromatography. J Chromatogr A 762:135–146

Minakuchi H, Nakanishi K, Soga N, Ishizuka N, Tanaka N (1998) Effect of domain size on the performance of octadecylsilylated continuous porous silica columns in reversed-phase liquid chromatography. J Chromatogr A 797:121–131

Motokawa M, Kobayashi H, Ishizuka N, Minakuchi H, Nakanishi K, Jinnai H, Hosoya K, Ikegami T, Tanaka N (2002) Monolithic silica columns with various skeleton sizes and through-pore sizes for capillary liquid chromatography. J Chromatogr A 961:53–63

Mougous JD, Leavell MD, Senaratne RH, Leigh CD, Williams SJ, Riley LW, Leary JA, Bertozzi CR (2002) Discovery of sulfated metabolites in mycobacteria with a genetic and mass spectrometric approach. Proc Natl Acad Sci USA 99:17037–17042

Müller C, Schäfer P, Störtzel M, Vogt S, Weinmann W (2002) Ion suppression effects in liquid chromatography-electrospray-ionisation transport-region collision induced dissociation mass spectrometry with different serum extraction methods for systematic toxicological analysis with mass spectra libraries. J Chromatogr B 773:47–52

Novotny M (1988) Recent advances in microcolumn liquid chromatography. Anal Chem 60:500A–510A

Schmeer K, Behnke B, Bayer E (1995) Capillary electrochromatography – electrospray mass spectrometry: a microanalysis technique. Anal Chem 67:3656–3658

Schmidt A, Karas M, Dülcks T (2003) Effect of different solution flow rates on analyte ion signals in nano-ESI MS, or: when does ESI turn into nano-ESI. J Am Soc Mass Spectr 14:492–500

Shen Y, Lee ML (1998) General equation for peak capacity in column chromatography. Anal Chem 70:3853–3856

Svec F (2004) Preparation and HPLC applications of rigid macroporous organic polymer monoliths. J Sep Sci 27:747–766

Svec F, Peters EC, Sýkora D, Yu C, Fréchet JMJ (2000) Monolithic stationary phases for capillary electrochromatography based on synthetic polymers: designs and applications. J High Resolut Chromatogr 23:3–18

Tanaka N, Nagayama H, Kobayashi H, Ikegami T, Hosoya K, Ishizuka N, Minakuchi H, Nakanishi K, Cabrera K, Lubda D (2000) Monolithic silica columns for HPLC, micro-HPLC, and CEC. J High Resolut Chromatogr 23:111–116

Tanaka N, Kobayashi H, Nakanishi K, Minakuchi H, Ishizuka N (2001) Monolithic LC columns. Anal Chem 73:420A–429A

Tanaka N, Kimura H, Tokuda D, Hosoya K, Ikegami T, Ishizuka N, Minakuchi H, Nakanishi K, Shintani Y, Furuno M, Cabrera K (2004) Simple and comprehensive two-dimensional reversed-phase HPLC using monolithic silica columns. Anal Chem 76:1273–1281

Taniguchi H, Murata Y (2002) The newest protocol of proteomics 9, Capillary HPLC. Cell Tech 21:1332–1343

Tolstikov VV, Fiehn O (2002) Analysis of highly polar compounds of plant origin: combination of hydrophilic interaction chromatography and electrospray ion trap mass spectrometry. Anal Biochem 301:298–307

Tolstikov VV, Lommen A, Nakanishi K, Tanaka N, Fiehn O (2003) Monolithic silica-based capillary reversed-phase liquid chromatography/electrospray mass spectrometry for plant metabolomics. Anal Chem 75:6737–6740

Tomita M, Nishioka T (2003) Frontier of metabolomics. Springer, Berlin Heidelberg New York

Venkatramani CJ, Zelechonok Y (2003) An automated orthogonal two-dimensional liquid chromatograph. Anal Chem 75:3484–3494

Wagner K, Miliotis T, Marko-Varga G, Bischoff R, Unger KK (2002) An automated on-line multidimensional HPLC system for protein and peptide mapping with integrated sample preparation. Anal Chem 74:809–820

Yoshida T (1997) Peptide separation in normal phase liquid chromatography. Anal Chem 69:3038–3043

Zhang R, Sioma CS, Wang S, Regnier FE (2001) Fractionation of isotopically labeled peptides in quantitative proteomics. Anal Chem 73:5142–5149

I.5 Capillary HPLC Coupled to Electrospray Ionization Quadrupole Time-of-flight Mass Spectrometry

S. Clemens, C. Böttcher, M. Franz, E. Willscher, E. v. Roepenack-Lahaye, and D. Scheel[1]

1 Introduction

Metabolite profiling in the pre-metabolomics era of the early 1970s to the late 1990s as well as the pioneering metabolomics projects since the late 1990s have been predominantly GC-MS based. GC-MS techniques are robust and well-established. Many primary metabolites (e. g. organic acids, sugars, amino acids, sugar alcohols) can easily be derivatized and are therefore amenable to GC-MS analysis. Also, spectral databases and deconvolution algorithms are available, which help extracting meaningful information. Early on, however, it was obvious that no single analytical technique would be sufficient to achieve comprehensive coverage of the metabolome (Sumner et al. 2003). As stated from the beginning and reiterated since, the chemical diversity of metabolites makes it virtually impossible to detect all compound classes in one "catch" (Goodacre et al. 2004; Dunn et al. 2005). That is why already the first reports describing GC-MS-based metabolomics platforms emphasized the need to develop complementing LC-MS platforms (Roessner et al. 2000). LC-MS covers in principle a much wider mass range and should allow one to target many compound classes not detectable by GC-MS. Furthermore, there is usually no need for derivatization and LC-MS offers superior options to elucidate unknown metabolites structurally. Particular fractions can easily be collected for NMR analysis and metabolites/molecular ions can be further analyzed by tandem-MS or even MS^n. Hampering the adoption of LC-MS approaches for metabolomics, however, was the fact that LC-MS has only rather recently (i. e. in the 1990s) developed into a routine technology (Niessen 1999a).

One might argue that the need for LC-MS-based profiling is even more pressing in plant science. A highly rich and diverse secondary metabolism is a hallmark of plant biology. Lacking the ability to avoid or to retreat from unfavorable conditions or potential foes, plants have evolved an enormous metabolic plasticity, which allows them to respond dynamically to environmental changes through the synthesis and/or degradation of particular compounds. This is complemented by the accumulation of various pre-formed defenses against microbial attack and other threats (Dixon 2001). Furthermore, many so-called secondary metabolites also apparently play major roles in primary developmental processes and as signaling molecules. Flavonoids

[1] Leibniz Institute of Plant Biochemistry, Weinberg 3, 06120 Halle/Saale, Germany, e-mail: sclemens@ipb-halle.de

Biotechnology in Agriculture and Forestry, Vol. 57
Plant Metabolomics (ed. by K. Saito, R.A. Dixon, and L. Willmitzer)

and their biosynthesis, for instance, have long been investigated because of their role in flower pigmentation, UV protection, or pathogen defense (Winkel-Shirley 2001). More recent work demonstrated that flavonoids negatively regulate auxin transport and are required for pollen germination (Taylor and Grotewold 2005).

A large fraction of plant secondary metabolites has been classically analyzed by LC techniques, predominantly through separation on reversed phase material. Thus, it is a straightforward concept to combine this with state-of-the-art mass spectrometry in order to develop powerful metabolomics platforms that cover important compound classes such as phenylpropanoids or alkaloids. A look at *Arabidopsis thaliana*, the most important plant model species, can illustrate the need for and the potential of LC-MS profiling. Because *A. thaliana* has no history of use as a medicinal plant, it initially did not attract the attention of too many natural product chemists. As a consequence, few secondary metabolites were identified 10 years ago. In the course of the genome sequencing, however, it became increasingly clear, that *A. thaliana* should produce thousands of different compounds. The *Arabidopsis* genome encodes a myriad of proteins likely to be involved in secondary metabolism (d'Auria and Gershenzon 2005). There are more than 270 cytochrome P450 genes, more than 100 glycosyl transferase genes, about 50 glutathione *S*-transferase genes, to name a few. For most of the encoded enzymes we do not know substrates or products.

The first major challenge for metabolomics is the huge chemical diversity of the metabolome. The second lies in the fact that – as indicated above for *Arabidopsis thaliana* – most of the metabolites in any given higher eukaryote are unknown. Current estimates are in the range of 4000–20,000 metabolites for a given species (Fernie et al. 2004). Unlike for proteins, genome sequences do not allow one to deduce the structure of the metabolites. Instead, the structure has to be elucidated because for only a very minor portion of the metabolites are standards available. Thus, the future success of metabolomics will also be determined by the ability to identify reliably metabolites and to establish the metabolomes of the important model species. Again, this is a particularly daunting task for plants and filamentous fungi, organisms that synthesize huge numbers of secondary metabolites, many of which might only be synthesized in certain cell types or at particular developmental stages. LC-MS, especially in the combination of quadrupole and time-of-flight analysis in modern hybrid instruments, holds the promise to meet this challenge as well. Structural information can in principal be obtained in three different ways: (i) by determining the elemental composition through the accurate mass, (ii) by exploiting the information provided by in-source fragmentation, and (iii) by performing targeted CID-MS (collision-induced dissociation). In contrast, GC-MS-based profiling faces severe limitations when it comes to de novo identification of unknown compounds (Fiehn 2002). Molecular ions are rarely detected because most analytes are derivatized and molecules are fragmented by the electron impact ionization.

We will in the following discuss the principles of capillary LC-MS-based profiling, describe the current state and present new data from our own laboratory on the optimization and the potential of capillary LC coupled to electrospray ionization quadrupole time-of-flight mass spectrometry (CapLC-ESI-QTOF-MS) (von Roepenack-Lahaye et al. 2004).

2 Extraction, Chromatography and Mass Spectrometry

When optimizing extraction and chromatographic separation of low molecular weight compounds, there are various considerations which are commonplace in analytical chemistry (Niessen 1999b) and which will therefore only be touched upon very briefly. The extraction of biological material with aqueous methanol has so far been the most widely used option for GC-MS as well as LC-MS metabolite profiling schemes (Roessner et al. 2000; Fiehn et al. 2000; Tolstikov et al. 2003). For the sake of stability of compounds and reproducibility of the analysis, cold extraction is preferred in most cases. Obviously, the choice of solvent greatly influences scope and range of the profiling. We tested, for instance, acetonitrile-water and methanol-water mixtures for the extraction of *Arabidopsis thaliana* seeds and counted simply the number of mass signals with a signal to noise ratio > 5 by analyzing subsets of the resulting LC-MS chromatograms with MetAlign (www.metalign.nl) (see below). We detected 1680 mass peaks in an 80% methanolic seed extract and 1771 signals in a 50% methanolic seed extract. Of these signals, 973 were found in both extracts. Utilization of acetonitrile-water gave comparable results: upon extraction with 80% and 50% acetonitrile, 2070 and 1771 mass peaks, respectively, were detected and 1029 were found in both extracts. Only 532 mass peaks were detected in all extracts.

There are several classical analytical options to enrich selectively certain compound classes by modifying the extraction. Solid-phase extraction can be used to remove problematic compounds such as lipids or to concentrate others that are of interest but give low signal intensity. These different options and their effect on the metabolome coverage of LC-MS approaches have not been systematically evaluated yet. A way of selectively targeting specific classes of molecules is derivatization. This can permit analysis of compounds with inadequate stability and results in better chromatographic behaviour as well as enhanced signal intensity. Also, derivatization has been proposed as a means to make the ionization of diverse analytes more uniform by adding a particular chemical group (Halket et al. 2005).

A major obstacle in the development of LC-MS was the general incompatibility of flow rates between LC and MS, i. e. the need to introduce a column effluent of about 1 mL/min into a high vacuum (Niessen 1999a). One solution to this problem was to reduce the flow rate by miniaturization of the LC column, a second to split the column effluent so that only a fraction reaches the mass spectrometer. Often these two options are combined. In capillary liquid chromatography the flow rate is reduced to meet the optimum flow rate range characteristic for many ESI interfaces. Splitting occurs – if at all – prior to chromatography between the pump and the column. Chromatography is performed at low flow rates of 2–20 μL/min (Abian et al. 1999). Column diameters are typically between 80 and 800 μm. In principle, MS is a mass flow sensitive detection because the response is proportional to the actual number of molecules reaching the detector. However, at a constant flow rate under atmospheric pressure ionization conditions, MS acts as a concentration sensitive detector, i. e. the signal is proportional to the analyte concentration in the eluent (Niessen 1999a). The smaller diameter of a capillary column as compared to a regular 4.6-mm analytical column combined with a lower flow rate allows the use of much smaller sample volumes and lower sample concentrations. Furthermore, depending on the design of the ESI interface a reduced flow-rate can result in higher sensitivity due to the enhanced ionization yield of the smaller primary droplet formation (Wilm and Mann 1994). Thus, since the mid 1990s there has been a trend towards miniaturization of the LC (Abian et al. 1999), although the better sensitivity – i. e. lower concentration detection limits – is partly offset by the need to reduce the injection volume and by the lower capacity of the column.

It is advisable to inject as small a volume as possible (and reproducible) in a solvent of low eluotropic strength. Otherwise, retention on the stationary phase is incomplete and many compounds will elute partly in the flow-through. Furthermore, separation could be seriously disturbed, which results in unsymmetrical peak shapes and altered retention times. Figure 1 shows the extracted ion chromatograms, which correspond to the molecular ion of 4-glucopyranosyloxybenzoyl choline, a secondary metabolite identified in methanolic seed extracts, injected in either 2 μL 80% methanol (a) or 2 μL 10%

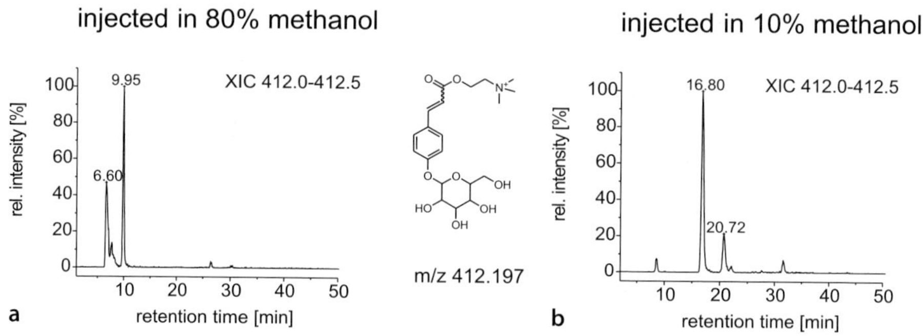

Fig. 1. Influence of solvent on the retention and separation. Extracted ion chromatograms (XIC 412.0–412.5) showing the altered retention behaviour of 4-glucopyranosyloxybenzoyl choline from a seed extract upon injection in different injection solvent mixtures: **a** 80% methanol – a fraction of the metabolite elutes in the flow-through (t_R = 6.60 min); **b** 10% methanol – diastereomers are retained on the column and baseline-separated (t_R = 16.80, t_R = 20.73 min)

methanol (b) following separation on C18 phase with hydrophilic end-capping. In case of an injection in 80% methanol most of the compound is eluted without any retention, whereas upon injection in 10% methanol the compound is retained on the stationary phase and both diastereomeres (probably *cis*/*trans*-isomers) are baseline separated.

A second major problem for the coupling of LC to MS was initially the incompatibility of commonly used mobile phases with MS. Therefore, non-volatiles such as phosphate ions or the frequently used ion-pairing agent trifluoroacetic acid (TFA), which causes ion suppression due to the extreme ionization capacity of the mother ion, had to be replaced with formate or acetate buffers. The organic component of the mobile phases is most frequently acetonitrile, sometimes methanol. Using classical reversed phase material (RP-18, 3 or 4 μm) as stationary phase, acceptable peak shapes for most of the compounds in leaf and root extracts could be achieved. In general peak widths of about 0.20–0.35 min for a 15-cm column with 3 μm particle size were observed. Application of a C18 phase with hydrophilic end-capping provides better separation for early eluting analytes. In particular, aromatic amino acids and biogenic amines show a considerably improved retention behaviour.

Concerns about the feasibility and reliability of LC-MS-based metabolite profiling have been raised repeatedly (Fiehn 2002; Fernie et al. 2004; Kell 2004). These concerns are mostly referring to the fact that electrospray ionization is prone to matrix effects. This term summarizes two phenomena potentially compromising quantification: (i) reduction or enhancement of ion signals caused by the sample matrix, and (ii) interferences from co-eluting molecules (Matuszewski et al. 2003). Matrix components that are non-volatile can have dramatic effects on the ion signal of an analyte. Mechanistically this effect is not fully understood. It is likely caused by competition between an analyte and non-volatile matrix components for access to the droplet surface in the spray and for reaction with ions formed during the ionization process (Niessen 1999a; Matuszewski et al. 2003; Manini et al. 2004). Thus, reproducibility of a quantification can be compromised, a potential problem that is further aggravated when diverse samples (= matrices) are analyzed. It is important to note, however, that matrix effects have predominantly been observed in cases where there was little chromatographic separation. Run times and column lengths were reduced because the coupling to MS/MS supposedly guaranteed highly selective detection (Matuszewski et al. 2003). It is obvious, however, that good separation prior to ionization is essential to reduce the impact of matrix effects and to minimize ion suppression – especially when highly complex metabolite mixtures are analyzed. The fewer analytes elute from the column simultaneously, the better the chances are of efficient and reproducible electrospray ionization and detection of a particular analyte. Thus, optimal separation is of paramount importance and, consequently, the use of very long monolithic columns for the liquid chromatography has been proposed (Tolstikov et al. 2003). In capillary LC, particle size and column diameter severely restrict the length of the column because of the backpressure build-up. At the same time, a certain minimum flow rate has to be maintained in order to obtain a stable electrospray. In our experience, therefore, it is not feasible to use columns much longer than 20 cm unless larger particles are used. We found, however, that the 3 μm–15 cm design is sufficient to achieve very good separation. Figure 1b shows as a selected example for the CapLC resolution the base-line separation of the two diastereoisomers of 4-glucopyranosyloxybenzoyl choline (below we will present and discuss an assessment of matrix effects in our metabolite profiling scheme).

A great advantage of ESI is its ability to provide soft ionization. Nevertheless, fragmentation can easily be induced in one of the higher-pressure regions of the ion passageway from the source into the mass analyzer (Fig. 2a). Three potentials which determine the opposite processes of declustering and focusing vs collision induced dissociation (Fig. 2b) can be varied on a QSTAR Pulsar and have to be optimized for the profiling in terms of mass signal yield and appropriate distribution. First we analyzed two model compounds, namely hirsutin (Fig. 2c) and rutin (Fig. 2d), which are prone to give in-source fragments and tried to optimize the intensities of the quasi-molecular ions by systematically ramping both declustering potentials and the focusing potential. For both hirsutin and rutin the quasi-molecular ions $[M+H]^+$ reached their maximum between 40 and 50 V for DP1 and 10 and 15 V for DP2. The effects of the focusing potential on the maxima of the breakdown curves were of minor importance. However, optimization was necessary (FP = 220 V, data not shown). It should be clearly stated that, in principle, for every analyte, such an optimization has to be done to get the full sensitivity, but since a highly complex mixture of mostly unknown compounds is analyzed, compromises have to be made. To get the optimal value of DP1 for a profiling experiment, we measured the same methanolic leaf extract (n = 4) with different DP1 values between 15 and 60 V and analyzed the mass signals with regard to its mass-to-charge ratio and signal-to-noise ratio distribution (Fig. 3, left panel and right panel, respectively). We found that in such a simplified analysis the density functions between 30 and 60 V show only minor differences and, thus, a value of about DP1 = 45 V appears to yield the best results concerning high signal-to-noise ratios as well as high mass-to-charge ratios.

In our metabolite profiling platform (von Roepenack-Lahaye et al. 2004) ions are injected into a QTOF system, a hybrid mass spectrometer. Practically, the third quadrupole in a triple quad instrument is replaced in these instruments with a time-of-flight mass analyzer (Chernushevich et al. 2001). Other mass analysis options for coupling to LC are discussed in detail in another chapter (Sumner et al.; this book, Chap. I.2). Likewise, the QTOF system is dealt with by Bino et al. (this book, Chap. I.3). Thus, we will only briefly summarize our experience with QTOF-MS. We routinely use an acquisition period ("scan time") of 2 s. For the deconvolution of data, which is even more vital for LC due to the lower resolution as compared to GC, there are currently two options available to us. The software MetaboliteID (Applied Biosystems) allows one to extract the mass spectra and to generate an output that lists mass peaks with

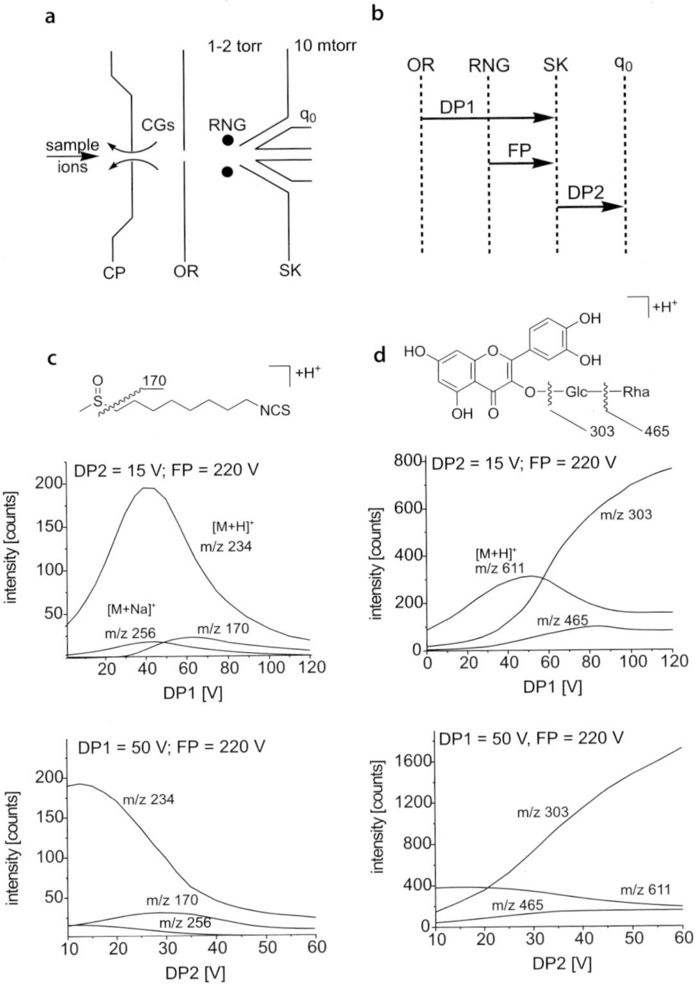

Fig. 2. Effects of ion source potentials on sensitivity and degree of in-source fragmentation: **a** schematic overview of the differentially pumped (evacuated) interface between ion source and mass spectrometer of an API QSTAR Pulsar Hybrid LC/MS system: curtain plate (CP), curtain gas (CGs), orifice (OR), ring (RNG), skimmer (SK); **b** definition of electrical potentials applied in the interfacial region: declustering potential (DP), focusing potential (FP); **c,d** breakdown curves for hirsutin (**c**) and rutin (**d**) obtained in DP1 and DP2 ramping experiments

retention times, accurate mass and intensity. Self-made macros are then needed to normalize, to align peak list and to compare intensities (von Roepenack-Lahaye et al. 2004). These latter steps are covered by the software MetAlign, developed by Arjen Lommen (www.metalign.nl; Tolstikov et al. 2003), which aligns and compares sets of chromatograms to identify differentially abundant mass signals. Reproducibility of the retention times of capillary LC is, in our experience, high enough to allow accurate alignment of chromatograms.

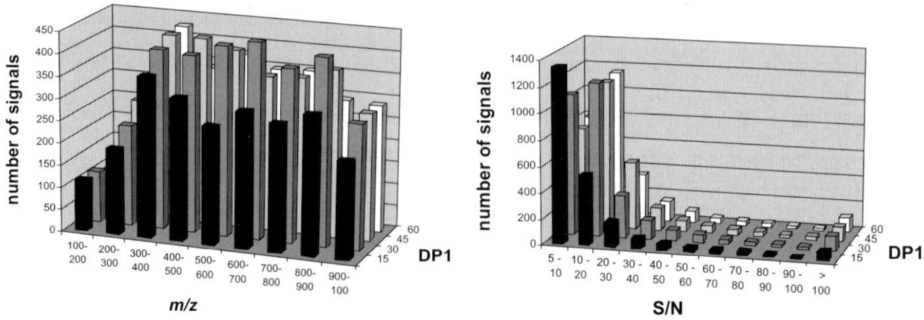

Fig. 3. Effect of declustering potential modulation (DP1 variable, DP2 = 15 V, FP = 220 V) on mass-to-charge ratio (*left panel*) and signal-to-noise ratio (*right panel*) distribution. A methanolic leaf extract was analyzed by CapLC-ESI(+)-QTOF-MS with varying declustering potential 1 settings between 15 and 60 V. Frequencies for different mass-to-charge ratio (m/z) and signal-to-noise ratio (S/N) classes were plotted against DP1 values (n = 4)

3 Potential and Limitations

3.1 Scope of the Analysis

Given the idealistic goal of metabolomics to achieve comprehensive coverage of the metabolome (Oliver et al. 1998), the number of detectable metabolites is an important feature of a metabolite profiling platform. CapLC-ESI-QTOF-MS has great potential because of its sensitivity (Chernushevich et al. 2001). In a single CapLC-MS run analyzed with the deconvolution software MetaboliteID we routinely detect about 1000–2000 mass signals, depending on the extracted material. Running the data through the MetAlign software and applying a signal-to-noise ratio cutoff of 5 gives comparable figures. Similar, albeit somewhat lower numbers (around 700) have been obtained for methanolic *Arabidopsis thaliana* leaf extracts in pilot experiments with monolithic silica columns, coupling of the LC to ion trap MS and deconvolution through MetAlign (Tolstikov et al. 2003).

A mass signal or *m*/*z*, however, does not necessarily represent a metabolite in all cases. Though isotope peaks should be eliminated through the deconvolution, many of the signals are likely to be fragments or adducts so that any given metabolite can in theory give rise to several mass signals. Thus, the number of detectable metabolites is certainly smaller than the number of mass signals and cannot reliably be estimated at this point in time. It is safe to state, however, that several hundred metabolites are routinely detectable in a methanolic extract using the combination of capillary LC and QTOF mass spectrometry. Formation of sodium or potassium adducts, for instance, is not too frequent because of the use of formic acid in the mobile phase.

How large a fraction of the *Arabidopsis* metabolome is covered using this technique? The range of secondary metabolite compound classes detectable by CapLC-ESI-QTOF-MS in its current state can be assessed by searching in the profiles for members of the various groups of metabolites known to occur in *Arabidopsis thaliana*. A recent compilation listed six biosynthetic classes (d'Auria and Gershenzon 2005): nitrogen-containing compounds, phenylpropanoids, benzenoids, polyketides such as flavonoids, terpenes and fatty acid derivatives. Metabolites of five of these classes can clearly be detected

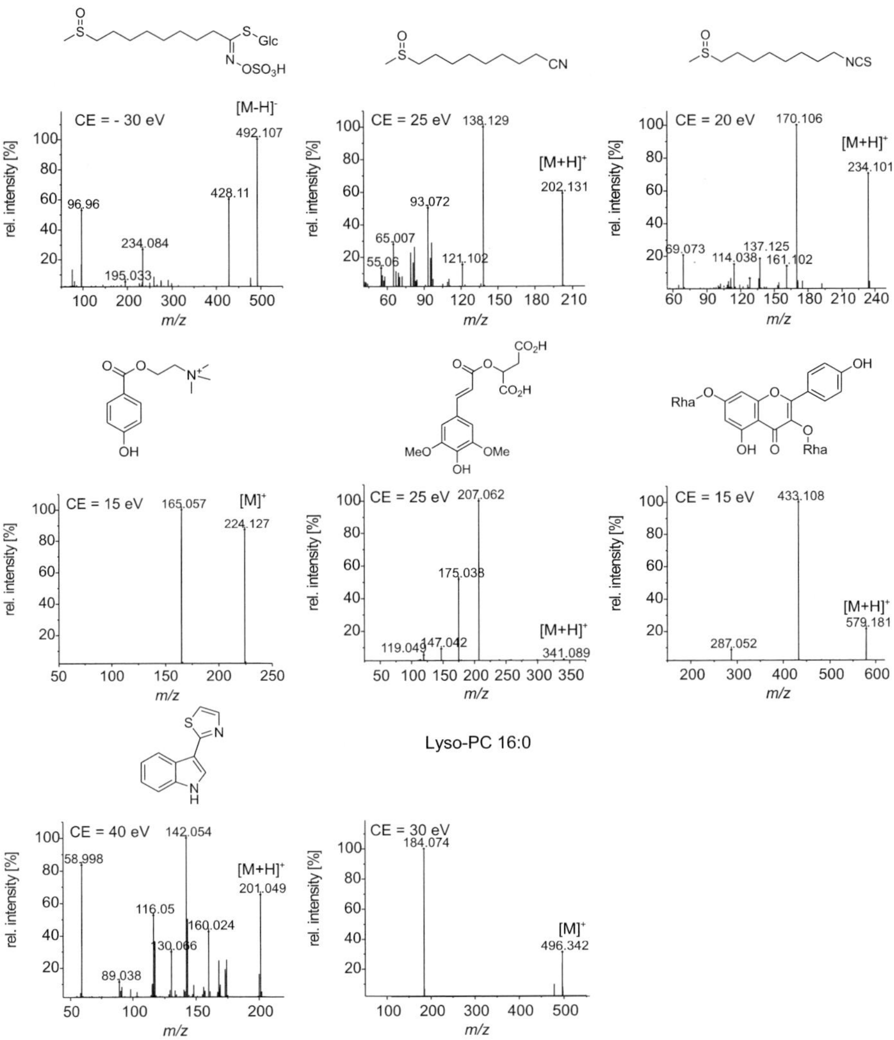

Fig. 4. Most of the known biosynthetic classes of *A. thaliana* secondary metabolites are detectable by CapLC-ESI-QTOF-MS. CapLC-ESI(+/-)-CID-MS spectra of representative metabolites detected in methanolic extracts of different *A. thaliana* tissues such as leaves, seeds and roots (for details on the biosynthetic classes and representative metabolites see text)

by CapLC-ESI-QTOF-MS. Figure 4 displays CID-MS spectra of representatives that we identified in *Arabidopsis thaliana* extracts. Intact glucosinolates can easily be detected by ESI in negative ion mode in nearly all tissues of *Arabidopsis thaliana.* Typical hydrolysis products of glucosinolates like isothiocyanates and nitriles can be detected by ESI(+). Examples are 8-methylsulfinyloctylisothiocyanate (hirsutin) and 8-methylthiononanitrile as well as several homologs. Furthermore, biosynthetic precursors of glucosinolates, like desulfoglucosinolates and thiohydroxamic acids, have been identified in certain cases. Indole-derived secondary metabolites such as the phytoalexin camalexin represent further nitrogen containing compounds that can be detected. Notably, methanolic root extracts contain a huge variety of indole derivatives. Furthermore, ascorbigens and glutathione-indole conjugates, which result from trapping reactions of the hydrolysis products of indole glucosinolates with several nucleophiles, could also be detected. As representatives of the phenylpropanoids, typical esters such as sinapoyl malate in leaves and sinapoyl choline in seeds could be specified. In particular, methanolic seed extracts show a wealth of different choline esters (unpublished observations). Besides the corresponding substituted cinnamoyl cholines various hydroxylated/methoxylated benzoyl cholines could be detected. Other choline containing compounds like differentially substituted phosphatidyl cholines have been identified in seed extracts, too. From the class of flavonoids the major flavonols kaempferol and quercetin and their glycosides could be detected either in positive or negative ion mode. Saccharide composition and aglycon structure can be determined by means of MS^2 and pseudo-MS^3 (product ion spectra derived from in-source CID fragments) experiments.

In conclusion, of the biosynthetic classes known to occur in *Arabidopsis thaliana*, all but one (terpenes) can be detected. Judging from this comparison, CapLC-ESI-QTOF-MS achieves a very good coverage of secondary metabolism. In addition, many primary metabolites such as amino a cids and oligopeptides can be analyzed. This assessment is based on data obtained in positive-ion mode. There is a potential to improve further the reach by also measuring in the negative-ion mode. We found that it is also possible to acquire reliable data in the negative mode without changing the mobile phase, albeit with a significantly reduced mass signal yield.

3.2 Quantification

Proper quantification of an analyte requires optimization of extraction, sample preparation, chromatography and detection, as well as the availability of a pure standard – which would ideally be isotopically labeled – that can be used to calibrate the signal. A standard also allows one to determine the dynamic range of a metabolite in question and to search for possible matrix effects by performing recovery experiments (Birkemeyer et al. 2005). Obviously it is extremely difficult for an unbiased profiling of mostly unknown compounds to meet the criteria for accurate and reproducible quantification. Standards are available for only a subset of the detected metabolites and this subset is particularly small for capillary LC-MS which targets hundreds of low abundance and species-specific metabolites. At the same time, it is inconceivable that all the required standards can be synthesized. Thus, any metabolomics approach comes at a cost of reduced precision (Trethewey et al. 1999) and has to be assessed critically and improved continuously with respect to accuracy of quantification.

For electrospray ionization there is inherently no correlation between signal strength and abundance when comparing different analytes because ionization efficiency is molecule-dependent. Decisive for quantitative profiling is, however, whether such a correlation exists for any given analyte and what the boundary minimum and maximum detector signals are, i. e. how wide the dynamic range is. In the absence of appropriate standards, serial dilution experiments are a way to assess linearity at least for the mass signals that are sufficiently strong. Our initial results demonstrated for a subset analyzed in detail that there is indeed a good correlation over the almost two orders of magnitude that were tested (von Roepenack-Lahaye et al. 2004). Still, because of the analyte dependency, there is a need to gather information continuously for more and more metabolites and to use these data for the quantification. As a first approximation an extensive set of reference compounds for different compound classes – if at all available – should be used. Eventually, the fascinating concept of mass isotopomer ratio analysis through in vivo labeling with ^{13}C (Birkemeyer et al. 2005) might at some point in the future allow internal standardization of profiling even with multicellular organisms.

The above-mentioned potential of LC-MS for severe matrix effects makes rigorous validation a necessity (Fernie et al. 2004). Matrix effects are again dependent on the analyte and the extract or the nature and origin of the biological sample. Ways to assess these indirectly for an analyte in question are spiking experiments with different matrices (Matuszewski et al. 2003). In order to obtain an estimate of matrix effects at the profiling scale, we performed, for instance, a series of mixing experiments. From previous data we know that leaf and root extracts are fundamentally different in their composition (von Roepenack-Lahaye et al. 2004). Leaf/root extracts were either diluted with 80% methanol or with equal amounts of root/leaf extracts and analyzed ($n = 4$). We focused attention on ions with a strong signal so that they would also be reliably detectable in dilutions. Also, we made sure that signal intensities were in the dynamic range; 45 m/z values eluting between 15 and 45 min were selected. Figure 5 shows the ratios of "signal after dilution with methanol" to "signal after dilution with root/leaf extract". A ratio above 1 is indicative of signal enhancement through the other extract, a ratio below 1 shows signal suppression. One can see in Fig. 5 that ion suppression occurs more frequently than enhancement. Of the 90 m/z measured in different dilutions, 76 showed ratios of 1 ± 0.4 (equals about two times the technical variation) and are therefore considered to be reliably quantifiable. In most experimental set-ups

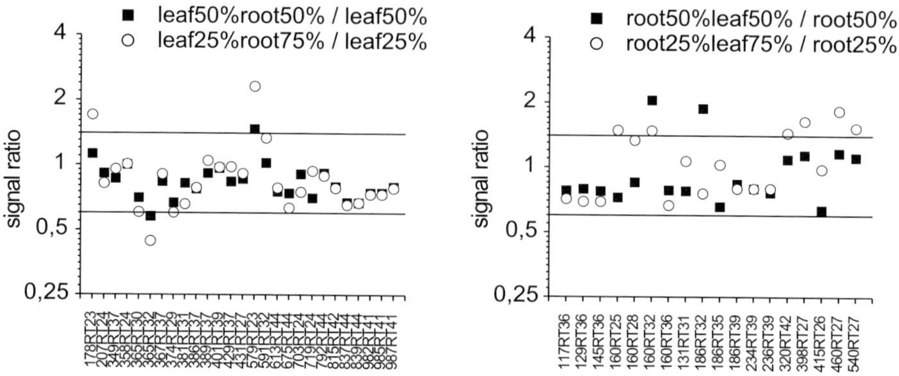

Fig. 5. Evaluation of matrix effects through mixing of extracts of different origin. Signal ratios for mass signals in crosswise matrix-diluted and solvent-diluted leaf and root extracts obtained by CapLC-ESI(+)-TOF-MS measurements. A total of 45 mass signals were analyzed in two different mixtures each. A ratio of "signal after dilution with methanol" to "signal after dilution with root/leaf extract" above 1 is indicative of signal enhancement through the other extract, a ratio below 1 of signal suppression. Most of the mass signals showed a ratio of 1 ± 0.4 (equals about two times the technical variation). This threshold of ±40% is indicated by *vertical lines*

matrix effects will probably be smaller than in this pilot experiment because matrices will be less diverse than a root and leaf extract. Information obtained by analyses such as these can be used to weigh the data obtained in a profiling experiment and to add "confidence tags" to each metabolite. Factored into such "confidence tags" should also be the results on the dynamic range and the degree of variability observed over many experiments. We conclude that the sensitivity and range of CapLC-ESI-QTOF-MS clearly make it a powerful approach for the identification of qualitative differences between samples but that – as predicted by Chernushevich et al. (2001) – quantitative analysis is also feasible provided analyte-dependent effects can be detected and corrected for. All available data suggest that overall the analytical variation is smaller than the biological variation – as is the case for the more established metabolomics approaches (Dunn et al. 2005).

3.3 Identification of Unknown Metabolites

As stated in the introduction, a major potential advantage of LC-MS-based profiling is the multitude of options to obtain structural information on unknown compounds. This is of paramount importance given, for instance, the conservatively estimated 5000 metabolites in *Arabidopsis thaliana* of which maybe 500 are annotated today (Bino et al. 2004). The first piece of information on unknowns is the accurate mass that can be obtained by TOF-MS with a deviation of only 5–10 ppm even in complex matrices (von Roepenack-Lahaye et al. 2004; Ibanez et al. 2005). Based on this, potential elemental compositions can be calculated. As recently proposed by Ibanez et al. (2005), the usually large number of possibilities can be reduced by calculating theoretical isotopic percentages for all possible elemental compositions and by comparing these to the experimental data. A further significant reduction of formulae can then be achieved through the second and third layer of structural information, in-source fragments and CID-MS spectra (see Fig. 4). The very high mass accuracy of QTOF instruments also applies to product ion scans (Chernushevich et al. 2001). With the information on accurate masses of precursor and product ions, databases can be searched. Obviously, the success rate of this approach is determined not only by the performance of the analysis but also by the availability of databases. In this respect the current situation is far from being satisfactory and future joint efforts will hopefully result in a significant improvement (Bino et al. 2004). One should add, however, that in the end identification will in most cases be tentative and further validation will be required (Bino et al. 2004). Also, discrimination between isomers is not possible without standards.

4 Conclusions and Outlook

Our experience with respect to the potential of this technique for metabolomics can in part be validated by taking a look at recent applications of LC-MS in general and of CapLC-ESI-QTOF-MS in particular. In occupational toxicology the superiority of LC-MS-MS with respect to sensitivity is now being exploited for the determination of trace and ultra-trace amounts of biomarkers of exposure (Manini et al. 2004). Quantification of low-abundance molecules in highly variable complex matrices is considered feasible, provided that precautions such as those outlined above are taken. Also, many novel metabolites have been identified and minor metabolic routes for well-known occupational hazards have been uncovered (Manini et al. 2004). Similarly, the mass accuracy and sensitivity of QTOF-MS coupled to liquid chromatography is now being applied to the elucidation of unknown environmental micro-contaminants in, for instance, water samples (Ibanez et al. 2005). Studies of this kind face challenges similar to those of metabolomics experiments. The emerging picture is that CapLC-ESI-QTOF-MS can be routinely applied (von Roepenack-Lahaye et al. 2004; Bino et al. 2005) and has high potential not only for the identification of selected molecules but for a highly sensitive, robust metabolite profiling that achieves very good coverage of the metabolome. Obviously this technology will be undergoing continuous validation and improvement. Developing the profiling is an iterative process. Any progress made with respect to the availability of standards or reference compounds, the identification of metabolites, the linear range or the possible matrix effects for a particular mass signal has to be used to increase further the accuracy of quantification. Also, protocols for extraction, selective enrichment of metabolites and chromatographic separation have to be tailored for specific questions so that a toolbox of different profiling schemes becomes available to fully exploit the power of CapLC-ESI-QTOF-MS.

Despite the current limitations – comparatively high cost and the lack of LC-MS spectral databases – this profiling approach most likely will contribute substantially to cataloguing the metabolome of *Arabidopsis thaliana* and other systems that are under investigation as models or economically important species. Also, it will help to elucidate biological functions of metabolites and will greatly facilitate the identification of enzyme substrates and products through the systematic analysis of mutants and the correlation with transcript and protein data (Hirai et al. 2005). Extensive data matrices will allow one to unravel metabolic and regulatory networks, especially in secondary metabolism.

References

Abian J, Oosterkamp AJ, Gelpi E (1999) Comparison of conventional, narrow-bore and capillary liquid chromatography mass spectrometry for electrospray ionization mass spectrometry: Practical considerations. J Mass Spectrometry 34:244–254

Bino RJ, Hall RD, Fiehn O, Kopka J, Saito K, Draper J, Nikolau BJ, Mendes P, Roessner-Tunali U, Beale MH, Trethewey RN, Lange BM, Wurtele ES, Sumner LW (2004) Potential of metabolomics as a functional genomics tool. Trends Plant Sci 9:418–425

Bino RJ, Ric de Vos CH, Lieberman M, Hall RD, Bovy A, Jonker HH, Tikunov Y, Lommen A, Moco S, Levin I (2005) The light hyperresponsive high pigment-2dg mutation of tomato: alterations in the fruit metabolome. New Phytol 166:427–438

Birkemeyer C, Luedemann A, Wagner C, Erban A, Kopka J (2005) Metabolome analysis: the potential of in vivo labeling with stable isotopes for metabolite profiling. Trends Biotechnol 23:28–33

Chernushevich IV, Loboda AV, Thomson BA (2001) An introduction to quadrupole-time-of-flight mass spectrometry. J Mass Spectrom 36:849–865

D'Auria JC, Gershenzon J (2005) The secondary metabolism of *Arabidopsis thaliana*: growing like a weed. Curr Opin Plant Biol 8:308–316

Dixon RA (2001) Natural products and plant disease resistance. Nature 411:843–847

Dunn WB, Bailey NJ, Johnson HE (2005) Measuring the metabolome: current analytical technologies. Analyst 130:606–625

Fernie AR, Trethewey RN, Krotzky AJ, Willmitzer L (2004) Metabolite profiling: from diagnostics to systems biology. Nat Rev Mol Cell Biol 5:763–769

Fiehn O (2002) Metabolomics – the link between genotypes and phenotypes. Plant Mol Biol 48:155–171

Fiehn O, Kopka J, Dormann P, Altmann T, Trethewey R, Willmitzer L (2000) Metabolite profiling for plant functional genomics. Nature Biotechnol 18:1157–1161

Goodacre R, Vaidyanathan S, Dunn WB, Harrigan GG, Kell DB (2004) Metabolomics by numbers: acquiring and understanding global metabolite data. Trends Biotechnol 22:245–252

Halket JM, Waterman D, Przyborowska AM, Patel RK, Fraser PD, Bramley PM (2005) Chemical derivatization and mass spectral libraries in metabolic profiling by GC/MS and LC/MS/MS. J Exp Bot 56:219–243

Hirai MY, Klein M, Fujikawa Y, Yano M, Goodenowe DB, Yamazaki Y, Kanaya S, Nakamura Y, Kitayama M, Suzuki H, Sakurai N, Shibata D, Tokuhisa J, Reichelt M, Gershenzon J, Papenbrock J, Saito K (2005) Elucidation of gene-to-gene and metabolite-to-gene networks in

Arabidopsis by integration of metabolomics and transcriptomics. J Biol Chem (epub ahead of print)

Ibanez M, Sancho JV, Pozo OJ, Niessen W, Hernandez F (2005) Use of quadrupole time-of-flight mass spectrometry in the elucidation of unknown compounds present in environmental water. Rapid Commun Mass Spectrom 19:169–178

Kell DB (2004) Metabolomics and systems biology: making sense of the soup. Curr Opin Microbiol 7:296–307

Manini P, Andreoli R, Niessen WM (2004) Liquid chromatography-mass spectrometry in occupational toxicology: a novel approach to the study of biotransformation of industrial chemicals. J Chromatogr A 1058:21–37

Matuszewski BK, Constanzer ML, Chavez-Eng CM (2003) Strategies for the assessment of matrix effect in quantitative bioanalytical methods based on HPLC-MS/MS. Anal Chem 75:3019–3030

Niessen WM (1999a) State-of-the-art in liquid chromatography-mass spectrometry. J Chromatogr A 856:179–197

Niessen WM (1999b) Liquid chromatography-mass spectrometry, 2nd edn. Dekker, New York

Oliver SG, Winson MK, Kell DB, Baganz F (1998) Systematic functional analysis of the yeast genome. Trends Biotechnol 16:373–378

Roessner U, Wagner C, Kopka J, Trethewey RN, Willmitzer L (2000) Technical advance: simultaneous analysis of metabolites in potato tuber by gas chromatography-mass spectrometry. Plant J 23:131–142

Sumner LW, Mendes P, Dixon RA (2003) Plant metabolomics: large-scale phytochemistry in the functional genomics era. Phytochemistry 62:817–836

Taylor LP, Grotewold E (2005) Flavonoids as developmental regulators. Curr Opin Plant Biol 8:317–323

Tolstikov VV, Lommen A, Nakanishi K, Tanaka N, Fiehn O (2003) Monolithic silica-based capillary reversed-phase liquid chromatography/electrospray mass spectrometry for plant metabolomics. Anal Chem 75:6737–6740

Trethewey RN, Krotzky AJ, Willmitzer L (1999) Metabolic profiling: a Rosetta Stone for genomics? Curr Opin Plant Biol 2:83–85

Von Roepenack-Lahaye E, Degenkolb T, Zerjeski M, Franz M, Roth U, Wessjohann L, Schmidt J, Scheel D, Clemens S (2004) Profiling of *Arabidopsis* secondary metabolites by capillary liquid chromatography coupled to electrospray ionization quadrupole time-of-flight mass spectrometry. Plant Physiol 134:548–559

Wilm MS, Mann M (1994) Electrospray and taylor-cone theory, Dole's beam of macro-molecules at last? Int J Mass Spectrom Ion Processes 136:167–180

Winkel-Shirley B (2001) Flavonoid biosynthesis. A colorful model for genetics, biochemistry, cell biology, and biotechnology. Plant Physiol 126:485–493

I.6 NMR Spectroscopy in Plant Metabolomics

J.L. Ward and M.H. Beale[1]

1 Introduction

Nuclear magnetic resonance (NMR) spectroscopy is one of the most powerful and widely used structural analysis techniques available to the analytical phytochemist and continues to be the technique of choice for unknown structure determination. As a technique for plant metabolomics it benefits from the fact that it is non-compound class selective and non-sample destructive. NMR spectra contain a wealth of accurate qualitative and quantitative information regarding the components of a sample. Whilst measurements of the ^{1}Hs are the most commonly used for metabolomic studies, analysis of the ^{13}Cs has also been employed. However, the low sensitivity of ^{13}C-NMR (due to its lower natural abundance and magnetogyric ratio) prevents its routine use for large numbers of complex extracts. General disadvantages of the sensitivity of NMR (relative to mass spectroscopy) and overlapping signals can be largely overcome, for single compounds and partially fractionated mixtures, by use of instruments with higher field strength magnets (600 MHz or greater) or by the use of modern cryoprobes. In very complex mixtures, such as crude plant extracts, that contain compounds at widely differing concentrations, sensitivity and overlapping signals are problematic for traditional 1D-NMR spectral interpretation. Techniques to deal with overlapping signals, such as 2D-J-resolved spectroscopy, can provide reconstructed 1D spectra that are simplified by the absence of proton–proton coupling (Viant 2003). However, in this review we concentrate on high-throughput 1D ^{1}H-NMR and demonstrate how the technique in combination with chemometrics has become well-established as a plant metabolomic screen. We also discuss how application of 2D- and hyphenated NMR techniques can provide further solutions to the sensitivity and deconvolution problems associated with analysis of complex natural product mixtures. For more general reviews in the use of NMR in plant sciences the reader is directed to Roberts (2000), Ratcliffe et al. (2001) and Ratcliffe and Shachar-Hill (2001, 2005).

[1] The National Centre for Plant and Microbial Metabolomics, Rothamsted Research, West Common, Harpenden, Herts. AL5 2JQ, UK, e-mail: Jane.ward@bbsrc.ac.uk, mike.beale@bbsrc.ac.uk

Biotechnology in Agriculture and Forestry, Vol. 57
Plant Metabolomics (ed. by K. Saito, R.A. Dixon, and L. Willmitzer)

2 High-throughput Screening by 1D ^{1}H-NMR

A key advantage in the use of ^{1}H-NMR spectroscopy in metabolomic screens is the robustness of the technique such that any compound that is soluble in the solvent of choice will be detected, providing that it contains hydrogen atoms. Furthermore, integration of signals from different compounds is absolutely quantitative and is truly representative of the relative concentrations of those compounds. The fact that NMR reliably detects most compounds present gives the technique a clear advantage in screening and fingerprinting applications over mass spectroscopic techniques, which are beset by problems caused by variable ionisation of different types of compounds.

Methodologies for metabolomic screening of plant extracts by NMR spectroscopy are based on the large body of work carried out in the biomedical area, particularly on plasma and urine in relation to disease biomarkers and drug metabolism. Much of this work was done by the prolific research group of Nicholson, Lindon and Holmes at Imperial College, London (Nicholson et al. 1999; Lindon et al. 2000, 2001, 2004; Bollard et al. 2005).

In plant metabolomics, solvent extraction of metabolites from tissue is necessary. Apart from the experimental design aspects where decisions on the plant numbers, growth, and tissue type have to be made, key choices influencing the range of metabolites detected must be made. These are (i) whether to use fresh or freeze-dried tissue and (ii) the polarity of the solvent to be used. On the whole, better quality NMR spectra are obtained by directly extracting freeze-dried tissue with deuterated NMR solvents. Most published work utilises polar solvents (aqueous buffers, perchloric acid, or methanol-water mixtures), although chloroform has also been utilised (Choi et al 2004a). Figure 1a depicts the ^{1}H-NMR spectrum of a deuterated water-methanol extract of *Arabidopsis thaliana.* The spectrum is typical for most green tissue polar extracts (Charlton et al. 2003; Ward et al. 2003; Choi et al. 2004b,c), and is dominated by signals arising from carbohydrates, amino-acids and organic acids. Polar extracts of other plant tissues such as potato tubers (Defernez et al. 2004), wheat flour (Lewis et al. 2003) and tomato fruits (Le Gall et al. 2003) have similar NMR spectra but contain other features reflecting the concentration of certain metabolites associated with these storage tissues. Exudates from plant roots also contain distinctive metabolites (Fan et al. 1997) and represent an area that has been neglected in the recent upsurge in metabolomic studies.

Classic interpretation of complex NMR spectra such as Fig. 1a is difficult although some 30 or 40 metabolites can be definitively identified by virtue of signals that appear in non-overlapped regions of the spectrum and quantified by integration against the internal reference standard (usually trimethylsilyl-*d*4-propionate). Use of libraries of standard spectra run in the same solvent on the same instrument (or at least on an instrument of the same field strength) is an aid to this interpretation, as is 'spiking' of extracts with pure compounds. In the example shown in Fig. 1b, clear differences in the anomeric proton region in the spectrum of wild-type *Arabidopsis thaliana* and that of two different

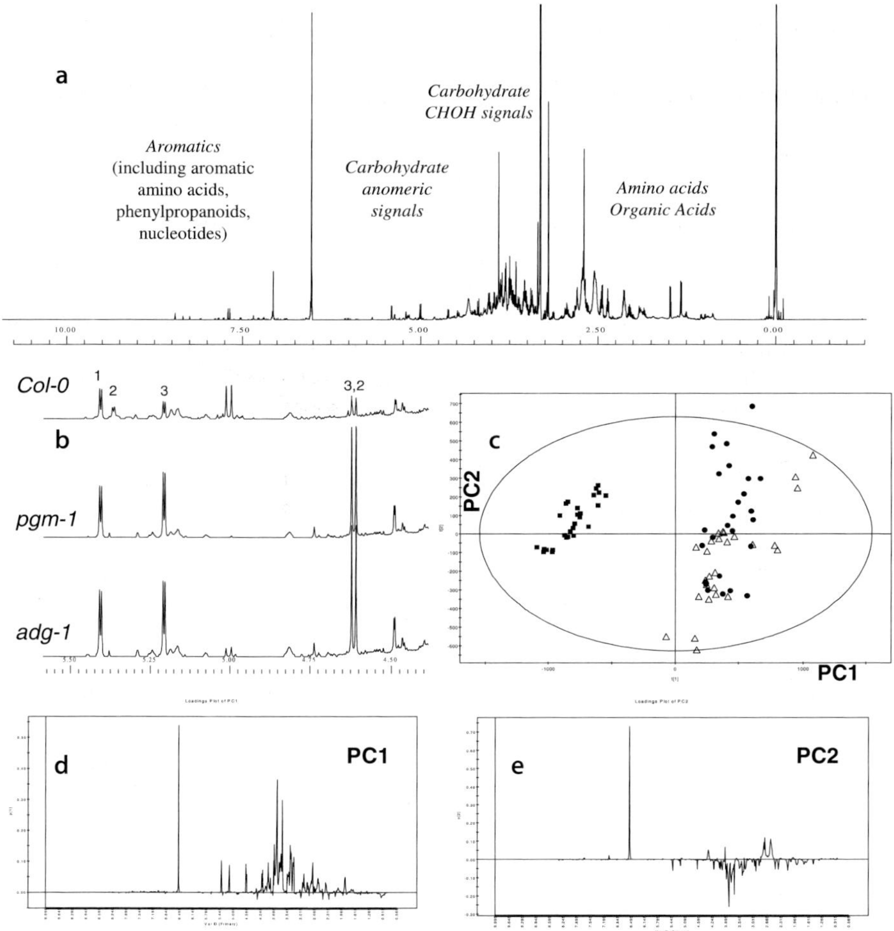

Fig. 1. **a** 600 MHz ^{1}H NMR spectrum of 4:1 (D_2O:CD_3OD) extract of freeze-dried *Arabidopsis thaliana* Col-0 tissue. **b** Expanded portion of the spectrum featuring the carbohydrate anomeric proton region highlighting differences in carbohydrate concentrations between Col-0 and the starch biosynthesis mutants *pgm-1* and *adg-1*. Labelled peaks: *1*-sucrose, *2*-maltose, *3*-glucose. **c** PCA scores plot illustrating differences observed between Col-0 and the *adg-1/pgm-1* mutants (*filled squares* – Col-0, *open triangles* – *pgm-1*, *filled circles* – *adg-1*). **d** Loadings plot of PC1 depicting the 'spectra' of compounds responsible for differences between Col-0 and the mutants. **e** Loadings plot of PC2 depicting the differences between *adg-1* and *pgm-1*

mutants in the starch biosynthesis pathway (*pgm-1* and *adg-1*) can be seen. These differences can be interpreted in relation to the known function of the enzymes missing in the mutants. However, in most plant metabolomic applications many hundreds of similar spectra are collected. Simultaneous classical interpretation of these numbers of individual spectra is not possible, but the use of chemometrics (see next section) allows the spectroscopist to focus on compounds that are responsible for differences between individual plants, or populations of plants, and target more detailed analysis to particular compounds or biochemical pathways.

3 Data Analysis

NMR-based metabolomic datasets are very large, both in terms of the number of datapoints per sample (typically 32 k or 64 k), and also the number of samples and resulting spectra acquired (from dozens to thousands, often including replicates). In order to draw conclusions and make comparisons between large numbers of spectra, automated strategies must be employed for the analysis and interpretation of such data once they have been acquired. The literature on data analysis is extensive and will only be briefly discussed here. Interested readers are directed to the review on pattern recognition methods (Lindon et al. 2001) for further coverage of some of the issues.

Data manipulation typically starts with some form of 'bucketing' or 'binning' whereby the spectrum is split into discrete regions (typically between 0.01 and 0.04 ppm in width), which are then integrated to return a list of integral values for each spectrum. Whilst this reduces the resolution of the data, it has the advantage of removing small chemical shift changes due to slight pH variation between samples. Increasingly, work is being carried out using all of the datapoints in the spectrum by employing an algorithm to align the peaks, eliminating any unwanted variation (Stoyanova et al. 2004). NMR data is usually analysed initially using multivariate statistical methods such as Principal Component Analysis (PCA). PCA is a data visualisation method, useful for observing groupings within large datasets. There are a number of commercially available software products that carry out PCA and other related multivariate analyses. One that has been widely used for NMR data is SIMCA-P (Umetrics, Sweden). A PCA model can be displayed in a graphical fashion as a "scores" plot as shown in Fig. 1c. This example compares ^{1}H-spectra collected from polar extracts of wild-type *Arabidopsis thaliana* Col-0 with those from the two mutants in starch biosynthesis (*pgm-1* and *adg-1*). This plot is useful for observing any groupings in the data set and in addition will highlight outliers that may be due to errors in sample preparation or instrumentation parameters etc. Coefficients by which the original variables must be multiplied to obtain the score are called "loadings". Thus, "loading plots" [e. g. Fig. 1d,e] can be used to detect and display the spectral areas responsible for the separation in the data, and can be interpreted as positive and negative NMR spectra of the compounds responsible for the differences between the clusters. The numerical value of the loading of a given variable on a PC indicates how much the variable has in common with that component (Massart et al. 1988). In Fig. 1d, PC1 represents the NMR spectra of compounds differing between wild-types and both mutants, whilst PC2, Fig. 1e, represents the (smaller) difference between the mutants.

When carrying out PCA it is necessary to apply scaling methods to the bucketed data matrices. In NMR spectroscopic data, although the integral values across a spectrum are proportional to concentration and the number of resonances present, the largest resonances would, without scaling, have a dominant effect in multivariate analysis. Before PCA the data can be scaled in different ways. In the covariance matrix method the data are just mean-centred. In the correlation matrix method the data are mean-centred and then the columns (variables) of the data matrix are scaled to unit variance. Covariance matrices are most widely used for NMR data because they have the advantage that the loadings plots retain the scale of the original data and can be compared back to libraries of spectra for assignment. Variable stability scaling (VAST) has recently been described and offers advantages over previously employed scaling methods in terms of the downstream multivariate modelling (Keun et al. 2003). This method weights each variable according to a metric of its stability and can unearth subtle differences between lines against backgrounds of biological variation. However, in the plant research arena, where growth of many identical clones under controlled environment is easily achieved, biological variation can be minimised, for example by pooling many individuals, and thus presents less of a problem.

An alternative method of highlighting differences between sample sets against backgrounds of biological or experimental variation is Orthogonal Signal Correction (OSC) (Gavaghan et al. 2002). This data filtering method can be applied to the scaled data matrix before multivariate analysis, and can, if used carefully, yield insights that are not evident from PCA. Such deeper mining of large data sets for differences against a background of noise can also be obtained by supervised modelling methods such as Partial Least Squares-Discriminant Analysis (Lindon et al. 2001). The use of neural networks to classify spectra has been applied to the study of the herbicide mode of action on maize seedlings (Aranìbar et al. 2001).

4 Two-dimensional NMR

Two-dimensional (2D) NMR experiments, which make use of interactions between NMR-detectable nuclei within a molecule, can be used to increase the spectral resolution and highlight which peaks belong to the same molecule. These experiments generally have much longer acquisition times, posing problems to those researchers wanting to carry out high throughput data collection. Nevertheless, these experiments can be run in automation and generate data useful to the metabolomics researcher and are particularly useful in the assignment of identities to unknown peaks.

The 2D experiments can be split into homonuclear and heteronuclear experiments. Homonuclear experiments examine the correlations between nuclei of the same type (commonly ^{1}H). The TOCSY experiment (Total Correlation

Spectroscopy) describes all interactions in a spin system and therefore is one of the most informative experiments available. The experiment has been used to examine complex matrices such as root exudates and cell extracts (Fan et al. 1997). Heteronuclear correlation experiments – as the name suggests – examine the correlation between two different types of nuclei within a molecule. Indirect experiments such as HMQC (heteronuclear multiple quantum coherence) and HSQC (heteronuclear single quantum coherence) spectroscopy are particularly useful in structure assignment as is HMBC (heteronuclear multiple bond coherence) which can examine long range correlations. A relatively new technique DOSY (diffusion ordered spectroscopy), that is not dependent on analysis of spin–spin coupling, has been applied to analysis of complex mixtures such as liquid foods (Gil et al. 2004). Here resonances are separated in the second dimension by virtue of their diffusion coefficient. This coefficient is governed by molecular size and thus DOSY represents a novel tool to deconvolute overlapping signals and may be particularly useful in plant spectra to assign carbohydrate signals to mono-, di- or higher saccharides.

5 Stable Isotope Labelling

NMR spectroscopy is not restricted to the analysis of ^{1}H signals. ^{2}H, ^{13}C and ^{15}N isotopes, however, have a very low natural abundance, making the detection of these signals difficult. Stable isotope labelling leads to the selective enhancement of some of these signals, providing a powerful method for the scrutiny of metabolic pathways in many organisms including plants (Roberts 2000; Roscher et al. 2000). Stable isotope labelling in plants has been extensively reviewed in recent years (Ratcliffe et al. 2001; Ratcliffe and Shachar-Hill 2001, 2005). ^{13}C is the most useful isotope and there are many suitable precursors including acetate, amino acids, carbohydrates and carbon dioxide. ^{15}N labelling can be achieved using labelled nitrate or ammonium ions whilst D_2O can be used to supply plant tissue with ^{2}H. In the majority of cases, ^{13}C labelling studies involve the use of singularly labelled precursors although multiply labelled precursors can also be used. One of the first applications of this technique in plant metabolism concerned the measurement of ^{13}C–^{15}N bonds using a solid state cross polarisation technique (Schaefer et al. 1981). An important example that demonstrates the use of isotope labelling in plant metabolonics has been published by Kikuchi et al. (2004). Using wild type and ethanol-insensitive mutants of *Arabidopsis thaliana,* labelled with ^{13}C, they were able, using subtractive 2D-HSQC, to isolate and assign only those metabolites that were affected by ethanol treatment.

Isotopic labelling techniques are often used for measuring the fluxes of metabolites through metabolic pathways. The information can be used to establish the identity of the biochemical pathways involved. This type of study has been carried out to investigate metabolic pathways in plants using ^{2}H,

^{13}C and ^{15}N-labelling (Fox et al. 1995; Prabhu et al. 1996; Schleucher et al. 1998). The analysis of stable isotope labelling in pulse-chase and time-course experiments can also provide quantitative information on metabolic fluxes although this is restricted to simple linear pathways that are reasonably close to the entry point of the label into metabolism. For complicated pathways it may be useful to examine the distribution of the label once the system has reached a steady state.

6 Hyphenated NMR

Liquid chromatography-NMR-mass spectrometry (LC-NMR-MS) is arguably the most powerful of the hyphenated techniques available to the phytochemical researcher (Hostettmann and Wolfender 2001; Wolfender et al. 2001). Metabolite profiling using such hyphenated techniques can help to provide clean spectral information on components of a mixture of unknown metabolites in an extract or fraction, leading to a partial or a complete structure determination in a single online experiment. One of the disadvantages of LC-NMR is its lack of sensitivity, which hampers the on-flow measurement of minor metabolites. Another problem is the need to suppress solvent signals from the mobile phase which if left unsuppressed would dominate the spectrum. Signals residing near these solvent peaks may be suppressed together with the solvent signal. This can be a major drawback when dealing with unknown constituents. In these cases the use of sequential analysis using different solvent systems for the LC is necessary.

LC-NMR hyphenation has been a reality for over 20 years (Buddrus and Herzog 1980). However it is only since the improvements in solvent suppression, NMR sensitivity and the use of shielded magnets that the technique has received more widespread recognition. The technique has been successful when applied to plant extracts rich in natural products of relatively low molecular mass. Recent studies have emphasised the value of LC-NMR as a technique for obtaining detailed chemical profiles of species for taxonomic work. For example, using LC-NMR in both on-flow and stop-flow modes, flavones, xanthones and secoiridoids of several *Gentianaceae* taxa have been identified (Wolfender et al. 1997). Vogler et al. (1998) also used LC-NMR in the on-flow mode to identify nine anti-bacterial sesquiterpene lactones from a partially purified extract of *Vernonia fastigiata* (Asteraceae) without the need for isolation of the individual compounds.

LC-SPE-NMR is a relatively new concept with incorporation of an online SPE (solid phase extraction) cartridge to trap an analyte peak prior to introduction into the NMR flow probe. This can be done in automation without interruption of the column flow. An additional advantage of this system is that after drying the cartridge, analytes can be eluted in fully deuterated solvents, reducing the need for solvent suppression. Furthermore multiple trapping on the same cartridge concentrates the analytes. There are still relatively few publications using this technology although, an application of LC-SPE-NMR to the detection of compounds from oregano was recently reported (Exarchou et al. 2003). Very recently the technology was used for the rapid identification of antioxidants in complex commercial rosemary extracts (Pukalskas et al. 2005). In this work, all major compounds present in the extract were collected on SPE cartridges after their separation and analysed by both NMR and ESI-MS. LC-SPE-NMR using post column solid-phase extraction was also applied to the direct analysis of phenolic compounds in the polar fraction of olive oil (Christophoridou et al. 2005). As well as the identification of simple phenolic acids, lignans and flavonoids the technique enabled the identification of several new phenolic compounds not previously reported as constituents of olive oil.

7 Discussion: Applying NMR to Plant Metabolomics

NMR has proven to be an exceptionally useful tool in animal metabolomics. In plant metabolomics 1D-NMR coupled with multivariate pattern matching serves as an excellent screen to cluster plant lines/treatments by global analysis of the total extractable metabolome. This type of analysis serves as a first pass screen that also gives quantitative data on important abundant metabolites. The clustering information and preliminary metabolite data can then be used to guide more detailed analysis by other techniques. These more targeted techniques include the GC-, LC- and CE-mass spectroscopic techniques discussed elsewhere in this volume, but also include the 2D-NMR, hyphenated-NMR and isotope-labelling techniques described above.

Examples of 1D-NMR-PCA application can be found in several areas, for example, in functional genomics (Cornah et al. 2004; Le Gall et al. 2005), in analysis of ecotypic and cultivar variation (Ward et al. 2003; Frederich et al. 2004), in safety evaluation of GM crops (Noteborn et al. 2000; Charlton et al. 2003; Le Gall et al. 2003; Lewis et al. 2003; Defernez et al. 2004; Manetti et al. 2004), in analysis of the affects of infection (Choi et al. 2004b,c), in classifying mode of action of chemicals (Aranibar et al. 2001) and of course in quality control of food and herbal products (Vogels et al. 1996; Charlton et al. 2002). The scale of this type of screening will increase and new challenges facing researchers in this area concern the construction of databases of fingerprints from mutants and the interfacing of these with similar databases of spectra of pure standards, such that automated interpretation of complex spectra can be performed. Generic plant metabolomic problems such as temporal batch to batch variation caused by machine or chromatographic drift are not generally seen for NMR as instrument drift is minimal. Furthermore, for *Arabidopsis thaliana* at least, biological batch to batch variation in NMR spectra has been eliminated by careful control of growth and experimental procedures (Lewis et al. 2003). Thus the functional genomic goal of a database of electronically comparable profiles of large collections of gene knockout mutants and other genetic resources may now be achievable.

The potential of 2D- and hyphenated NMR to increase the number of metabolites that can be observed and quantified is yet to be realised. Although throughput will be decreased, technology platforms where selected samples from first pass 1D-NMR-MS-PCA screens are selected for further fractionation by SPE-NMR-MS and LC-SPE-NMR-MS are being put in place. Success with these will inevitably broaden the metabolome coverage, especially when used in parallel with GC-, CE- and LC-MS^n methods.

References

Aranìbar N, Singh BJ, Stockton GW, Ott KH (2001) Automated mode-of-action detection by metabolic profiling. Biochem Biophysl Res Commun 286:150–155

Bollard ME, Stanley EG, Lindon JC, Nicholson JK, Holmes E (2005) NMR-based metabonomic approaches for evaluating physiological influences on biofluid composition. NMR Biomed 18:143–162

Buddrus J, Herzog H (1980) Coupling of HPLC and NMR.1. Analysis of flowing liquid-chromatographic fractions by proton magnetic-resonance. Org Magn Reson 13:153–155

Charlton AJ, Farrington WHH, Brereton P (2002) Application of ^{1}H-NMR and multivariate statistics for screening complex mixtures: quality control and authenticity of instant coffee. J Agric Food Chem 50:3098–3103

Charlton A, Allnutt T, Holmes S, Chisholm J, Bean S, Ellis N, Mullineaux P, Oehlschlager S (2003) NMR profiling of transgenic peas. Plant Biotech J 2:27–35

Choi HY, Kim HK, Hazekamp A, Erkelens C, Lefeber AWM, Verpoorte R (2004a) Metabolomic Differentation of *Cannabis sativa* Cultivars using ^{1}NMR spectroscopy and principal component analysis. J Nat Prod 67:953–957

Choi H, Choi HY, Verberne M, Lefeber AMW, Erkelens C, Verpoorte R (2004b) Metabolic fingerprinting of wild type and transgenic tobacco plants by ^{1}H NMR and multivariate analysis technique. Phytochemistry 65:857–864

Choi HY, Tapias Casas E, Kim KH, Lefeber AMW, Erkelens C, Verhoeven JTH, Brzin J, Zel J, Verpoorte R (2004c) Metabolic discrimination of *Catharanthus roseus* leaves infected by *Phytoplasma* using ^{1}H-NMR spectroscopy and multivariate data analysis. Plant Physiol 135:398–2410

Christophoridou S, Dais P, Tseng L-H, Spraul M (2005) Separation and identification of phenolic compounds in olive oil by coupling high-performance liquid chromatorgraphy with post-column solid-phase extraction to nuclear magnetic resonance spectroscopy (LC-SPE-NMR). J Agric Food Chem 53:4667–4679

Cornah JE, Germain V, Ward JL, Beale MH, Smith SM (2004) Lipid utilisation, gluconeogenesis and seedling growth in Arabidopsis mutants lacking the glyoxylate cycle enzyme malate synthase. J Biol Chem 279:42916–42923

Defernez M, Gunning YM, Parr AJ, Shepherd LVT, Davies HV, Colquhoun IJ (2004) NMR and HPLC-UV profiling of potatoes with genetic modifications to metabolic pathways. J Agric Food Chem 52:6075–6085

Exarchou V, Godejohann M, van Beek TA, Gerothanassis IP, Vervoort J (2003) LC-UV-solid-phase extraction-NMR-MS combined with a cryogenic flow probe and its application to the identification of compounds present in Greek oregano. Anal Chem 75:6288–6294

Fan TMW, Lane AN, Pedler J, Crowley D, Higashi RM (1997) Comprehensive analysis of organic ligands in whole root exudates using nuclear magnetic resonance and gas chromatography-mass spectrometry. Anal Biochem 251:57–68

Fox GG, Ratcliffe RG, Robinson SA, Stewart GR (1995) Evidence for deamination by glutamate-dehydrogenase in higher-plants – commentary. Can J Bot 73:1112–1115

Frederich M, Choiu YH, Angenot L, Harnischfeger G, Lefeber AWM, Verpoorte R (2004) Metabolomic analysis of *Strychnos nux-vomica, Strychnos icaja* and *Strychnos ignatii* extracts by ^{1}H nuclear magnetic resonance spectrometry and multivariate analysis techniques. Phytochemistry 65:1993–2001

Gavaghan CL, Wilson ID, Nicholson JK (2002) Physiological variation in metabolic phenotyping and functional genomic studies: use of orthogonal signal correction and PLS-DA. FEBS Lett 530:191–196

Gil AM, Duarte I, Cabrita E, Goodfellow BJ, Spraul M, Kerssebaum R (2004) Exploratory applications of diffusion ordered spectroscopy to liquid foods: an aid towards spectral assignment. Anal Chim Acta 506:215–223

Hostettmann K, Wolfender JL (2001) Applications of liquid chromatography/UV/MS and liquid chromatography/NMR for the online identification of plant metabolites. In: Tringali C (ed) Bioactive compounds from natural products-isolation, characterisation and biological properties. Taylor and Francis, London, pp 31–68

Keun HC, Ebbels TMD, Antti H, Bollard ME, Beckonert O, Holmes E, Lindon JC, Nicholson JK (2003) Improved analysis of multivariate data by variable stability scaling: application to NMR-based metabolic profiling. Anal Chim Acta 490:265–276

Kikuchi J, Shinozaki K, Hirayama T (2004) Stable isotope labelling of *Arabidopsis thaliana* for an NMR-based metabolomics approach. Plant Cell Physiol 45:1099–1104

Le Gall G, Colquhoun IJ, Davis AL, Collins GJ, Verhoeyen ME (2003) Metabolite profiling of tomato (*Lycopersicon esculentum*) using ^{1}H NMR spectroscopy as a tool to detect potential unintended effects following a genetic modification. J Agric Food Chem 51:2447–2456

Le Gall G, Metzdorff SB, Pedersen J, Bennett RN, Colquhoun IJ (2005) Metabolite profiling of *Arabidopsis thaliana* (L.) plants transformed with an antisense chalcone synthase gene. Metabolomics 1:181–198

Lewis J, Baker JM, Beale MH, Ward JL (2003) Metabolite profiling of GM plants: the importance of robust experimental design and execution. In: Nap JP, Atanassov A, Stiekema WJ (eds) Genomics for biosafety in plant biotechnology. NATO science series I, 359. IOS Press, Amsterdam, pp 47–57

Lindon JC, Nicholson JK, Holmes E, Everett JR (2000) Metabonomics: metabolic processes studied by NMR spectroscopy. Concepts Magn Reson 12:289–320

Lindon JC, Holmes E, Nicholson JK (2001) Pattern recognition methods and applications in biomedical magnetic resonance. Prog Nucl Magn Reson Spectrosc 39:1–40

Lindon JC, Holmes E, Nicholson JK (2004) Toxological applications of magnetic resonance. Prog Nucl Magn Reson Spectrosc 45:109–143

Manetti C, Bianchetti C, Bizzari M, Casciani L, Castro C, d'Ascenzo G, Delfini M, di Cocco ME, Lagana A, Miccheli A, Motto M, Conti F (2004) NMR-based metabonomic study of transgenic maize. Phytochemistry 65:3187–3198

Massart DL, Vandeginste BGM, Deming SN, Michotte Y, Kauffman L (1988) Chemometrics: a textbook. Elsevier, Amsterdam

Nicholson JK, Lindon JC, Holmes E (1999) 'Metabonomics': understanding the metabolic responses of living systems to pathophysiological stimuli via multivariate statistical analysis of biological NMR spectroscopic data. Xenobiotica 29:1181–1189

Noteborn HPJM, Lommen A, van der Jagt RC, Weseman JM (2000) Chemical fingerprinting for the evaluation of unintended secondaty metabolic changes in transgenic food crops. J Biotech 77:103–114

Prabhu V, Chatson KB, Abrams GD, King J (1996) C-13 nuclear magnetic resonance detection of interactions of serine hydroxymethyltransferase with C1-tetrahydrofolate synthase and glycine decarboxylase complex activities in Arabidopsis. Plant Physiol 112:207–216

Pukalskas A, van Beek TA, de Waard P (2005) Development of a triple hyphenated HPLC-radical scavenging detection-DAD-SPE-NMR system for the rapid identification of antioxidants in complex plant extracts. J Chromatography A 1074:81–88

Ratcliffe RG, Shachar-Hill Y (2001) Probing plant metabolism with NMR. Annu Rev Physiol Plant Mol Biol 52:499–526
Ratcliffe RG, Shachar-Hill Y (2005) Revealing metabolic phenotypes in plants: inputs from NMR analysis. Biol Rev 80:27–43
Ratcliffe RG, Roscher A, Shachar-Hill Y (2001) Plant NMR spectroscopy. Prog Nucl Magn Reson Spectrosc 39:267–300
Roberts JKM (2000) NMR adventures in the metabolic labyrinth within plants. Trends Plant Sci 5:30–34
Roscher NJ, Kruger NJ, Ratcliffe RG (2000) Strategies for metabolic flux analysis in plants using isotope labelling. J Biotechnol 77:81–102
Schaefer J, Skokut TA, Stejskal EO, McKay RA, Varner JE (1981) Estimation of protein-turnover in soybean leaves using magic angle double cross-polarization N-15 nuclear magnetic-resonance. J Biol Chem 256:1574–1579
Schleucher J, Vanderveer PJ, Sharkey TD (1998) Export of carbon from chloroplasts at night. Plant Physiol 118:1439–1445
Stoyanova R, Nicholls AW, Nicholson JK, Lindon JC, Brown TR (2004) Automatic alignment of individual peaks in large high-resolution of spectral data sets. J Magn Reson 170:329–335
Viant MR (2003) Improved mthods for the acquisition and interpretation of NMR metabolomic data. Biochem Biophys Res Commun 310:943–948
Vogels JTWE, Terwel L, Tas AC, van den Berg F, Dukel F, van der Greef J (1996) Detection of adulteration in orange juices by a new screening method using proton NMR spectroscopy in combination with pattern recognition techniques. J Agric Food Chem 44:175–180
Vogler B, Klaiber I, Roos G, Walter CU, Hiller W, Sandor P, Kraus W (1998) Combination of LC-MS and LC-NMR as a tool for the structure determination of natural products. J Nat Prod 61:175–178
Ward JL, Harris C, Lewis J, Beale MH (2003) Assessment of ^{1}H NMR spectroscopy and multivariate analysis as a technique for metabolite fingerprinting of *Arabidopsis thaliana*. Phytochemistry 62:949–957
Wolfender JL, Rodriguez S, Hostettmann K, Hiller W (1997) Liquid chromatography/ultra-violet/mass spectrometric and liquid chromatography/nuclear magnetic resonance spectroscopic analysis of crude extracts of *Gentianaceae* species. Phytochem Anal 8:97–104
Wolfender JL, Ndjoko K, Hostettmann K (2001) The potential of LC-NMR in phytochemical analysis. Phytochem Anal 12:2–22

I.7 Hetero-nuclear NMR-based Metabolomics

J. Kikuchi[1,2,3] and T. Hirayama[2,3,4,5]

1 Introduction

Novel methods for measurement of living systems are making new breakthroughs in life science. In the era of the metabolome (analysis of all measurable metabolites), a mass spectrometry (MS)-based approach is considered to be the major technology (Aharoni et al. 2002; Fiehn 2002; Sumner et al. 2003), whereas a nuclear magnetic resonance (NMR)-based method is frequently regarded as a minor technology due to its low sensitivity. However, we intend to strengthen the NMR-based approach, using advantages of NMR measurement, such as high quantification, non-invasive measurements, localized in vivo spectroscopy, selectivity of nuclear environments, and validity of structure analysis of diverse biomolecules including stereo-isomers. Attractive NMR-based metabolic analyses can be achieved by uniform stable isotope labeling of organisms allowing the application of multi-dimensional NMR experiments that have been used in protein structure determination (Kikuchi et al. 2004; Kikuchi and Hirayama 2005). Using these novel methods, the dynamic molecular networks inside cells and tissues will be dissected.

2 Historical Aspects of NMR Studies of Plant Metabolism

The history of NMR has been sharpened by a succession of major technological and methodological advances, including greatly enhanced sensitivity due to improvements in electronic devices, probe design, high-field superconducting magnets, the field/frequency stability to allow multi-scan averaging, and also the development of pulse Fourier-transform methods, significant progress in data handling facilities, and the development of multi-dimensional NMR

[1] RIKEN Plant Science Center, 1-7-22 Suehiro-cho, Tsurumi-ku, Yokohama, 230-0045 Japan, e-mail: kikuchi@psc.riken.jp
[2] International Graduate School of Arts and Sciences, Yokohama City University, 1-7-29 Suehiro, Tsurumi-ku, Yokohama, 230-0045 Japan
[3] CREST, Japan Science and Technology Agency, 4-1-8 Hon-cho, Kawaguchi, 332-0012 Japan
[4] Genomic Sciences Center, RIKEN Yokohama Institute, 1-7-22 Suehiro, Tsurumi-ku, Yokohama, 230-0045 Japan
[5] Laboratory of Plant Molecular Biology, RIKEN Tsukuba Institute, 3-1-1 Koyadai, Tsukuba, 305-0074 Japan

Biotechnology in Agriculture and Forestry, Vol. 57
Plant Metabolomics (ed. by K. Saito, R.A. Dixon, and L. Willmitzer)

(Ernst 1992; Claridge 1999). NMR spectroscopy provides many new insights into the physiology of higher plants. The evolution of this particular application of NMR can be traced back to the ground-breaking ^{13}CNMR studies using magic-angle spinning methods (Schaefer and Stejskal 1976). The subsequent developments of the technique and its applications have been charted at regular intervals in the review literature, and although not as widely exploited as its proponents might wish, NMR is now becoming an established technique in the armory of plant biochemists.

3 ^{1}H-NMR-based Metabolomics

NMR signals are highly reproducible, and quantitative assessment of each metabolite in a sample is therefore guaranteed. In contrast, MS-signals are sometimes less quantitative due to problems of matrix effects ("ion suppression" or "ion enhancement") (Mei et al. 2003; Mallet et al. 2004). Because NMR is a nondestructive technique, it is easy to combine NMR analysis with a complementary technique such as gas chromatography/MS or liquid chromatography/MS (Corcoran and Spraul 2003; Ott et al. 2003). In contrast to these applications in which numerous specific metabolites can be identified in complex mixtures, other investigators have addressed the question of whether computer-aided comparisons of the ^{1}H NMR spectra of partially fractionated extracts can yield statistically meaningful metabolic fingerprints of the extracted tissue. Using this approach, it was possible to show that there were minimal compositional differences between certain transgenic and non-transgenic tobacco varieties, but only after accounting for the substantial effects of external factors (Choi et al. 2004).

4 Use of Stable Isotope Labeling Technique to Enable Monitoring of the Dynamic Movement of Metabolites

NMR signals can be detected from the nuclei of many isotopes; ^{1}H, ^{13}C, ^{15}N and ^{31}P are the most widely used for biological NMR spectroscopy (Ratcliffe et al. 2001). For carbon the relevant magnetic isotope is ^{13}C. Its natural abundance is only 1.11%, contributing to the considerably lower sensitivity for ^{13}C NMR than for ^{1}H NMR. Accordingly, the application of ^{13}C NMR in unlabeled systems is largely confined to the detection of the most abundant metabolites, such as the organic solutes that accumulate in response to salt stress or certain secondary metabolites (Ratcliffe and Sachar-Hill 2001). Indirect detection techniques such as ^{13}C-hetero-nuclear single quantum coherence (HSQC) pulse sequence increases the sensitivity of the experiments (Vuister and Bax 1992). An example of this approach can be found in an analysis of alkaloid biosynthesis in vivo (Hinse et al. 2003).

The nitrogen atom has two magnetic isotopes, ^{14}N and ^{15}N, and both can be useful for the detection of metabolites in vivo and in extracts. The practicality of detecting the naturally abundant (99.63%) ^{14}N isotope was first demonstrated in root tissues and subsequently in vivo ^{14}N NMR has mainly been used for the analysis of ammonium and nitrate. The extremely low natural abundance of the ^{15}N isotope (0.037%) rules out the detection of unlabeled metabolites, but after labeling with [^{15}N]ammonium or [^{15}N]nitrate it is possible to use in vivo ^{15}N NMR to detect amino acids, as well as certain secondary products. NMR methods are relatively insensitive, so only signals from compounds present at relatively high levels (concentrations of at least 10 µmol/L) can be detected in spectra (Krishnan et al. 2005). Since metabolic engineering often results in the accumulation of relatively high concentrations of metabolites, this insensitivity is often not as restrictive for compound detection and identification as it is in other areas of biochemistry.

5 Approach for Hetero-nuclear NMR-based Metabolomics

In recent years, hetero-nuclear NMR methods and their spectral editing technologies have developed rapidly. For example, careful selection of window functions and base-line corrections of two dimensional (2D)-spectra yielded improved signal dispersion and line shapes of cross peaks permitting clear subtraction 2D-spectra, a technically difficult and time-consuming procedure using conventional 1D-NMR technology (Deferenz and Colquhoun 2003). With the methodology used in recent protein NMR studies, differences in the molecular composition between wild type and mutant strains can be easily quantified. Therefore, we think advanced technologies in NMR analysis combined with stable isotope labeling are useful tool for metabolomic analysis. We report here stable isotope labeling experiments in *Arabidopsis* using carbon or nitrogen, two of the largest components of all organic compounds (Kikuchi et al. 2004; Kikuchi and Hirayama 2005). Figure 1 shows the basic concept of NMR-based plant metabolomics proposed in this study. The NMR-based approach has an advantage when comparing different samples. Spectral subtraction between different mutants or stimuli enables metabolite levels between different samples to be quantified.

5.1 Example of Hetero-nuclear NMR Experiments *In Vitro*

To demonstrate the usefulness of the NMR method (metabolomics), the metabolite profile of an ethanol-hypersensitive mutant of *Arabidopsis, gek1* (Hirayama et al. 2004), was analyzed (Fig. 2a). *Arabidopsis* seedlings were grown for two weeks on agar plates (see above) and treated with 0.5% ethanol or water for 10 h. Figure 2b shows the ^{1}H–^{13}C HSQC spectra of ethanol or water-treated wild type *Arabidopsis* extracts. The subtraction spectra were obtained

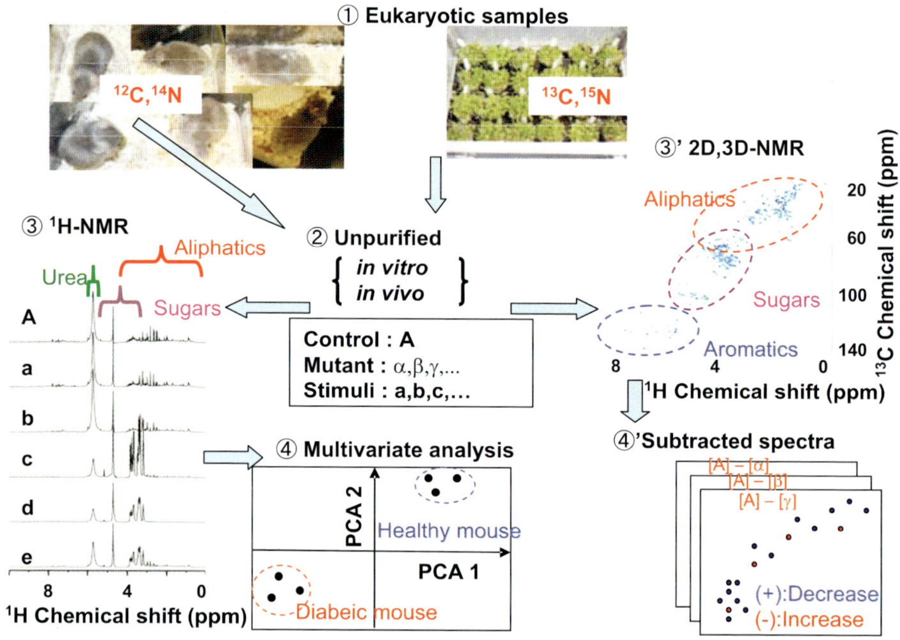

Fig. 1. Comparison of ordinal metabolomics approach (*left*: PCA-based) and our hetero-nuclear NMR metabolomics approach (*right*: multi-dimensional NMR-based)

by subtracting the spectrum of an ethanol-treated sample from that of a water-treated sample. Figure 2c shows the subtraction spectra from the wild-type (WT) (left) or the *gek1* (right) samples. The subtraction of measured spectra generates virtual NMR spectra that highlight compounds that are different between samples. In the present case, the subtraction spectra clearly show that upon ethanol treatment, glutamic acid is synthesized de novo in both the WT and the mutant, consistent with the previous observation that ethanol is converted into amino acids and lipids rapidly in plant tissues (Mellema et al. 2002; Rawyler et al. 2002). In addition, the ethanol-hypersensitive *gek1* mutant synthesized proline and γ-amino butylic acid (GABA) de novo, two compounds that have been reported to accumulate in cells under abiotic stresses such as drought and salinity (for reviews, Hare et al. 1998; Shelp et al. 1999). The assignments of these compounds were possible by comparing both ^{1}H and ^{13}C chemical shifts independently obtained with corresponding commercial reagents. Since ^{13}C-NMR chemical shifts are sensitive to differences in chemical structure but insensitive to the surrounding environment such as solvent effects (Kikuchi and Asakura 1999), the 2D-HSQC type spectra offer exceptionally useful information for assignment of individual chemical groups. From this point of view, construction of a database of 2D-HSQC spectra of main metabolites will enhance the NMR metabolomics.

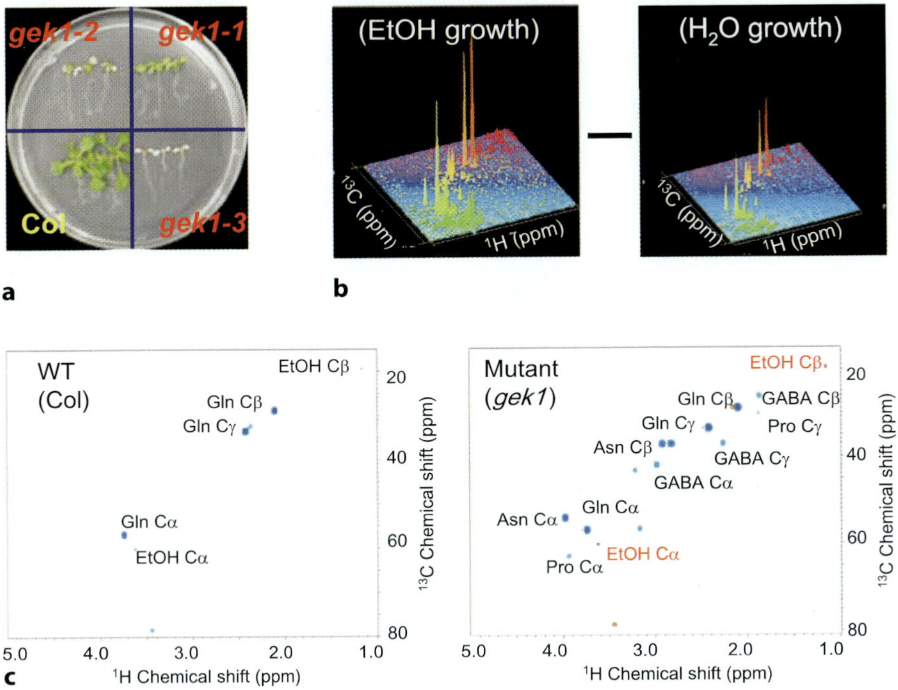

Fig. 2. a Example of how spectral subtraction can be used to differentiate environmental stress responses between WT and *gek1* (Hirayama et al. 2004) mutant. **b** For NMR spectroscopy, 5 mg of freshly frozen samples were heated with 0.5 mL H_2O and centrifuged at 15,000 *g* for 5 min to remove insoluble fractions. After adding 50 μL of ^2H_2O for NMR lock, supernatants were transferred into 5-mm NMR tubes. The spectra were measured on a Bruker DRX-500 spectrometer equipped with a ^1H inverse probe with triple axis gradient. A total of 200 complex f1 (^{13}C) and 1024 complex f2 (^1H) points were recorded with 64 scans per f1 increment. The spectral widths were 12,000 Hz and 8400 Hz for f1 and f2, respectively. **c** To quantify the signal intensities, a Lorentzian-to-Gaussian window with a Lorentzian line width of 10 Hz and a Gaussian line width of 15 Hz was applied in both dimensions, prior to Fourier transformation. A fifth order polynomial baseline correction was subsequently applied in the f1 dimension (Kikuchi et al. 2002). The indirect dimension was zero-filled to 2048 points in the final data matrix. NMR spectra were processed using NMRPipe software (Delaglio et al. 1995). Quantitative 2D-spectral subtraction was accomplished by editing a macro program of the NMRPipe software. Signal assignments are highlighted next to the corresponding cross peaks

5.2 Example of Hetero-nuclear NMR Experiments In Vivo

^{15}N uniformly labeled Arabidopsis seeds can be obtained from plants fed with a nutrient solution containing $^{15}NO_3$ as the sole nitrogen source. Using such seeds, the first ^1H–^{15}N HSQC-type NMR (Bodenhausen and Ruben 1980; Grzesiek and Bax 1993) in vivo experiments in plants were performed (Kikuchi et al. 2004). Figure 3 shows the development of the ^1H–^{15}N HSQC spectrum measured in living ^{15}N-labeled seeds that was induced by soaking the dry

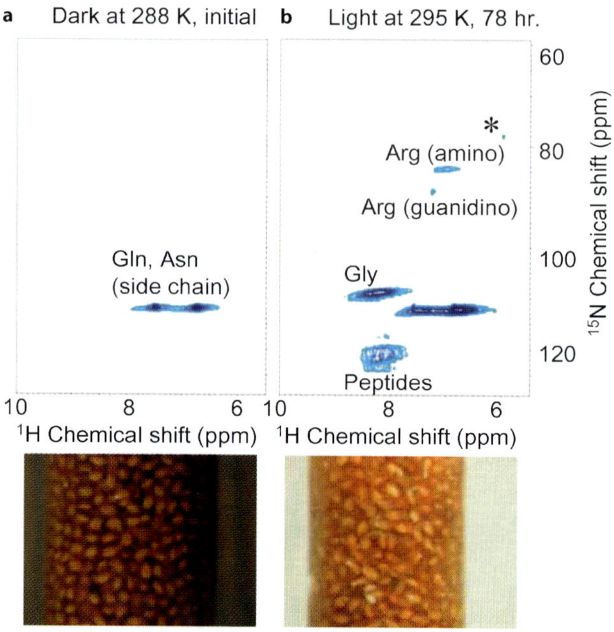

Fig. 3. Development of the ^{1}H–^{15}N HSQC spectrum measured in living ^{15}N-labeled seeds that was induced by soaking the dry seeds in water (*pictures shown at the bottom*). A total of 128 complex f1 (^{15}N) and 1024 complex f2 (^{1}H) points were recorded with 96 scans per f1 increment. The spectral widths were 4500 Hz and 8400 Hz for f1 and f2, respectively. Two spectra: **a** dark at 0 h; **b** light at 78 h are shown for comparison. Signal assignments *are highlighted* next to the corresponding cross peaks

seeds in water. Using in vivo measurement, dynamic movement of metabolites can be observed. In this case, at the initial stage just after water absorption into dried ^{15}N seeds, all cross peaks (especially those corresponding to peptide backbones) were very broad due to slow molecular motion in the dried seeds (Fig. 3a). After 12 h of imbibition, the line shapes of cross peaks, especially those corresponding to the glycine backbone and the side-chains of glutamic acid and asparatic acid, started to sharpen due to the enhanced molecular motion caused by the increasing water content. The temperature shift from 4 to 22 °C at 72 h of imbibition, and light illumination which started at 78 h of imbibition, both of which accelerate germination, further sharpened all cross peaks and enhanced the peptide backbone signal dramatically (Fig. 3b).

6 Prospects for the Future

As described above, NMR techniques possess an advantage over common analytical methods because they simultaneously provide information on the concentrations of numerous compounds as well as their spatial distribution. Therefore, NMR offers useful methodology for metabolomics. At this moment, however, there are several issues to be resolved before we can utilize the full power of NMR measurement in metabolomics. First, the sensitivity of NMR is rather lower. This disadvantage is being overcome with the progress in NMR technology. The sensitivity of a spectrometer scales as the 7/4th power of the static magnetic-field, and our group has developed the highest magnetic field (21.2 Tesla) used in biomolecular studies (Kiyoshi et al. 2004). In addition, NMR signal-to-noise (S/N) ratios can be substantially improved by cooling the NMR radio frequency detector and preamplifier . We are currently developing a 4.5-K cryogenetically cooled probe for the 920-MHz NMR spectrometer (Yokota et al. 2004). The increase of S/N gain is expected to be 8-fold, corresponding to a 64-fold reduction of the NMR acquisition time. The ^{1}H-^{13}C HSQC spectrum (shown in Fig. 1) recorded by the 500-MHz spectrometer exhibited 477 cross peaks identified by the NMRPipe software (Delaglio et al. 1995), corresponding to 100–200 metabolites at concentrations over 10^{-6} mol/L. However, theoretically the 920-MHz spectrometer equipped with a 4.5-K cryogenically cooled probe will be able to detect metabolites at concentrations as low as 10^{-8} to 10^{-9} mol/L. Furthermore, S/N gain by the cryogenetically cooled probe is significantly enhanced in low dielectric solvents (Horiuchi et al. 2005). In other words, the ^{1}H–^{13}C HSQC spectra recorded by 64 scans with the 500-MHz spectrometer (shown in Fig. 2) will be taken by only one scan with equivalent S/N ratio for the same sample but with higher resolution due to the higher magnetic field. Second, to facilitate the identification of metabolites in samples, a database of 2D HSQC spectra of known metabolites is required. We have just started to construct such a database. Once developed, metabolite analyses will be conducted with simultaneous quantification and metabolite identification. Furthermore, recent solid-state NMR methods will facilitate the study of insoluble metabolites such as starch, cell-wall components, and biomembranes (Kikuchi et al. 2000). Thus, the use of isotope labeling together with newly developed NMR technologies open a new avenue for plant metabolomics.

Acknowledgements. This work was supported in part by RIKEN GSC Internal Collaborations (No. 830-56625), by CREST (No. A88-54366), Japan Science and Technology Agency to J.K., T.H. We also acknowledge Grants-in-Aid for Scientific Research (No. 15710171, to J.K.; No. 15570045, to T.H.) from the Ministry of Education, Science, Sports and Culture of Japan.

References

Aharoni A, Ric de Vos CH, Verhoeven HA, Maliepaard CA, Kruppa G, Bino R, Goodenowe DB (2002) Nontargeted metabolome analysis by use of Fourier Transform Ion Cyclotron Mass Spectrometry. OMICS 6:217–234

Bodenhausen G, Ruben DJ (1980) Natural abundance nitrogen-15 NMR by enhanced heteronuclear spectroscopy. Chem Phys Lett 69:185–189

Choi H-K, Choi YH, Verberne M, Lefeber AWM, Erkelens C, Verpoorte R (2004) Metabolic fingerprinting of wild type and transgenic tobacco plants by ^{1}H NMR and multivariate analysis technique. Phytochemistry 65:857–864

Claridge TDW (1999) High-resolution NMR techniques in organic chemistry. Elsevier Science, London, UK

Corcoran O, Spraul M (2003) LC-NMR-MS in drug discovery. Drug Discov Today 8:624–631

Defernez M, Colquhoun IJ (2003) Factors affecting the robustness of metabolite fingerprinting using ^{1}H NMR spectra. Phytochemistry 62:1009–1017

Delaglio F, Grzesiek S, Vuister G W, Zhu G, Pfeifer J, Bax A (1995) NMRPipe: a multidimensional spectral processing system based on UNIX pipes. J Biomol NMR 6:277–293

Ernst RR (1992) Nuclear magnetic resonance Fourier transform spectroscopy. Angew Chem Int Ed Engl 31:805–823

Fiehn O (2002) Metabolomics – the link between genotypes and phenotypes. Plant Mol Biol 48:155–171

Grzesiek S, Bax A (1993) The importance of not saturating H_2O in protein NMR. Application to sensitivity enhancement and NOE measurements. J Am Chem Soc 115:12593–12594

Hare PD, Cress WA, van Staden J (1998) Dissecting the roles of osmolyte accumulation during stress. Plant Cell Environ 21:535–553

Hinse C, Richter A, Provenzani J, Stöckgt J (2003) In vivo monitoring of alkaloid metabolism in hybrid plant cell cultures by 2D cryo-probe NMR without labeling. Bioorg Med Chem 11:3913–3919

Hirayama T, Fujishige N, Kunii N, Iuchi S, Shinozaki K (2004) A novel ethanol hypersensitive mutant of *Arabidopsis*. Plant Cell Physiol 45:703–711

Horiuchi T, Takahashi M, Kikuchi J, Yokoyama S, Maeda H (2005) Effect of dielectric properties of solvents on the quality factor for a beyond 900 MHz cryogenic probe model. J Magn Reson 174:34–42

Kikuchi J, Asakura T (1999) Use of ^{13}C conformation-dependent chemical shifts to elucidate the local structure of a large protein with homologous domains in solution and solid state. J Biochem Biophys Method 38:203–208

Kikuchi J, Hirayama T (2005) Novel methods for uniform stable isotope labeling in plant and animal systems for a hetero-nuclear NMR based metabomics. 1st Int Metabol Meeting

Kikuchi J, Williamson MP, Shimada K, Asakura T (2000) Structure and dynamics of photosynthetic membrane-bound proteins in *Rhodobacter sphaeroides*, studied with solid-state NMR spectroscopy. Photosyn Res 63:259–267

Kikuchi J, Iwahara J, Kigawa T, Murakami T, Okazaki T, Yokoyama S (2002) Solution structure determination of the two DNA-binding domains in the *Shizosaccharomyces pombe* Abp1 protein by a combination of dipolar coupling and diffusion anisotropy restraints. J Biomol NMR 22:333–347

Kikuchi J, Shinozaki K, Hirayama T (2004) Stable isotope labeling of Arabidopsis thaliana for a hetero-nuclear NMR-based metabolomics approach. Plant Cell Physiol 45:1099–1104

Kiyoshi T, Maeda H, Kikuchi J, Ito Y, Hirota H, Yokoyama S, Ito S, Miki T, Hamada M, Ozaki O, Hayashi S, Kurihara N, Suematsu H, Yoshikawa M, Matsumoto S, Sato A, Wada H (2004) Present status of 920 MHz high-resolution NMR spectrometers. IEEE Trans Appl Supercond 14:1608–1612

Krishnan P, Kruger NJ, Ratcliffe RJ (2005) Metabolic finger printing and profiling in plants by NMR. J Exp Bot 56:255–265

Mallet CR, Lu A, Mazzeo JR (2004) A study of ion suppression effects in electrospray ionization from mobile phase additives and solid-phase extracts. Rapid Commun Mass Spectrom 18:49–58

Mei H, Hsieh Y, Nardo C, Xu X, Wang S, Ng K, Korfmacher WA (2003) Investigation of matrix effects in bioanalytical high-performance liquid chromatography/tandem mass spectrometric assays: application to drug discovery. Rapid Commun Mass Spectrom 17:97–103

Mellema S, Eichenberger W, Rawyler A, Suter M, Tadege M, Kuhlemeier C (2002) The ethanolic fermentation pathway supports respiration and lipid biosynthesis in tobacco pollen. Plant J 30:329–336

Ott K-H, Aranibar N, Singh B, Stockton GW (2003) Metabolomics classifies pathways affected by bioactive compounds. Artificial neural network classification of NMR spectra of plant extracts. Phytochemistry 62:971–985

Ratcliffe RJ, Shachar-Hill Y (2001) Probing plant metabolism with NMR. Annu Rev Plant Physiol Plant Mol Biol 52:499–526

Ratcliffe RJ, Roscher A, Shachar-Hill Y (2001) Plant NMR spectroscopy. Prog NMR Spect 39:267–300

Rawyler A, Arpagaus S, Braendle R (2002) Impact of oxygen stress and energy availability on membrane stability of plant cells. Ann Bot 90:499–507

Schaefer J, Stejskal EO (1976) C-13 Nuclear magnetic resonance of polymers spinning at magic angle. J Am Chem Soc 98:1031–1032

Shelp BJ, Bown AW, McLean MD (1999) Metabolism and functions of gamma-aminobutyric acid. Trends Plant Sci 4:446–452

Sumner LW, Mendes P, Dixon RA (2003) Plant metabolomics: large-scale phytochemistry in the functional genomics era. Phytochemistry 62:817–836

Vuister GW, Bax A (1992) Resolution enhancement and spectral editing of uniformly ^{13}C enriched proteins by homonuclear broadband ^{13}C–^{13}C decoupling. J Magn Reson 98:428–435

Yokota H, Okamura T, Ohtani Y, Kuriyama T, Takahashi M, Horiuchi T, Kikuchi J, Yokoyama S, Maeda H (2004) 4.5 K cooling system for a cryogenically cooled probe of a 920 MHz NMR. Adv Cryo Eng 49:1826–1833

Section II Bioinformatics

II.1 Bioinformatics Approaches to Integrate Metabolomics and Other Systems Biology Data

B. Mehrotra and P. Mendes[1]

1 Introduction

To understand the functioning of cells fully it is important to unravel the roles of genes and their products. The study of gene transcripts (transcriptomics) and proteins (proteomics) is progressing rapidly through the use of microarrays and mass spectrometry. Additionally, cells contain numerous other organic molecules not directly encoded in the DNA, the metabolites, which are critical for cell function. Knowledge about metabolites is crucial for an understanding of most cellular phenomena (Weckwerth 2003; Fernie et al. 2004; Kell 2004). Metabolomics is an emerging field consisting of the study of metabolites at a systems scale. It is similar in objectives to transcriptomics and proteomics; two major goals of metabolomics are the identification of all metabolites in each organism (their metabolomes) and measurements of their dynamics under many different challenges. Integrated approaches combining metabolomics with transcriptomics and proteomics are now underway (e. g. Verhoeckx et al. 2004; Broeckling et al. 2005) and are expected to result in much deeper insights than any of these techniques alone.

Metabolomics shares several characteristics with proteomics and transcriptomics. Like these, it is a technique where large numbers of molecules are profiled simultaneously (though current methods identify only hundreds of metabolites, vs thousands of proteins, and tens of thousands of transcripts). Metabolite profiles are, like transcript and protein profiles, snapshots of the state of a biological sample. Experimental data of all three types are usually dominated by a number of variables (molecules) much larger than samples, posing a hard challenge for data analysis and interpretation. Like proteomics, most metabolomics methods rely on spectroscopies to identify molecules; however, this is done through comparison against standards, rather than by mass fingerprints or sequence. The lack of a concept of "sequence" for metabolites is a major difference between metabolomics and the other two methodologies. Sequence is the key for identification of proteins and nucleic acids in large-scale profiles, but alternative methods must be used for metabolite profiles. De novo identification of metabolites can be done with 2D NMR, but requires considerable amounts of highly purified material, a major obstacle. Tandem mass spectrometry requires smaller amounts of material, but alone is often

[1] Virginia Bioinformatics Institute, Virginia Polytechnic Institute and State University, Washington St., MC 0477, Blacksburg, Virginia 24061, USA, e-mail: mendes@vt.edu

Biotechnology in Agriculture and Forestry, Vol. 57
Plant Metabolomics (ed. by K. Saito, R.A. Dixon, and L. Willmitzer)

insufficient to identify completely unknown metabolites. The alternative to de novo metabolite identification is to construct a library of standard profiles, created with purified metabolites (e. g. Wagner et al. 2003). While large spectral libraries for NMR, IR and mass spectrometry have existed for decades, these are very incomplete compared to the nearly 200,000 known natural products (Buckingham 1994). Additionally, these published spectral libraries are often not accurate enough for clear identifications. In the case of chromatography-mass spectrometry techniques, much better results are obtained if one has constructed a library with one's own equipment, which requires a large investment of time and finance. Thus, metabolite identification is one of the fundamental differences between metabolomics and proteomics/transcriptomics, and it is common to find two thirds or more unidentified metabolites in non-targeted profiles. The second major difference between these techniques is that metabolites have widely different chemical properties, such as polarity, volatility, molecular mass, and chemical reactivity. Comparatively, nucleic acids are very uniform in their properties; proteins, while more diverse than the former, are still approachable by their common properties (such as the amide bonds that can be used to sequence them). Because metabolites vary in their composition and structure, they require many different methods for extraction and separation, and no single existing technique is able to profile all metabolites in a biological sample (Sumner et al. 2003). Comprehensive coverage of the metabolome requires parallel analyses carried out with several different techniques.

Since the turn of the century, systems approaches have regained popularity with biologists. Perhaps this is because the analysis of purified molecules is rapidly approaching its limit, or simply because global experimental analyses have become possible. Either way, it is now recognized that systems biology studies of complex cellular phenomena are sorely needed (Kitano 2002). An increasingly appealing approach consists of experiments that simultaneously monitor the levels of transcripts, proteins, and metabolites, and combine their data to make inferences about the structure and dynamics of the underlying biochemical networks (Mendes 2001; Mendes et al. 2002; Oliver et al. 2002). In order to integrate the diverse and large amount of data generated by such experiments, several statistical and computational methods are required. In particular, it is important that all data be managed in a single database which also keeps track of the intricate details about how the experiments have been designed and the data generated. Ultimately, the data must be traceable backwards to samples and experiments, allowing not only for their interpretation but also for enabling others to replicate the experiments. These data about data are usually known as *metadata* and there have been several attempts at standardizing them. The MIAME standard was proposed for transcriptomics data (Brazma et al. 2001) and received widespread support, including support by prominent journals requiring data to conform to that standard. Similar proposals for proteomics data have been put forward, e. g. PEDRo (Taylor et al. 2003) and MIAPI (Orchard et al. 2004), but these are still under development and have not yet received the crucial support from the publishing world. Metabolomics is no different, and recently two proposals have been published to define standards for plant metabolomic data and metadata, ArMet (Jenkins et al. 2004) and MIAMET (Bino et al. 2004). Invariably, these attempts build upon the MIAME standard and hopefully will soon allow unequivocal specification of systems biology data. (The existing Systems Biology Markup Language, SBML (Hucka et al. 2003), is a standard for specifying systems biology models, rather than data.)

In the remainder of this chapter we will delineate ongoing efforts in our laboratory pertaining to integrating metabolomics and other functional genomics data. These efforts have arisen from our participation in plant systems biology studies of *Medicago truncatula* and *Vitis vinifera*, and also of the yeast *Saccharomyces cerevisiae*. All of these are large team efforts and we acknowledge all our collaborators for their vital role in these projects (see below).

2 Databases

We are developing a database system, DOME (database for OMEs), which stores functional genomics data originated from microarray measurements of transcripts, 2D-PAGE-QTOFMS protein assays, and GC-MS, LC-MS, CE-MS, and CE-LIF assays of metabolites. Early on it became evident that, in order to analyze all these data sets in an integrated way, they should be stored in a single database. This avoids, by design, the infamous data integration problem of bioinformatics (Davidson et al. 1995), as all data reside in the same schema, and queries can be made across all of them irrespective of their nature. This integration on a single schema was only possible by assuring that all necessary metadata was included and structured in an appropriate way. Figure 1 depicts a high-level overview of the DOME schema.

The main skeleton of the metadata schema consists of a hierarchy of experiment sets, experiments, sampling points, samples, and extracts, following exactly the way in which the biological material is manipulated in the experiments. This metadata schema also allows for easy export of transcript data in a format compatible with MIAME, and hopefully with the future proteomics and metabolomics standards as they stabilize (it is already compatible with PEDRo and ArMet). Linking through metadata is one of the main ways in which one can put together data from metabolomics with microarray and proteomics. For example, one single sampling point (a sample collection time) is attached to a set of perturbed and control replicate transcript levels, protein levels, and metabolite levels, as well as to their respective average values, comparisons and statistical significance. Thus one can relate the metabolite levels at one time in the experiment with the transcript and protein levels and, because the sampling points reflect experiment time, also to the molecular levels of previous or subsequent sampling points.

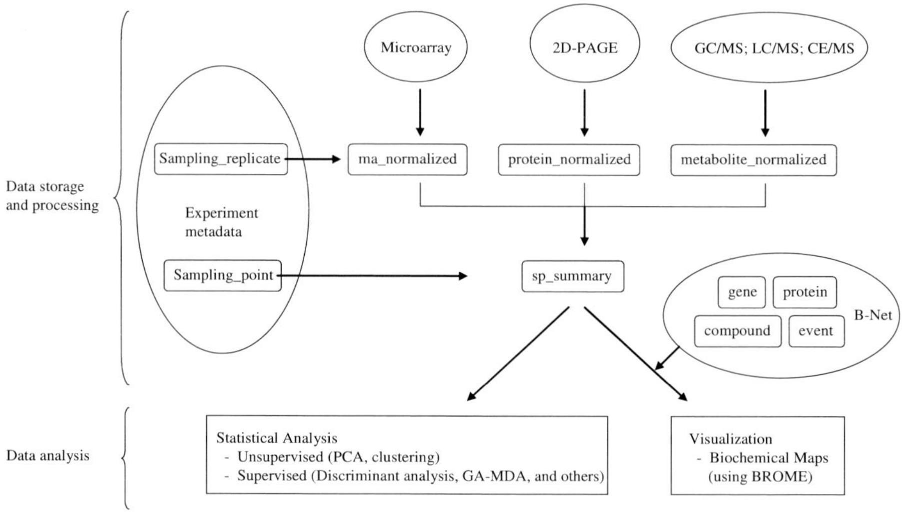

Fig. 1. High level schema of the integrative functional genomics database DOME. Metadata tables are used to provide context to the actual experimental data. Raw data from microarray, 2D-PAGE-MS, and various metabolomics technologies are stored in separate tables. These data are transformed by their appropriate normalization methods, keeping intermediate values, and finally arriving at numerical summaries of each sample (such as means and standard deviations), which are then comparable across all technologies and stored in the sp_summary table. It is the data in sp_summary that is then processed with higher-level statistical analyses or visualizations. It is also at this level that background information about the known molecular biology of the system (B-Net) can be integrated

Another feature that makes the comparison of metabolite with transcript and protein data possible is that each of these data types is first processed in appropriate ways, resulting in data of a similar type. Presently these data are reduced to ratios of levels (usually the level of a molecule in the perturbed state over its level in the control). Before data are available in this state they need to be processed through methods that are different for each technique. For example, the microarray data goes through a series of standardization and normalization procedures that take into account the technical details of microarray technology (Quackenbush 2001). The metabolite GC-MS data are first corrected for sample size, then deconvoluted into a series of peaks, which are identified by the software AMDIS (Davies 1998) relative to an internal standard that was included in the samples. An important issue is that the data should all be represented either in linear or in logarithmic space, not a mixture of the two. It is also important to preserve raw data, because there is always the possibility that in the future better methods to process it will appear; however, because the raw data are not commonly used in the analysis, it is enough to store these offline as long as the database keeps appropriate track of their location.

A third way to relate metabolite data with transcript and protein data is through the use of existing biochemical knowledge. This is, of course, the traditional way to analyze such comparisons; however, it is usually done by experts in an ad-hoc way. In order for this to be automated into computational procedures, it is necessary to represent this domain knowledge in appropriate schemas. Several biochemical databases exist that partially fulfill our needs (Kanehisa et al. 2004; Krieger et al. 2004), though they fall short in a number of ways (Wittig and de Beuckelaer 2001; Xing Li et al. 2002). In particular, they are rather poor in terms of their coverage of plant secondary metabolism, even AraCyc (Mueller et al. 2003), which is specific for *Arabidopsis*. Thus we created a sub-schema in our database to represent the existing knowledge about the biochemistry of the species in question. Because this can be useful on its own (i. e. independent of the experimental data), it was designed in a manner that allows it to be an autonomous database, and has been named B-Net. B-Net has been populated with gene/transcript information from TIGR's gene indices (Lee et al. 2005) and SGD (Dwight et al. 2004), with protein information from UniProt (Bairoch et al. 2005), metabolite information from LIGAND (Kanehisa et al. 2004) and AraCyc (Mueller et al. 2003), and supplemented with data collected directly from the literature by a team of curators in our laboratory. All of the facts in B-Net are documented for the type of evidence that supports them, using a method generalized from the Gene Ontology's evidence codes (Ashburner et al. 2000). Note that the information imported from external databases is filtered to remove entries that do not represent specific molecular entities. This was particularly important in the case of LIGAND, where groups of molecules are represented alongside individual molecules (e. g. "amino-acids", instead of the specific ones); however only the latter were imported to B-Net. B-Net also classifies entries with gene ontology terms wherever possible.

It is relevant here to highlight two problems that pervade metabolomics databases. The first is the issue of metabolites that are detected but not clearly identified, already mentioned above. These metabolites are sometimes referred to as "unknowns". Despite their identity not being known, a database must distinguish between them, and so these are usually named by the analytical chemists through some ad-hoc scheme. Such names are often attributed in ways that prevent comparison of data between different labs, for example by choosing identifiers that are used in different contexts meaning different things. A negative consequence would be that in two separate experiments two unknown metabolites might receive a common name, even though no one had intended to mean that they were the same molecular entity. In order to overcome this problem, and because our database contains data originated from several laboratories, we have developed a naming scheme that assures that unknown metabolites from different experiments and laboratories are not accidentally named the same thing. This naming convention has been proposed to the community in a recent Opinion article (Bino et al. 2004) and we hope that it becomes adopted by many laboratories, as this is the only way in which it would become useful. Essentially, the name for each unknown metabolite is composed of an identifier for the laboratory, one for the extraction method, another one for the type of analysis carried out, and at least two coordinates from the analysis. These coordinates are context specific, and could be retention time of a separation, mass-to-charge ratio, chemical shift, wavenumber, etc. Another objective with this naming convention is to allow for future analyses that attempt to establish identity between these unknown metabolites. It is expected that many of these are observed in different studies in separate laboratories, and through the inclusion of the analysis coordinates in their names it becomes easier to recognize that two unknowns may actually be the same molecule. For example, if several studies consistently identified a peak in GC-MS (using the same extraction and analysis parameters) with the same retention time and main ion mass, then it may be that the two are the same metabolite. By assigning names derived from this scheme, it then also becomes possible to create lists of molecular entities that have been observed and not yet identified (a kind of "orphan" list for metabolomics).

Another unresolved issue that is being encountered in our projects is that the same metabolites in a sample might have been observed by more than one technique. The problem that is posed then is which quantification should one chose if they do not agree. This is complicated by the fact that when metabolites appear in an analysis, they may not have been present in the original sample, but instead result from an artifact of the extraction method. Another reason could be that the same metabolite might be present in different locations in the cell, leading to the metabolite being isolated in two separate pools. In the latter case the two pools should both be represented in the database, while in the case of artifacts, one should use only the more accurate quantification. This issue results in a need for careful annotation of metabolomics results, but also requires special structures in the database schema that are capable of representing several pools of a single metabolite.

3 Data Visualization

Scientific data visualization is the activity of displaying properties of a data set that help the human scientist to identify quickly its most important characteristics. This is not a simple problem because it is very hard to identify what would be important for each scientist, and it is as much an issue of the scientific domain as it is of psychology. Nevertheless, there are data properties that are generally sought by a large class of researchers, and visualization software is focused on them. For metabolomics, a frequent way in which biochemists like to visualize data is through the use of maps that depict portions of the biochemical network. Several packages exist that allow for this map-based data visualization (Luyf et al. 2002; Shannon et al. 2003; Thimm et al. 2004; Lange and Ghassemian 2005), and we have also developed one (BROME, BRowser for OMEs) that is coupled to our database system DOME for visualization of metabolomics data with transcriptomics and/or proteomics data. This allows our database to select only a small set of metabolites, enzymes and genes that are present in a certain map, such that the researcher can then quickly observe their levels organized according to how we believe the biochemistry is organized. This could help in understanding how changes in mRNA or protein levels affect the level of metabolites in a certain pathway or network. However this is not as straightforward as it may seem: the changes in level of mRNA are likely very different from changes in protein levels or changes in metabolite levels. Cells cannot tolerate large changes of many metabolites, while mRNA levels can change widely without much toxicity. Thus, in order to visualize the expression of metabolites, mRNA, and proteins in the same biochemical map, they need to be expressed on different scales, or otherwise normalized to some comparable scale. A problem with thinking about data as part of some biochemical map ("pathway") is that it is likely that molecules in the map are also involved in other interactions not depicted there. Therefore, looking at a particular slice of a network could be highly misleading. It has been shown that the concentrations of metabolites next to each other in a metabolic map do not necessarily have high correlation (Steuer et al. 2003; Camacho et al. 2005), strengthening this point. In order to understand a change in the level of a particular metabolite, it may be more useful to view the expression changes of *all* enzymes (i. e. their protein and mRNA levels) linked with that metabolite. For this we have developed the concept of metabolite neighborhood maps (Xing Li et al. 2002), which are local views of the biochemical network and consist of all the reactions that affect the metabolite of interest, including all the metabolites and enzymes that take part in those reactions. BROME has a large number of maps available, from these neighborhood maps to the nice pathway maps of the KEGG system (Kanehisa et al. 2004).

4 Data Analysis

Analysis of metabolomic data can use the same multivariate statistical methods that are widely used in microarray data analysis. These methods can be either supervised, where each sample or variable (molecule) is associated to an already known class, or unsupervised, where there is no pre-classification of the data (Mendes 2002; Sumner et al. 2003; Goodacre et al. 2004). Unsupervised methods are widely popular, and the most used are principal component analysis (PCA), hierarchical clustering (HCA), k-means clustering, and self-organizing maps (SOM). Unsupervised analyses are mostly guided by the variance and covariance (or correlation) in the data sets, so they are good at finding patterns therein; however nothing guarantees, other than a careful experimental design, that the largest variance is indeed a result of the perturbation rather than other unwanted effects. On the other hand, supervised analyses are guided by the pre-existing knowledge provided by the researcher and so are usually based on discrimination, a property that is more related to the consistency of the members in a class, and the differences between classes. Supervised methods already demonstrated for metabolomics data are linear discriminant analysis (Raamsdonk et al. 2001; Bundy et al. 2005), discriminant partial least squares (PLS-DA) (Gavaghan et al. 2002; Jonsson et al. 2004), genetic algorithms (Johnson et al. 2003; Goodacre 2005), genetic programming (Allen et al. 2003; Goodacre 2005) and other methods (Goodacre et al. 2000; Shi et al. 2004).

Obviously, much more information could be extracted from systems biology experiments if metabolomics data were analyzed coupled with transcriptomics and proteomics data, but this requires much attention to ensure that the data be comparable, as has been discussed in the previous section (see also Purohit et al. 2004). For example, multidimensional scaling should be preferred to principal component analysis since the former takes into account the different scales of each variable, but the latter does not (at least in its original incarnation that is based on covariance).

We have recently tried to understand how to interpret correlation between metabolites in metabolomics data sets (Camacho et al. 2005) following results from an analysis by Steuer et al. (2003). Briefly, we have found that strong correlation arises when two or more metabolites are in chemical equilibrium, when they conserve a common chemical moiety, when their concentrations are mostly controlled by a single enzyme, or when one of the enzymes that affects their concentration changes with greater magnitude than the other enzymes that also affect them (Camacho et al. 2005). We confirmed that two metabolites next to each other in a biochemical network (e. g. substrate and product of a single enzyme) are not expected to be strongly correlated, in general. These conclusions put the results that have been obtained demonstrating high correlations between metabolite pairs into a systems biology context; often such correlations are only explained at a global level, and through the action of proteins.

Since metabolomics data are best understood when accompanied by other systems biology data, there seems to be a great need for methods that specifically address data integration. As demonstrated by the study of metabolite correlations, methods that take into consideration pre-existing biochemical knowledge are likely to be more effective. Since many metabolomics experiments are carried out through time intervals, methods that consider the time dimension explicitly are likely to be more productive than those that do not.

5 Conclusion

In this chapter we have addressed some of the bioinformatic issues related to metabolomics and its integration within the systems biology framework. We believe that metabolite, transcript, and protein analyses are much more powerful combined than individually. In order to extract maximal benefit from such combined studies, specific bioinformatics support is necessary in the form of databases, visualization, and data analysis. Ultimately, a full understanding of the underlying phenomena will require an additional layer of computational and theoretical tools, supporting the formulation and evaluation of dynamic models that attempt to represent the biological system. Such models will need to be predictive, but we believe that, much more than that, they need to be explanatory. Within our laboratory we are pursuing several projects in this direction and have a strong interest in combining that approach with the data and informatics systems described here, as have others. This will be a topic of much discussion in the near future and we await it with excitement.

Acknowledgements. We thank our collaborators John Cushman, Grant Cramer, Rick Dixon, Greg May, David Schooley, Vladimir Shulaev, and Lloyd Sumner for the excellent experimental and analytical data collection work in their laboratories. We also thank our colleagues Xing Li and Aejaaz Kamal for their work on the DOME and BROME systems, and Diogo Camacho and Alberto de la Fuente for the metabolite correlation analysis. We acknowledge the generous support of the National Science Foundation's Plant Genome Research Program (awards DBI-0109732 and DBI-0217653).

References

Allen J, Davey HM, Broadhurst D et al. (2003) High-throughput classification of yeast mutants for functional genomics using metabolic footprinting. Nat Biotechnol 21:692–696

Ashburner M, Ball CA, Blake JA et al. (2000) Gene ontology: tool for the unification of biology. The Gene Ontology Consortium. Nat Genet 25:25–29

Bairoch A, Apweiler R, Wu CH et al. (2005) The Universal Protein Resource (UniProt). Nucleic Acids Res 33:D154–D159

Bino RJ, Hall RD, Fiehn O et al. (2004) Potential of metabolomics as a functional genomics tool. Trends Plant Sci 9:418–425

Brazma A, Hingamp P, Quackenbush J et al. (2001) Minimum information about a microarray experiment (MIAME)-toward standards for microarray data. Nature Genet 29:365–371

Broeckling CD, Huhman DV, Farag MA et al. (2005) Metabolic profiling of *Medicago truncatula* cell cultures reveals the effects of biotic and abiotic elicitors on metabolism. J Exp Bot 56:323–336

Buckingham J (1994) Dictionary of natural products. Chapman and Hall/CRC, London

Bundy JG, Willey TL, Castell RS, Ellar DJ, Brindle KM (2005) Discrimination of pathogenic clinical isolates and laboratory strains of Bacillus cereus by NMR-based metabolomic profiling. FEMS Microbiol Lett 242:127–136

Camacho D, de la Fuente A, Mendes P (2005) The origin of correlations in metabolomics data. Metabolomics 1:53–63

Davidson SB, Overton C, Buneman P (1995) Challenges in integrating biological data sources. J Comput Biol 2:557–572

Davies T (1998) The new Automated Mass Spectrometry Deconvolution and Identification System (AMDIS). Spectroscopy, Europe 10:24–27

Dwight SS, Balakrishnan R, Christie KR et al. (2004) *Saccharomyces* genome database: underlying principles and organisation. Brief Bioinform 5:9–22

Fernie AR, Trethewey RN, Krotzky AJ, Willmitzer L (2004) Metabolite profiling: from diagnostics to systems biology. Nat Rev Mol Cell Biol 5:763–769

Gavaghan CL, Wilson ID, Nicholson JK (2002) Physiological variation in metabolic phenotyping and functional genomic studies: use of orthogonal signal correction and PLS-DA. FEBS Lett 530:191–196

Goodacre R (2005) Making sense of the metabolome using evolutionary computation: seeing the wood with the trees. J Exp Bot 56:245–254
Goodacre R, Shann B, Gilbert RJ et al. (2000) Detection of the dipicolinic acid biomarker in Bacillus spores using Curie-point pyrolysis mass spectrometry and Fourier transform infrared spectroscopy. Anal Chem 72:119–127
Goodacre R, Vaidyanathan S, Dunn WB, Harrigan GG, Kell DB (2004) Metabolomics by numbers: acquiring and understanding global metabolite data. Trends Biotechnol 22:245–252
Hucka M, Beale M, Fiehn O et al. (2003) The systems biology markup language (SBML): a medium for representation and exchange of biochemical network models. Bioinformatics 19:524–531
Jenkins H, Hardy N, Beckmann M et al. (2004) A proposed framework for the description of plant metabolomics experiments and their results. Nat Biotechnol 22:1601–1606
Johnson HE, Broadhurst D, Goodacre R, Smith AR (2003) Metabolic fingerprinting of salt-stressed tomatoes. Phytochem 62:919–928.
Jonsson P, Broadhurst D, Goodacre R et al. (2004) A strategy for identifying differences in large series of metabolomic samples analyzed by GC/MS. Anal Chem 76:1738–1745
Kanehisa M, Goto S, Kawashima S, Okuno Y, Hattori M (2004) The KEGG resource for deciphering the genome. Nucleic Acids Res 32:D277–D280
Kell DB (2004) Metabolomics and systems biology: making sense of the soup. Curr Opin Microbiol 7:296–307
Kitano H (2002) Systems biology: a brief overview. Science 295:1662–1664
Krieger CJ, Zhang P, Mueller LA et al. (2004) MetaCyc: a multiorganism database of metabolic pathways and enzymes. Nucleic Acids Res 32:D438–D442
Lange BM, Ghassemian M (2005) Comprehensive post-genomic data analysis approaches integrating biochemical pathway maps. Phytochemistry 66:413–451
Lee Y, Tsai J, Sunkara S et al. (2005) The TIGR Gene Indices: clustering and assembling EST and known genes and integration with eukaryotic genomes. Nucleic Acids Res 33:D71–D74
Luyf AC, de Gast J, van Kampen AH (2002) Visualizing metabolic activity on a genome-wide scale. Bioinformatics 18:813–818
Mendes P (2001) Modeling large scale biological systems from functional genomic data: parameter estimation. In: Kitano H (ed) Foundations of systems biology. MIT Press, Cambridge, MA, pp 163–186
Mendes P (2002) Emerging bioinformatics for the metabolome. Brief Bioinform 3:134–145
Mendes P, de la Fuente A, Hoops S (2002) Bioinformatics and computational biology for plant functional genomics. Rec Adv Phytochem 36:1–13
Mueller LA, Zhang P, Rhee SY (2003) AraCyc: a biochemical pathway database for *Arabidopsis*. Plant Physiol 132:453–460
Oliver DJ, Nikolau B, Wurtele ES (2002) Functional genomics: high-throughput mRNA, protein, and metabolite analyses. Metabolic Eng 4:98–106
Orchard S, Hermjakob H, Julian RK et al. (2004) Common interchange standards for proteomics data: Public availability of tools and schema. Proteomics 4:490–491
Purohit PV, Rocke DM, Viant MR, Woodruff DL (2004) Discrimination models using variance-stabilizing transformation of metabolomic NMR data. Omics 8:118–130
Quackenbush J (2001) Computational analysis of microarray data. Nat Rev Genet 2:418–427
Raamsdonk LM, Teusink B, Broadhurst D et al. (2001) A functional genomics strategy that uses metabolome data to reveal the phenotype of silent mutations. Nature Biotechnol 19:45–50
Shannon P, Markiel A, Ozier O et al. (2003) Cytoscape: a software environment for integrated models of biomolecular interaction networks. Genome Res 13:2498–2504
Shi H, Paolucci U, Vigneau-Callahan KE, Milbury PE, Matson WR, Kristal BS (2004) Development of biomarkers based on diet-dependent metabolic serotypes: practical issues in development of expert system-based classification models in metabolomic studies. Omics 8:197–208
Steuer R, Kurths J, Fiehn O, Weckwerth W (2003) Observing and interpreting correlations in metabolomic networks. Bioinformatics 19:1019–1026
Sumner LW, Mendes P, Dixon RA (2003) Plant metabolomics: large-scale phytochemistry in the functional genomics era. Phytochemistry 62:817–836

Taylor CF, Paton NW, Garwood KL et al. (2003) A systematic approach to modeling, capturing, and disseminating proteomics experimental data. Nat Biotechnol 21:247–254

Thimm O, Blasing O, Gibon Y et al. (2004) MAPMAN: a user-driven tool to display genomics data sets onto diagrams of metabolic pathways and other biological processes. Plant J 37:914–939

Verhoeckx KC, Bijlsma S, Jespersen S et al. (2004) Characterization of anti-inflammatory compounds using transcriptomics, proteomics, and metabolomics in combination with multivariate data analysis. Int Immunopharmacol 4:1499–1514

Wagner C, Sefkow M, Kopka J (2003) Construction and application of a mass spectral and retention time index database generated from plant GC/EI-TOF-MS metabolite profiles. Phytochemistry 62:887–900

Weckwerth W (2003) Metabolomics in systems biology. Annu Rev Plant Biol 54:669–689

Wittig U, de Beuckelaer A (2001) Analysis and comparison of metabolic pathway databases. Brief Bioinform 2:126–142

Xing Li X, Brazhnik O, Kamal A et al. (2002) Databases and visualization for metabolomics. In: Harrigan GG, Goodacre R (eds) Metabolic profiling: its role in biomarker discovery and gene function analysis. Kluwer Academic Publ, Boston, pp 293–309

II.2 Chemometrics in Metabolomics – An Introduction

J. Trygg[1], J. Gullberg[2], A.I. Johansson[2], P. Jonsson[1], and T. Moritz[2]

1 Introduction

In the post-genomics era, the use of methodologies that enable transcriptomic, proteomic and metabolomic data to be analysed in detail have revolutionized biological investigations. One of the major advantages with metabolomics investigations compared to traditional target metabolite analysis is that metabolomics data can give an unbiased view of changes in metabolism during environmental, genetic or developmental changes. Instead of tracking only a few metabolites, changes in relative amounts in 300 to 1000 or even more metabolites can be recorded and analysed, covering all major metabolic pathways. This development has accentuated the need to apply and further develop multivariate methodology. Chemometrics (see Eriksson et al. 2001) provides tools to make good use of measured data, enabling practitioners to make sense of measurements and to model quantitatively and produce visual representations of information. Today, chemometrics has grown into a well established data analysis tool in areas such as multivariate calibration, quantitative structure-activity modeling, pattern recognition and multivariate statistical process monitoring and control. Although seemingly diverse disciplines, the common denominators in these applications are that high complexity data tables are generated and that these data tables can be analysed and interpreted by means of chemometric methods.

In chemometrics, there are three basic categories of analysis (Fig. 1):

1. Exploratory analysis (Fig. 1A). This gives an overview of all the data in order to detect trends, patterns or clusters.
2. Classification analysis and discriminant analysis (Fig. 1B), which classifies samples into categories or classes, for example wild-type and mutant.
3. Regression analysis and prediction models (Fig. 1C) are used when a quantitative relationship between two blocks of data is sought. For example, when prediction of growth or fiber properties from mass spectrometry data.

However, in biology, chemometric methodology has still been largely overlooked in favour of traditional statistics. It is not until recently that the

[1] Research Group for Chemometrics; Organic Chemistry, Department of Chemistry, Umeå University, SE-901 87 Umeå, Sweden, e-mail: johan.trygg@chem.umu.se
[2] Umeå Plant Science Centre, Department of Forest Genetics and Plant Physiology, Swedish University of Agricultural Sciences, SE-901 87 Umeå, Sweden, e-mail: thomas.moritz@genfys.slu.se

Biotechnology in Agriculture and Forestry, Vol. 57
Plant Metabolomics (ed. by K. Saito, R.A. Dixon, and L. Willmitzer)

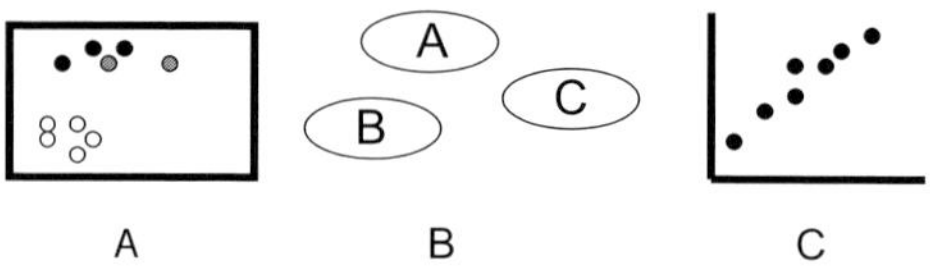

Fig. 1. Overview of the basic categories of chemometrics analysis: **A** overview of data structure; **B** classification and discriminant analysis; **C** regression analysis

overwhelming size and complexity of the 'omics' technologies has driven biology towards the adoption of chemometric methods. Here we will give an introduction to chemometrics and also give examples of why and when chemometrical methodologies should be used.

2 Theory and Methods

2.1 Making Data Contain Information – Design of Experiments

In experimental biology, e. g. when investigating how a number of different environmental factors (e. g. temperature, day length, nutrition) affect different responses such as growth, transcript profiles and metabolite profiles in plants, there is a need to carry out experiments in a systematic way. One way to investigate how the factors affect the plant's responses is to Change One Factor at a Time, i. e. the COST approach. This approach has severe problems: (1) finding optimal conditions for experiments (e. g. method development), (2) unnecessarily many experiments are needed (inefficiency), (3) ignores interaction among variables (lost information) and (4) provides no map over the experimental space.

Design of Experiments (DOE) (Lundstedt et al. 1998) is the methodology of how to conduct and plan experiments in order to extract the maximum amount of information in the fewest number of runs. The basic idea is to devise a small set of experiments, in which all pertinent factors are varied systematically. It is a fundamental tool for planning experiments and making data informative by simultaneously, albeit in a structured way, varying controllable factors (e. g. environmental conditions, instrument settings, experimental procedures) of the studied system. Today they comprise a tool box for virtually any experimental problem.

2.1.1 Stages in the DOE Process

Most of us can only grasp the effect of one factor at a time in our minds, and that often leads us into the inefficient COST approach. We need the mathematics (and the computer) to keep track of the factors and their combinations.

In summary, (1) all factors are varied together over a set of experimental runs, (2) noise is decreased by means of averaging, (3) the functional space is efficiently mapped, interactions and synergisms are seen.

1. What do I want? – formulate question(s) stating the objectives and goals of the investigation. For example identify factors (e. g. temperature, day length, nutrition) and factor ranges (e. g. 15–25 °C, 6–12 h, 1–10 mmol N/L) that affects flowering time.
2. Screening design – finding out a little about many factors. Which factors are the dominating ones in controlling flowering time? Screening designs provide simple models with information about dominating variables, and information about ranges. Pareto's principle states that 20% of the data (factors) account for 80% of the information. Different types of screening designs exist – which one to choose depends on the problem. The most common one is the fractional factorials design (Fig. 2). The full factorial design is a set of experimental runs where every level of a factor is investigated at both levels of all the other factors. It requires $N = 2^k$ number of runs for k factors. Investigating more than five factors with the full factorial design can in some cases become time consuming, i. e. $2^5 = 32, 2^6 = 64, 2^7 = 128$ experiments, etc. Instead, performing a *fractional factorial design* reduces

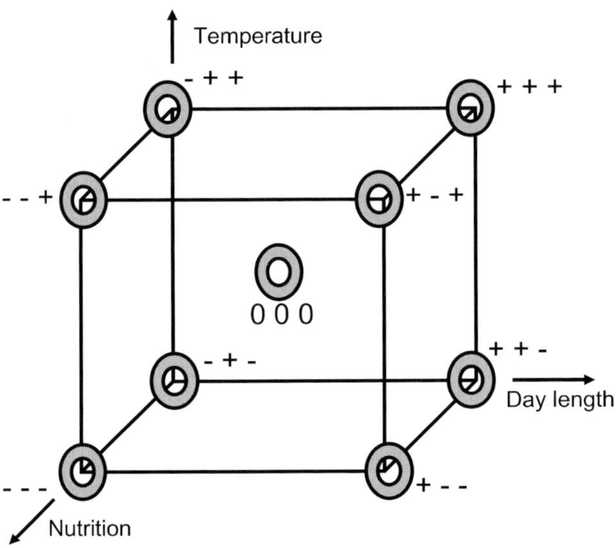

Fig. 2. Example of a full factorial design of experiments (DOE) for investigating how three factors (temperature, day length and nutrition) control flowering time. Varying the three factors at two levels (coded as +/-) requires $2^3 = 8$ experiments + center points. Each experiment according to the design set of experiments is *marked with a circle* in the figure. Evaluating the results from such an experimental design reveals the influence of each of the different factors separately and also any interactions between them. DOE is the only feasible approach to separate cause and effect from each other

that number quickly without the loss of too much information regarding the estimation of factors involved. Fractional factorial design takes advantage of the fact that three-way and higher interactions are seldom significant. It requires only $N = 2^{k-p}$ number of runs for k factors, where p is set manually. For example five factors can be run in only $2^{5-2} = 8$ experiments instead of $2^5 = 32$ experiments compared to the full factorial design. Fractional factorial design takes advantage of the fact that three-way and higher interactions are seldom significant. The downside, of course, for not performing all experiments, is that confounding patterns are present. In other words, the estimated effects are not "pure" but instead mixed with higher degree interaction effects. This loss of information is the prize we need to pay for the reduction of the number of experiments. The degree of confounding is determined by the choice of p.

3. Response surface modeling (RSM) and optimization (few factors) – after screening the factors involved in, e.g. determination of flowering time or derivatization of metabolites, the goal of the investigation is usually to create a valid map of the experimental domain (local space) given by the significant factors and their ranges. This is done with a quadratic polynomial model. The higher order models have an increased complexity, and therefore also require more experiments/factors than screening designs. Different types of RSM designs include Central composite designs, Box Behnken designs and D-optimal designs (see, e.g. Lundstedt et al. 1998 for more information).
4. Robustness testing – in robustness testing of, for instance, an analytical method, the aim is to explore how sensitive the responses are to small changes in the factor settings, e.g. temperature. Ideally, a robustness test should show that the responses are not sensitive to small fluctuations in the factors, that is, the results are the same for all experiments. Robustness testing is usually applied as the last test just before the release of a product or a method. The fractional factorial design is usually applied here.

Plant metabolomic studies typically constitute a set of samples from Arabidopsis wild types and mutants. Assume that these have been subjected to different external conditions such as variation in day length and temperature. Design of Experiments can then be used to select representative samples, related to the biological question we are investigating (how flowering time is affected by temperature, day length, nutrition). An experimental design in three factors can be setup, with factor 1 (temperature), factor 2 (day length), and factor 3 (nutrition). In total, only eight different experiments equal 2^k where $k = 3$ factors are required to explore the experimental space. In addition, a number of replicates, typically three experiments, are added to estimate the noise level. By adding extra experiments, one can investigate more thoroughly the day length and temperature dependence (increase the number of different day lengths and temperatures).

2.2 The Data Table, X-matrix

In plant metabolomics studies, typically a set of samples are characterised using modern instrumentation such as GC/MS, LC/MS or H^1-NMR spectroscopy. The choice of instrument (see Sumner et al. 2003; Dunn et al. 2005) and experimental procedure (Gullberg et al. 2004) are important and largely determined by the biological system and the scientific question. Design of Experiments can here be used to optimize the experimental protocol.

In contrast to a ^{1}H-NMR spectrum, GC/MS and LC/MS data must be processed before multivariate analysis. The reason is the two-dimensional nature (chromatogram/mass spectra) of the data for each sample. For GC/MS data, curve resolution or deconvolution methods are mainly applied for data processing (see, e. g. Halket et al. 1999; Jonsson et al. 2005a). This gives a resolved spectral and chromatographic profile for each detected compound. The 1D multivariate profile used to characterize each sample is made up of the integrated areas of all detected chromatographic peaks. The corresponding mass spectrum and retention index are used for identification purposes (Schauer et al. 2005). For LC/MS data, curve resolution can be applied (e. g. Idborg-Björkman et al. 2003) or a peak detection algorithm that identifies all chromatographic peaks and uses their integrated areas as the multivariate profile characterizing that sample (e. g. Andreev et al. 2003). Another alternative is to sum the chromatographic direction to create a 1D multivariate profile produced by the total intensity over all mass spectral channels (e. g. Allen et al. 2004). Recently, partly alternative methodologies have been applied to GC/MS data (Jonsson et al. 2004, 2005a) and LC/MS data (Jonsson et al. 2005b) where all samples are processed simultaneously and a common set of descriptor variables are extracted.

After, e. g. the GC/MS analysis, we now have a multivariate profile (300–1000 s of variables) for each sample that is a fingerprint of the inherent properties (e. g. phenotype) for each sample. For multiple samples we can therefore construct a two-dimensional data table, an X matrix, by stacking each sample on top of each other. The question is then, how do we go about analysing this multivariate, highly collinear and complex data set? The univariate approach (e. g. student's t-test [Jackson 1991]) is not recommended. It assumes independent variables in X (i. e. more samples than variables) and this creates problems with interpretation, spurious correlations (so called Type I, II errors) and the evident risk of missing information in combinations of variables. Traditional statistical methods (e. g. multiple linear regression, MLR) are also not recommended. They also assume independent variables and have difficulties with noisy data (Eriksson et al. 2001). Instead, multivariate analyses based on projection methods represent a number of efficient and useful methods for the analysis and modeling of these complex data. Projection methods convert the multi-dimensional data table into a low-dimensional model plane, usually consisting of two to five dimensions. Principal component analysis (PCA) (Jackson 1991) and partial least squares (PLS) (Wold et al. 1984) methods are two widely used methods that can handle incomplete, noisy and collinear data structures.

2.3 Geometrical Interpretation of a Data Table

An easy way to understand and appreciate projection based methods is to translate the data table into a swarm of points in a multi-dimensional space. For a data table or matrix X, with N rows (biological samples) and K columns (e. g. relative amounts of different metabolites), each row (individual sample) can be represented as a point in a K-dimensional space. Its position in this space is given by its coordinates, i. e. its values in each of the K columns. Repeating this for all N rows in a matrix, we have produced a swarm of points in K-dimensional space. Points (samples) that lie close to each other in this multi-dimensional space are more biologically similar to each other than points that lie far apart (dissimilar). Projection methods find a model hyperplanes of much lower dimensionality that closely approximates X, i. e. the swarm of points. Figure 3 gives an overview of how multivariate projection methods work.

2.4 Principal Component Analysis

Principal Component Analysis (PCA) is the workhorse in chemometrics. It is a multivariate projection method designed to extract and display the systematic variation in a data matrix X. The first two *principal components* define a plane, a window into the K-dimensional space. By projecting each of the sample points (in K-dimensional space) onto this two-dimensional sub-space, it is possible to visualize all the samples. The coordinates of each of these samples projected onto this plane are called *scores* T, and they are weighted averages of all X-variables (e. g. metabolites). Hence the visualization of these scores T is called a *score plot*. The score plot is very informative because it gives an overview of all samples in X and how they relate to each other. It may reveal groupings of samples (clusters), trends and outliers (deviating samples). e. g. two genotypes (wild type and mutant) would show up as two distinct clusters of samples, representing wild type and mutant samples respectively. In addition, an experiment that suffered from a broken GC-vial would translate into an unique point in the score plot, i. e. an outlier (Fig. 3).

The score plot allows us to investigate the relation among the samples, but once interesting patterns are found (groupings, outliers etc.), it is possible to understand the reason for this, i. e. what variables (e. g. metabolites) are responsible for this pattern found in the score plot. Hence, there also exists a corresponding plot related to the measured variables (metabolites), i. e. the columns in the X matrix. This plot is known as the *loading plot* P and describes the influence (weight) of the X-variables (metabolites) in the model. An important feature is that directions in the score plot correspond to directions in the

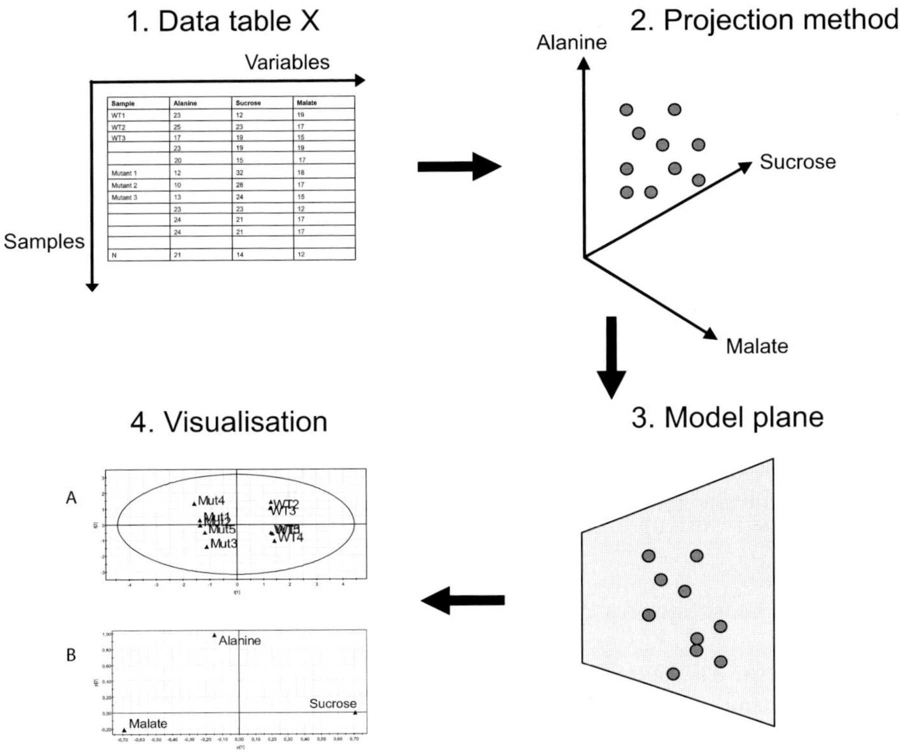

Fig. 3. (1) Each row (representing one biological sample) in a data table with K = 3 variables can be represented as one point in a K = 3 dimensional space. The position of that point is given by the coordinates given by the values in each of the K = 3 variables. (2) Repeating this for all rows (samples) in a data table produces a swarm of points in K = 3 dimensional space. Points (samples) that are close to each other have more similar biological properties than points that are far apart. (3) Projection methods such as PCA, finds a representative low-dimensional plane (here two-dimensional) that is a good summary of the variation in the X data table (swarm of points). (4) This model plane can then be visualised in scatter plots (A) and provides an overview, e. g. if there are any groupings, trends or outliers in the data. For example in the figure (A) there is a clear separation between the *Arabidopsis* wild type and mutant. It is also possible to understand the reason for this separation by looking at the direction of the model plane with respect to the original axes (original variables). These are summarized in the PCA model loadings, P (B)

loading plot (Fig. 3). This is a powerful tool for understanding the underlying patterns in the data.

The PCA model can be expressed as

$$\text{Model of X: } X = TP^T + E$$

where T are the scores, P defines the loadings, and E represent the residual matrix. The residual matrix E contains the residuals for each sample between

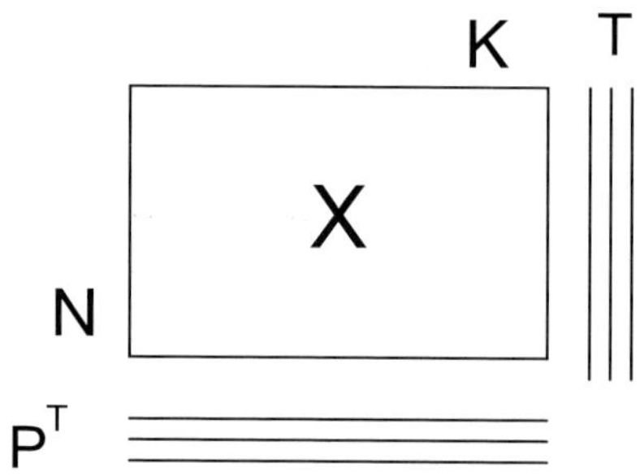

Fig. 4. PCA summarise all variation in X into a few new variables called scores T. These new variables are linearly weighted combinations of the original X-variables. The loadings P contain the weights used for each X-variable and thus reveal the influence of individual X-variables

its point in K-dimensional space and its point on the model plane. The residuals are important for detection of outliers and for defining the model boundaries (see Fig. 4).

2.5 Partial Least Squares Projections to Latent Structures (PLS)

The PLS method is used instead of the PCA method when additional knowledge about each sample exists, the Y matrix, e. g. genotype of each sample (wild type/mutant). The sample information according to the design matrix from the Design of Experiments (see Sect. 2.1) is often used as a Y matrix. Hence, PLS represents the regression analogy of PCA working with two matrices, X and Y (Wold et al. 1984). It is one of the most common methods when a quantitative relationship between a descriptor matrix X and a response matrix Y is sought. The Y matrix can contain both quantitative (e. g. glucose concentration) and qualitative (genotype) information. This additional sample information in Y is used by the PLS method to focus the model plane to capture the *Y-related variation* in X, e. g. separation between genotypes, rather than providing an overall view of *all variation* in the data as done by the PCA model. In addition, the PLS method can also be used to predict the properties (Y-values) of new unknown samples, e. g. predict the glucose concentration or genotype.

The Y matrix consists of the same number of rows as the X matrix. Each column in Y indicate a certain property, e. g. glucose concentration or genotype for each sample. When Y contains qualitative information such as genotype, the number of columns in Y equals the number of classes. Each row in Y describes the group membership for that sample where "1" indicates class belonging for that sample and "0" does not. When Y is qualitative, the PLS method is called PLS Discriminant Analysis (PLS-DA), to distinguish it from the situation when Y is quantitative.

3 Example: Metabolomics Study on Arabidopsis Mutants

We will work through a metabolomics example using GC/MS data from the analysis of *Arabidopsis* extracts. Shoots of higher plants are characterized by axillary branching, where the shoot branches develop from shoot meristems located between a leaf and the shoot stem. The control of axillary shoot growth (branching) is not well understood, but it is known that several internal factors such as the plant hormones IAA and cytokinins are involved (McSteen and Leyser 2005). Mutations screens in *Arabidopsis* have identified four loci involved in the repression of axillary bud growth, *MAX1–4*. Based on the mutants, it is now suggested that an unknown transmittable substance might be involved in controlling branching (see McSteen and Leyser 2005). The biosynthesis of this compound in *Arabidopsis* is catalyzed by a number of MAX (more-axillary growth) proteins.

We have used a metabolomics approach to classify and identify the metabolic differences between the MAX-mutants. Root samples from WT, max3 and max4 mutants were analysed by GC/TOFMS as described by Gullberg et al. (2004). The GC/MS data was processed by hierarchical multivariate curve resolution (Jonsson et al. 2005a), and the obtained X-matrix was thereafter subjected to PCA and PLS-DA analysis. The GC/MS processing resulted in 514 resolved peak areas. Log transformation, column centering and scaling to unit variance was done on the resolved peak areas (X-matrix) prior to modeling and two dummy Y-variables were constructed based on the class belonging of each sample to the

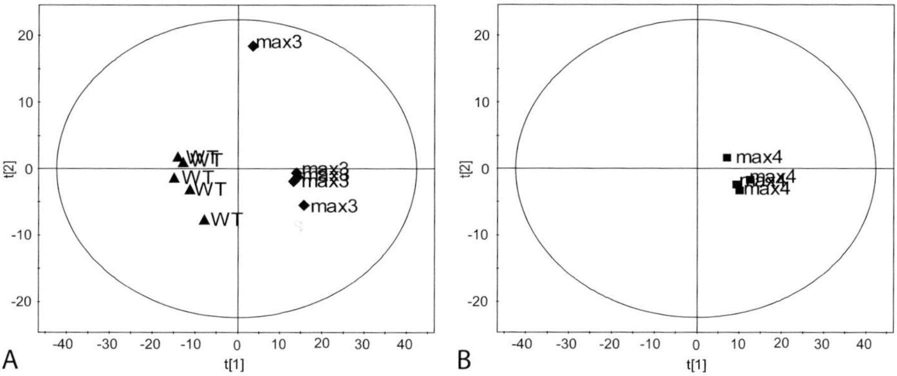

Fig. 5. A PLS-DA score-plot from the analysis of metabolite profiles in roots of *Arabidopsis* WT, max3 and max4. The PLS-DA model is based on WT and max3. The X-matrix was centered and scaled to unit variance. The explained variation in the X-matrix (R^2X) is 0.74, the explained variation in the Y matrix (R^2Y) is 0.99 and the predictive ability according to sevenfold cross-validation (Q2) is 0.84. R^2X is the cumulative modelled variation in X, R^2Y is the cumulative modelled variation in Y and Q^2Y is the cumulative predicted variation in Y, according to cross-validation. The range of these parameters is 0–1, where 1 indicates a perfect fit. **B** Based on the model max4 samples were predicted into the model showing that max3 and max4 are very similar regarding metabolic content (compare position score plot in **A**)

genotypes, WT and max3. The PLS-DA model score plot is shown in Fig. 5A. The score plot reveals the relationship among the samples. It is clear from the figure that the model plane displays a clear separation of the two genotypes.

To validate the model results, predictions were made for the genotype max4, using the calculated PLS-DA model based on the other sample-set (WT and max3). The results, shown in the obtained PLS-DA score plot (Fig. 5B) predicted that the max4 is closer to max3 than WT. This is consistent with the facts that max3 is very similar to the max4 genotype, where the MAX3 and MAX4 proteins use the same substrate (Schwarz et al. 2005). Interpretation of the first weight vector (w1) from the PLS-DA model, as described by Trygg and Wold (2002), together with the 99% confidence intervals calculated using jack-knifing (Martens and Martens 2000), highlighted 64 significant variables (metabolites) differing between WT and max4. The importance of these metabolites is a part of *biological validation* of the data set. The *statistical validation* was done by prediction of the max3 mutants into the WT/max4 model. Both type of validation is of importance for validating the multivariate data set.

4 Summary and Future Prospectives

Multivariate projection methods, e. g. PCA and PLS, represent a useful and versatile technology to modelling, monitoring and prediction of complex problems and data structures encountered within metabolomics and other 'omics' disciplines. The common denominator is that high complexity data tables are generated and that these data tables can be analysed and interpreted by means of chemometric methods. The principal component analysis (PCA) method summarizes the variation in a data table X into a model plane (the scores T). A scatter plot of these scores gives an overview of the samples (observations) and how they relate to each other, e. g. if there are groupings or trends or deviating samples and so on. In order to interpret the patterns found in a score plot one examines the corresponding loading plot (P). The loadings P reveal how each variable contributes to the separation among samples in the model plane and also gives insights into the relative importance of each variable.

However, one fundamental property is that the data does contain relevant information regarding our biological question. In other words, how to maximise the information content in the data? The traditional way to Change One Factor at a Time, i. e. the COST approach, is not recommended. Design of Experiments (DOE) is the methodology of how to conduct and plan experiments in order to maximize information in the data in the fewest number of runs. A proper experimental design will reveal the influence of each of the different factors separately and also any interactions between them. DOE is the only feasible approach to separate cause and effect from each other. Therefore is DOE in combination with chemometrical analysis a powerful way of planning, conducting and evaluating metabolomics experiments.

One common discussion point in the analysis of "omics" data is how to correlate several types of data, usually with different data structures. Systems biology seeks to integrate information from multiple parts of a biological system in a holistic attempt to understand the whole system. There are still many obstacles and hurdles to overcome in order to succeed. One of these relates to how the actual integration of the different types of data will be done. Hence, the advancement of systems biology depends heavily on the ability to integrate multiple profiling techniques (e. g. transcriptomics, proteomics, GC/MS, LC-NMR). The current multivariate statistical methods (e. g. the PLS method) lacks the proper model structure to describe these types of data structures, because they focus only on the *correlation pattern* among multiple data tables (e. g. X = microarrays vs Y = metabolomics data) and not on the *non-correlated variation* among these data tables which, in a biological sense, can be of equal interest. It has also been demonstrated that, because of this, the interpretation of these models are negatively affected (Trygg and Wold 2002), e. g. positive correlation patterns are interpreted as negligible or even flipped and become negative. This is a fundamental problem as we certainly cannot expect that all variation in transcript and metabolite levels co-vary. Fortunately, recent advances in chemometrics provide the ability to compare multiple data sets with each other. Novel extensions of the PLS method, called O-PLS (Trygg and Wold 2002) and O2-PLS (Trygg 2002) contain the model structure to support both these features. In addition, the O2-PLS method is bi-directional which means that the flow of information can go in both ways, from X (e. g. microarray) to Y (e. g. metabolomics) and vice versa. Hence, the O2-PLS methodology will be important in selecting what genes or metabolites are important to do further experimentation upon, e. g. understanding biomarker patterns and selecting genes for knockout studies. The O2-PLS methodology can also be extended to more than two data tables, hence it nicely fits into the framework of a combined profiling approach.

Acknowledgements. The Swedish Research Council, Wallenberg Consortium North (WCN), the Kempe foundation, EU strategic funding, Knut and Alice Wallenberg Foundation (JT) and Strategic Research Funding (SSF) are acknowledged for financial support. Professor Ottoline Leyser, York, UK, for allowing us to show data from the max-mutant project, and Dr. Miyako Kusano, RIKEN Plant Science Centre, Yokohama, Japan for the initial analysis of metabolites in the max-mutants.

References

Allen J, Davey HM, Broadhurst D, Heald JK, Rowland JJ, Oliver SG, Kell DB (2003) High-throughput classification of yeast mutants for functional genomics using metabolic footprinting. Nature Biotechnol 21:692–696

Andreev VP, Rejtar T, Chen HS, Moskovets EV, Ivanov AR, Karger BL (2003) A universal denoising and peak picking algorithm for LC-MS based on matched filtration in the chromatographic time domain. Anal Chem 75:6314–6326

Dunn WB, Bailey NJC, Johnson HE (2005) Measuring the metabolome: current analytical technologies. Analyst 130:606–625
Eriksson L, Johansson E, Kettaneh-Wold N, Wold S (2001) Multi and megavariate data analysis. Umetrics (www.umetrics.com), ISBN 91-973730-1-X
Gullberg J, Jonsson P, Nordström A, Sjöström, M, Moritz T (2004) Optimisation of preparation of plant samples for metabolic profiling by GC-MS. Anal Biochem 331:283–295
Halket JM, Przyborowska A, Stein SE, Mallard WG, Down S, Chalmers RA (1999) Deconvolution gas chromatography mass spectrometry of urinary organic acids - potential for pattern recognition and automated identification of metabolic disorders. Rapid Commun Mass Spectrom 13:279–284
Idborg-Björkman H, Edlund, PO, Kvalheim OM, Schuppe-Koistinen I, Jacobsson SP (2003) Screening of biomarkers in rat urine using LC/electrospray ionization-MS and two-way data analysis. Anal Chem 75:4784–4792
Jackson JE (1991) A users guide to principal components. Wiley, New York
Jonsson P, Gullberg J, Nordström A, Kowalczyk M, Sjöström M, Moritz T (2004) A strategy for extracting information from large series of non-processed complex GC/MS data. Anal Chem 76:1738–1745
Jonsson P, Johansson AI, Gullberg J, Trygg J, A J, Grung B, Marklund S, Sjöström M, Antti H, Moritz T (2005a) Highthroughput data analysis for detecting and identifying differences between samples in GC/MS-based metabolomic analyses. Anal Chem 77:5635–5642
Jonsson P, Bruce SJ, Moritz T, Trygg J, Sjostrom M, Plumb R, Granger J, Maibaum E, Nicholson JK, Holmes E, Antti H (2005b) Extraction, interpretation and validation of information for comparing samples in metabolic LC/MS data sets. Analyst 130:701–707
Lundstedt T, Seifert E, Abramo L, Thelin B, Nyström A, Pettersen J, Bergman R (1998) Experimental design and optimization. Chem Intel Lab Systems 42:3–40
Martens H, Martens M (2000) Modified Jack-knife estimation of parameter uncertainty in bilinear modelling by partial least squares regression (PLSR). Food Qual Pref 11:5–16
McSteen P, Leyser O (2005) Shoot branching. Annu Rev Plant Biol 56:353–374
Schauer N, Steinhauser D, Strelkov S, Schomburg D, Allison G, Moritz T, Lundgren K, Roessner-Tunali U, Forbes MG, Willmitzer L et al (2005) GC-MS libraries for the rapid identification of metabolites in complex biological samples. FEBS Lett 579:1332–1337
Schwartz S, Qin XQ, Loewen MC (2005) The biochemical characterization of two Carotenoid cleavage enzymes from *Arabidopsis* indicates that a carotenoid-derived compound inhibits lateral branching. J Biol Chem 279:46940–46945
Sumner LW, Mendes P, Dixon RA (2003) Plant metabolomics: large-scale phytochemistry in the functional genomics era. Phytochemistry 62:817–836
Trygg J (2002) O2-PLS for qualitative and quantitative analysis in multivariate calibration. J Chemometr 16:283–293
Trygg J, Wold S (2002) Orthogonal projections to latent structures (O-PLS). J Chemometrics 16:119–128
Wold S, Ruhe A, Wold H, Dunn WJ III (1984) The collinearity problem in linear regression. The partial least squares (PLS) approach to generalized inverses. SIAM J Sci Statist Comput 5:735–743

II.3 Map Editor for the Atomic Reconstruction of Metabolism (ARM)

M. Arita[1,2], Y. Fujiwara[1], and Y. Nakanishi[3]

1 Introduction

In the systems study of biological networks, computational analysis is expected to contribute in three phases by (1) *model selection*, the formal definition of each pathway's role in the manifestation of the biological aspect under analysis, (2) *model refinement*, the estimation of model parameters to refine constructed mathematical models, and (3) *simulation and feedback*, the computer simulation for the feedback of the predicted results for a better understanding of the target mechanism(s) (Hood 2003; Arita et al. 2005). Of these, the first phase is of prime importance because it determines the target of the analysis and its abstraction level. In other words, in the *model selection* phase, an appropriate model is searched and selected among all hypothetical candidates. The process of model selection is often performed intuitively by researchers. For example, glycolysis, the best-known pathway module in energy metabolism, contains a sequence of ten biochemical reactions from glucose to pyruvate (Berg et al. 2002). Since the pathway is linear, it can be summarized as if it were a single reaction:

$$\text{Glucose} + 2\text{NAD}^+ + 2\text{ADP} + 2\text{Pi} \rightarrow 2\text{Pyruvate} + 2\text{ATP} + 2\text{NADH} + 2\text{H}^+$$

The behavior of the pathway may be mathematically described either as a set of ten reactions, or as a single, abstract reaction. The abstract model is preferred in explaining net ATP generation, whereas the ten-reaction model is used for metabolic simulations. In choosing a model, we must be aware of the trade-off between model accuracy and its description length. In general, any model inevitably loses its fit to its corresponding natural mechanism as its description becomes simpler. In glycolysis, the nine intermediate molecules in the pathway can be eliminated to obtain the net ATP model in return for sacrificing biochemical details (i. e., the intermediates) of the pathway.

However, when glycolysis is placed in a global metabolic network, the description of the intermediate molecules is no less important than that of the gateway molecules such as glucose and pyruvate. The question arises as to

[1] Department of Computational Biology, Graduate School of Frontier Sciences, The University of Tokyo, 5-1-5 Kashiwanoha, Kashiwa, 277-8561 Japan, e-mail: arita@k.u-tokyo.ac.jp
[2] Institute for Advanced Biosciences, Keio University, and Precursory Research for Embryonic Science and Technology, Japan Science and Technology Agency
[3] Intec Web and Genome Informatics Corporation

Biotechnology in Agriculture and Forestry, Vol. 57
Plant Metabolomics (ed. by K. Saito, R.A. Dixon, and L. Willmitzer)

which criteria should be evaluated and chosen for an appropriate abstraction of the given network. In biology, the focus has been directed at modularity in terms of function and structure.

The introduction of modularity, i. e. the encapsulation of network details by specifying the input/output, yields many advantages. First, it simplifies the description of the given network and facilitates its understanding. In metabolic networks, biologists have used intuitive, functional concepts such as 'amino-acid biosynthesis' or 'nitrogen assimilation' without formal, logical definitions. Second, the introduction of modularity simplifies the static verification of the network, a required step that precedes quantitative analyses such as simulation. The stoichiometric balance is one such property that can be verified statically.

In glycolysis, the structural and functional modularity is clear, mainly because the pathway is not branched; the pathway structure is linear and all carbon atoms in glucose are mapped to pyruvate and the function is the decomposition (lysis) of glucose. In general, modularity is less straightforward in branching metabolic pathways. Since molecular moieties are split into or merged with multiple molecules, it is not easy to trace carbon and other atomic elements, let alone delineate modularity. In fact, the determination of molecules to be regarded as intermediates is context-dependent: focusing on different atoms changes the pathways to be traced and therefore the resulting modular decomposition. For multiply branching pathways there is no universally effective, definitive decomposition.

Can the modularity of metabolic networks be detected computationally? Many automated methods exploit the stoichiometry of biochemical reactions (Mavrovouniotis 1992); this is one solution for the formal decomposition into metabolic modules. Currently, however, the most widely accepted method for finding modularity is through network topology; verification is by visualization with functional annotations (Ravasz et al. 2002; Ma et al. 2004). This is why pathway modules remain intuitive. Although electronic circuits can be verified using Boolean logic, no such formal system has been developed for biological systems. Indeed, many software programs for genomic and proteomic networks address only visualization, and their formal analysis remains to be solved.

The basic concept of our metabolic map editor, introduced in this chapter, combines a formal description of metabolism (structural conversions and their stoichiometric conditions) with intuitive network visualization. Many visualization tools and databases provide a static view of a given metabolic network (Mendes 2002; Kanehisa et al. 2004; Keseler et al. 2005). Because a metabolic network is well investigated and has a traditionally accepted layout for metabolites and enzymatic reactions, fully automated layout algorithms that ignore such standard layouts do little to further our understanding of its properties. Rather, an interactive drawing tool that can edit and modify standard metabolic networks is needed. Only with interactive software can biologists derive species-specific pathway images and visualize their intuitive ideas.

2 Definition of Metabolic Information

In this section, we introduce the concept of handling metabolic information at the atomic scale, and show how atomic representation can advance the formal understanding of metabolism.

2.1 Definition of Metabolic Pathways

Functions that should be computationally supported by a metabolic map editor include (1) searching and adding alternative or new metabolic pathways, (2) superimposing genomic, proteomic, and metabolomic data onto the network, and (3) rearranging the network topology to accommodate the metabolism of a particular species of interest. To fulfill these criteria, a formal definition of metabolic pathways is required. Previous computational studies tended to present a metabolic network as a graph where nodes and edges correspond to metabolites and their biochemical reactions, and the pathway as a sequence of graph edges (Ravasz et al. 2002; Ma et al. 2004). However, from a biochemical perspective, edge sequences (or graph paths) defined in this manner do not necessarily correspond to metabolic pathways. Since molecular structures are transformed in the course of each reaction, adjacent graph edges may not share the common structural moiety that corresponds to the metabolic (or atomic) flux (Fig. 1).

To resolve the conflict between graph paths and biochemical pathways, an additional constraint must be introduced (Arita 2003). In this chapter, *a metabolic pathway* (pathway for short) from metabolite *X* to metabolite *Y* is defined as a sequence of reactions through which at least one atom (carbon, nitrogen, or sulfur) in *X* reaches *Y*. A metabolite *Y* is called *reachable from X* if there is a pathway from *X* to *Y*. This is a rather strict constraint because the

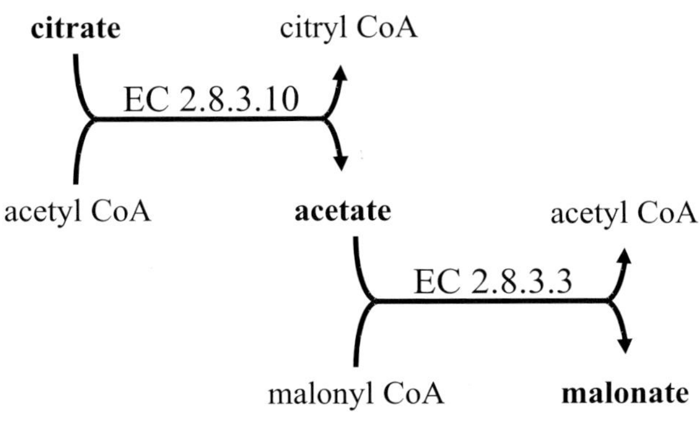

Fig. 1. Example of a biochemically inappropriate pathway from citrate to malonate. Analysis of molecular structures is required to enable the computer to detect that the citrate moiety is transferred to citryl CoA, rather than to acetate

conserved moiety throughout a pathway may not consist of carbon, nitrogen, or sulfur atoms; it may consist of oxygen or hydrogen atoms or even electrons. The map editor deals with only three types of elements because they can be computationally traced without ambiguity. The tracing of oxygen or hydrogen atoms is virtually impossible because the water molecule is involved in many reactions. The same is true for phosphates and metal ions that exist as free inorganics in a cell.

2.2 Representation of Pathways and Networks

In the atomic representation of a metabolism, each reaction is decomposed into a set of sub-structural correspondences called *atomic mappings* (Arita 2003). Each atomic mapping represents the transfer of a certain structural moiety in the course of biochemical reactions, and may be shared among multiple reactions. For example, atomic mapping between ATP and ADP, and between glutamate and α-keto-glutarate is prevalent in phosphate- and amino-transfer reactions, respectively (Fig. 2).

Conventionally, a metabolic pathway is thought of as a sequence of catalytic reactions classified by EC numbers. However, in the metabolic reconstruction from a whole genome sequence, a metabolic pathway is better viewed

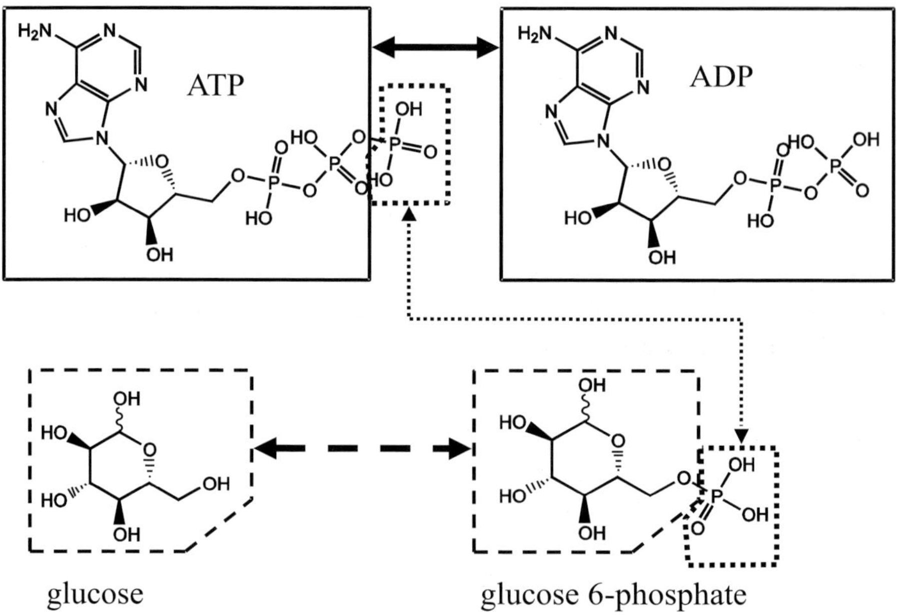

Fig. 2. Three atomic mappings in the reaction glucose + ATP = glucose 6-phospahte + ADP. The mapping between ATP and ADP (*shown in solid lines*) is common to all phosphorylation reactions with ATP and ADP

as a sequence of atomic mappings rather than as EC reactions. There are at least two reasons for this. First, the proposed atomic view can provide more candidate pathways in the reconstruction. Reconstruction based only on EC numbers often results in a set of incomplete pathways with multiple gaps. Such superficial gaps, however, may be filled with other reactions that share atomic mapping corresponding to the gap. Second, the proposed atomic view provides flexibility in choosing coenzymes. In theoretical analyses, the stoichiometric balance of reactions has been discussed as if coenzymes were fixed for all reactions (as in the traditional metabolic map) (Papin et al. 2004). In practice, however, their balance should be determined considering the net reaction balance in the network. For example, some NAD-dependent enzymes can also catalyze reactions using NADP, and the overall balance of their use depends on the total networking condition, not on individual reactions.

Thus, at least for the computational reconstruction of pathways, a metabolic network is better viewed as a set of atomic mappings rather than a set of EC-numbered reactions. In the biosynthesis of isoleucine and valine, for example, the same set of enzymes catalyzes 2-oxobutanoate and pyruvate to form isoleucine and valine, respectively. The only difference between the two pathways is a single alkyl group independent of the catalytic sequences in the biosynthesis. The decomposition into atomic mappings can explicitly describe such structural sharing between pathways.

3 Metabolic Map Editor

3.1 Overview

The design principle of the map editor is that users can flexibly integrate a sequence of atomic mappings (not reactions) into existing metabolic pathways to form metabolic maps (Fig. 3). First, users are expected to search metabolic pathways using the associated database that stores enzymatic reactions, their atomic mappings and molecular structures. The searched pathways (sequences of atomic mappings) are transferred to the main window where their layout can be freely edited as in a conventional graphical drawing editor. The advantage of our editor over conventional editors such as Microsoft PowerPoint is that users can import metabolic objects (e. g. compound structures and reactions) from the background database: although on-screen it appears as if only graphical objects for compound structures and reactions are imported, more information is processed in the background. For example, importing one enzymatic reaction on the screen implicitly invokes the integration of its associated atomic mappings into the already drawn metabolic map so that the route of any atom in the new reaction can be traced seamlessly on the resulting metabolic map.

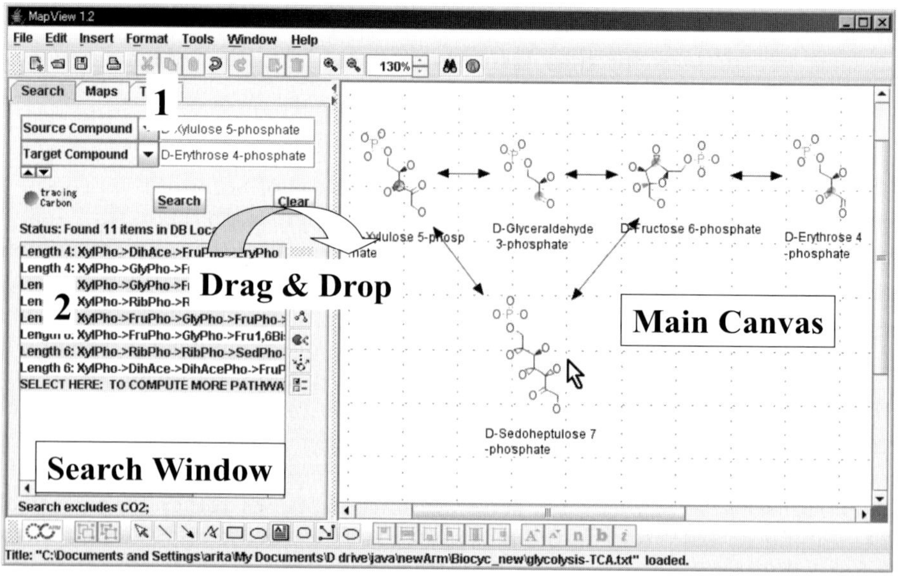

Fig. 3. Screen-shot of the Map Editor. Pathways are searched by typing molecular names in the input fields (*Number 1*). The search results are listed (*Number 2*). Users can drag pathways into the main canvas

3.2 Metabolic Database

The background database for atomic information stores three types of metabolic data: molecular structures, reaction formulas, and their atomic mappings. All data are freely available in text format from the website http://www.metabolome.jp/download.html.

Molecular structures are registered in the MOL-file format (MDL Information Systems; its description is downloadable from http://www.mdli.com/). The MOL-file format is the de facto standard to describe molecular structures; an example is shown in Fig. 4. Each MOL-file describes one molecular structure as a list of atoms with their XYZ coordinates and their chemical bondings. The chirality of carbon is specified using one integer value for each corresponding carbon atom. Information on the display of chirality (in thick and shaded lines) is specified using other integer values. The metabolic editor does not use the XYZ coordinates written in the format; rather, it applies the original drawing algorithm to assign XYZ positions (Arita 2005).

As in other metabolic databases, enzymatic reactions are described using compound names. Reaction formulas were obtained from the Enzyme Nomenclature of the International Union of Biochemistry and Molecular Biology (http://www.chem.qmw.ac.uk/iubmb/enzyme/). In each reaction, the order of molecules on the left- and right-hand side was manually rearranged so that the atomic mappings can be computationally detected by comparing molecu-

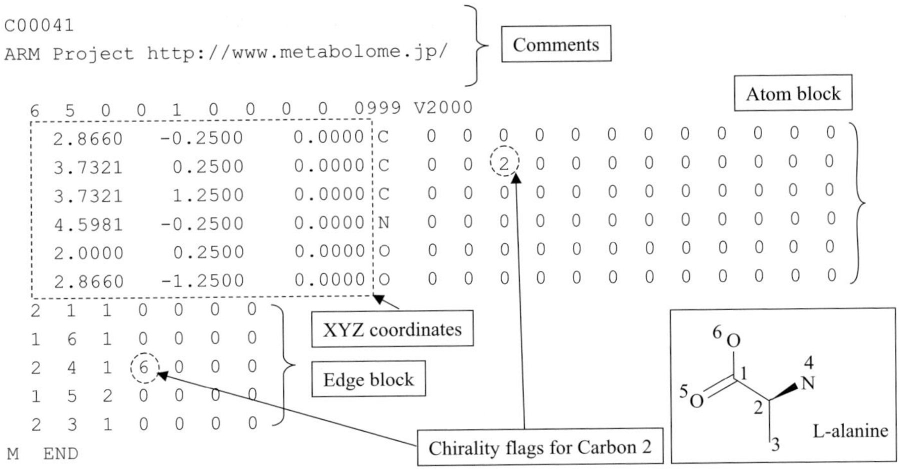

Fig. 4. MOL-file format for L-alanine. In the ARM database, carbon atoms in the atom-block are ordered according to the IUPAC positions for molecular structures. Only carbon atoms are correctly ordered

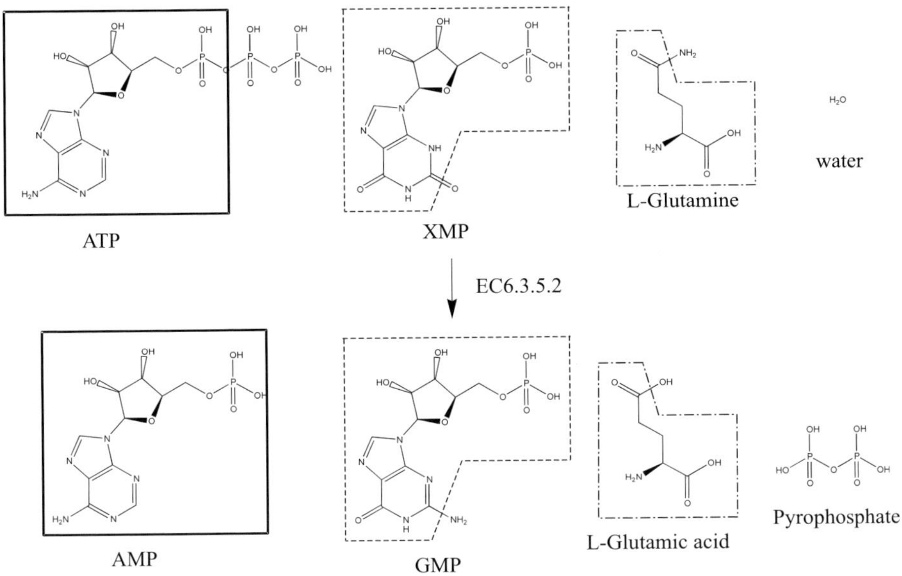

Fig. 5. Schematic view of reaction EC 6.3.5.2. The reaction formula is written as "ATP + XMP + L-glutamine + H_2O =>AMP + GMP + L-glutamic acid + pyrophosphate" in the database so that the molecular structures roughly correspond one-to-one (*top to bottom*)

lar structures sequentially from left to right (Fig. 5). For details on structure comparison to compute atomic mappings refer to Arita (2003).

Using molecular structures, atomic mappings were pre-computed for all registered reactions and, after manual verification, correct mapping results were stored in the database. Details, including the accuracy of the mapping computation, were described previously (Arita 2003). Although the mapping was computed for all atomic elements except hydrogen, the results were registered only for carbon, nitrogen, and sulfur atoms due to ambiguities in the mapping of the rest of the elements.

3.3 Drawing Maps from Pathways

The map editor is equipped with a search engine for metabolic pathways. Given a source and target metabolites, the engine computes logically possible pathways between these metabolites from the shortest- to pathways of any length. Although pathway length is measured by the number of reaction steps, an arbitrary value can be assigned in the algorithm used. In other words, the engine can compute any pathway throughout which at least one carbon (or nitrogen, sulfur) atom is conserved. An arbitrary combination of pathways can be visualized by dragging a searched pathway into the main canvas window (Fig. 3). When a pathway is dragged into the canvas, it is merged with the already drawn network (Fig. 6). Although its initial layout is automatically assigned, a user can freely rearrange the orientation or location of any metabolic object by using the mouse.

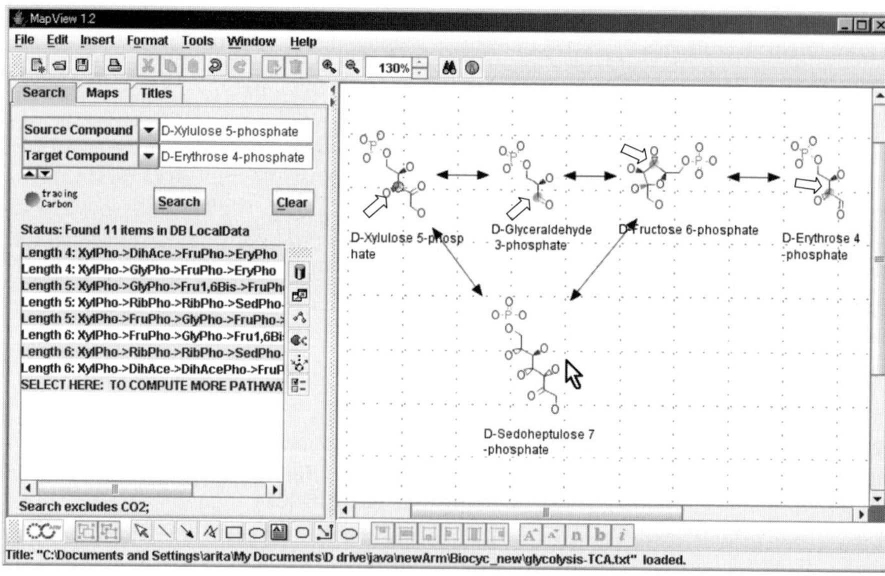

Fig. 6. The network generated by merging two pathways from xylulose 5-phosphate (X5P) to erythrose 4-phosphate. Every time a pathway is dropped, only the difference from the existing network is drawn. Carbon 3 in X5P is traced in this example (*shown with blank arrows*)

The unique function of the map editor is its ability to trace a particular atom on the map. Since each metabolic object on the map is linked with its atomic information in the database, the logical tracing of each atomic position is possible by transitive calculation of the atomic mappings in the network. A user needs only to mouse-click a particular atom on the network to see its traces (Fig. 6).

4 Applications

4.1 Carbon Flow in Cyclic Pathways

Metabolic pathways contain two types of cycles in terms of tracing atoms: cycles where carbon atoms are exchanged in each round (e. g. the tricarboxylic acid (TCA) cycle), and cycles where all carbon atoms are conserved (e. g. the urea cycle). If reversible, a single reaction catalyzing two identical molecules (e. g. 2 pyruvate = 2-acetolactate + CO_2) can form the former type of cycle by itself. Likewise, the latter type of cycle can be formed by any reversible reaction. Cyclic pathways may be biochemically meaningful, but in practice, their existence is problematic in searching metabolic pathways. Since pathways are searched and output according to the number of reaction steps in our system, short cycles drastically increase the number of spurious pathways with local loops. To eliminate such futile pathways, our pathway-search algorithm eliminates all pathways that visit the same molecule multiple times. However, this constraint is too strict for searching all possibly existing pathways. For example, users studying the TCA cycle may want to analyze carbon traces that go round the cycle multiple times.

To support the atomic analysis of cyclic pathways, a metabolic map editor is indispensable. First, users search the pathways of interest and paste them into the main window using the edit function (i. e., model selection). Then, a particular atom can be interactively traced within the selected set of reactions. Since the target model is highly constrained, it is feasible to compute pathways visiting the same compound multiple times.

4.2 Carbon Flow in Energy Metabolism

Because of the shared metabolites between the glycolytic and pentose phosphate pathways, the atomic traces of a particular carbon atom often become hard to follow. This is the case for the correspondence between the C-1 of glucose and the C-1 of pentose 5-phosphate. By clicking the corresponding atomic position in the map editor, the atomic trace can be grasped at a glance. Although the entry point of the pentose phosphate pathway decarboxylates the C-1 position of glucose, this position is identical to the C-1 of fructose 6-phosphate through glycolysis. Thus, in the pentose phosphate pathway, the C-1 of fructose 6-phosphate corresponds to the C-1 of sedoheptulose 7-phosphate, of xylulose 5-phosphate, and of ribose- and ribulose 5-phosphate. In fact, the C-1 of glucose corresponds to both the C-1 and C-6 positions of fructose 6-phosphate, and therefore to all C-1 and C-5 positions of pentose 5-phosphate. These positions are invariable throughout the glycolytic and pentose phosphate pathways.

4.3 Visualization of Lipid Metabolism

Recently, metabolome analysis has been facilitated due to the rapid technical progress made in mass spectrometry (MS). To detect lipid molecules, for example, an effective strategy is to couple MS with liquid chromatography-electrospray ionization (Houjou et al. 2004). More than 1000 glycerophospholipid species can be quantitated in a single assay in less than 2 h (R. Taguchi, personal communication). However, the efficient analysis of such large-scale data sets poses a vexing problem. Network visualization remains the first step for gaining an overview of the data; however, the traditional metabolic map is not suitable for visualization because it contains abstract notations. For example, the 'phosphatidyl group' contains two fatty acids of variable lengths (usually 12$\sim$24 carbon atoms) and degrees of unsaturation (usually 0$\sim$6 double bonds, depending on the length). Since experimentally confirmed fatty acids in a phosphatidyl group are comprised of more than 30 species, the number of actual phosphatidyl species may be as many as its square, i. e., $\approx$ 1000. To visualize the distribution of the spectrum of molecular species, our map editor supports an interactive instantiation of abstract moieties. For each abstract notation using 'R-group', a user can assign a list of molecules as its possible instantiation. The map editor can also display an integer value for each molecule (such as the concentration, mass, logP, etc.). Given the list of possible instantiations for each R-group and their corresponding concentrations (i. e., metabolomic data), the editor can display the percentage fraction of candidate molecules. When multiple R-groups exist, the amount to be displayed will be the integration of all possible assignments. In a phosphatidyl group, various fatty acids can be linked at R1 and R2 positions of glycerol phosphate. With a mouse click, the candidate list for the R1 (or R2) position is displayed together with the relevant percentage fractions. The percentage for a docosahexanoic acid (DHA) in the R1 group, for example, is calculated as the sum of all phosphatidyl molecules that have DHA at R1. When DHA is chosen for R1, the percentage list for R2 consists of fully instantiated molecular species that have DHA at R1 and another fatty acid at R2.

Due to the abstract notations for molecules, lipid metabolism is a particularly unspecific part in the traditional metabolic map. The computer-assisted metabolic map is indispensable to visualize the metabolomic data of such pathways.

5 Conclusions

The map editor is not only a tool for visualizing metabolic pathways, but is a necessary component for the systematic and modular understanding of species- and context-dependent metabolic networks. Since the software system is linked with atomic-level information in the background database, users can trace any atomic position on any metabolic network they draw. It is a desirable realization of a pathway database. Most web-based pathway databases do not support the users' own arrangement of networks, although in computer science, the definition of a database system is ' a collection of information organized in such a way that a computer program can quickly select desired data in a desired arrangement'. Our map editor compensates for this drawback, and represents a step forward to a more flexible analysis of large-scale biological information.

Acknowledgements. The ongoing analysis of lipid metabolism is a joint effort with Prof. Ryo Taguchi at The University of Tokyo. The authors thank Ursula Petralia for editing the manuscript. This work was supported by The Ministry of Education, Culture, Sports, Science and Technology (MEXT), Grant-in-Aid for Scientific Research on Priority Areas.

References

Arita M (2003) *In silico* atomic tracing by substrate-product relationships in *Escherichia coli* intermediary metabolism. Genome Res 13(11):2455–2466

Arita M (2005) Introduction to the ARM database: database on chemical transformations in metabolism for tracing pathways. In: Tomita M, Nishioka T (eds) Metabolomics: the frontier of systems biology. Springer, Berlin Heidelberg New York, pp 193–211

Arita M, Robert M, Tomita M (2005) All systems go: launching cell simulation fueled by integrated experimental biology data. Curr Opin Biotechnol 16(3):344–349

Berg JM, Tymoczka JL, Stryer L (2002) Biochemistry, 5th edn. Freeman, New York

Hood L (2003) Systems biology: integrating technology, biology, and computation. Mech Ageing Dev 124:9–16

Houjou T, Yamatani K, Nakanishi H, Imagawa M, Shimizu T, Taguchi R (2004) Rapid and selective identification of molecular species in phosphatidylcholine and sphingomyelin by conditional neutral loss scanning and MS3. Rapid Commun Mass Spectrom 18(24):3123–3130

Kanehisa M, Goto S, Kawashima S, Okuno Y, Hattori M (2004) The KEGG resource for deciphering the genome. Nucleic Acids Res 32 (Database Issue):D277–D280

Keseler IM, Collado-Vides J, Gama-Castro S, Ingraham J, Paley S, Paulsen IT, Peralta-Gil M, Karp PD (2005) EcoCyc: a comprehensive database resource for Escherichia coli. Nucleic Acids Res 33 (Database Issue):D334–337

Ma HW, Zhao XM, Yuan YJ, Zeng AP (2004) Decomposition of metabolic network into functional modules based on the global connectivity structure of reaction graph. Bioinformatics 20(12):1870–1876

Mavrovouniotis ML (1992) Computer-aided synthesis of biochemical pathways. Biotechnol Bioeng 36:1119–1132

Mendes P (2002) Emerging bioinformatics for the metabolome. Brief Bioinform 3(2):134–145

Papin JA, Reed JL, Palsson BO (2004) Hierarchical thinking in network biology: the unbiased modularization of biochemical networks. Trends Biochem Sci 29(12):641–647

Ravasz E, Somera AL, Mongru DA, Oltvai ZN, Barabasi AL (2002) Hierarchical organization of modularity in metabolic networks. Science 297(5586):1551–1555

II.4 AraCyc: Overview of an Arabidopsis Metabolism Database and its Applications for Plant Research

S.Y. Rhee, P. Zhang, H. Foerster[1], and C. Tissier

1 Introduction

Metabolism is one of the most fundamental processes of life. Each organism possesses an intricate network of metabolic pathways, whose elaborate regulatory circuitry may be developmentally programmed and hard-wired to respond to changes in the environment. With the release of the fully sequenced plant genomes of Arabidopsis and rice (AGI 2000; Goff et al. 2002; Yu et al. 2002), and the initiation of many sequencing projects of other plant species, there is a growing desire to place the sequenced and annotated genomes in a metabolic context. AraCyc (http://arabidopsis.org/tools/aracyc/) was the first plant organism-specific metabolism database to be computationally predicted by the PathoLogic component of the Pathway Tools software using MetaCyc as the reference database (Mueller et al. 2003). With continued manual curation, the goal of AraCyc is to describe the complete set of metabolic pathways for *Arabidopsis thaliana* whilst placing genes and enzymes within their metabolic context. Though many enzymes in AraCyc have yet to be manually curated, most of the pathways have been manually validated and it is so far the only genome-wide, comprehensive metabolic database for a single plant species (Zhang et al. 2005).

The benefits of a species-specific metabolic pathway database are substantial: (1) it depicts the biochemical components of an organism; (2) it aids in comparative studies of pathways across species to facilitate metabolic engineering to improve crop metabolic traits; (3) it can be used as a platform to integrate and analyze data from large-scale experiments such as gene expression, protein expression, or metabolite profiling; finally (4) by presenting pathway steps lacking assigned genes, or having genes assigned but solely based on computational prediction, it allows the identification of the biochemical steps that remain to be identified and experimentally characterized. The manual, de novo creation of a pathway database can be labor intensive and time consuming. SoyBase (http://www.soybase.org/) is the only other plant pathway database, specific to soybean, which was manually created and made publicly available. Alternatively, there are metabolic pathway databases that cover a wide range of organisms. Examples of comprehensive pathway databases include Kyoto Encyclopedia of Genes and Genomes (KEGG, http://www.genome.jp/kegg/) (Ogata

[1] Carnegie Institution, Department of Plant Biology, 260 Panama Street, Stanford, CA 94305, USA, e-mail: rhee@acoma.stanford.edu

Biotechnology in Agriculture and Forestry, Vol. 57
Plant Metabolomics (ed. by K. Saito, R.A. Dixon, and L. Willmitzer)

et al. 1999; Kanehisa 2002; Kanehisa et al. 2004), Enzymes and Metabolic Pathways (EMP, http://www.empproject.com/) (Selkov et al. 1996), and MetaCyc (www.metacyc.org) (Krieger et al. 2004). Each has its strengths and weaknesses, some of which have been reviewed (Maranas and Burgard 2001; Kanehisa 2002).

In this review, we describe the content and functionalities of AraCyc database as well as examples of applications that use the information contained in the database, in conjunction with functional genomics data to address systems-wide questions about metabolism. In addition, we discuss the current limitations and future directions of the database.

2 Database Content

AraCyc (version 2.5) currently features 197 pathways, comprising 979 unique reactions and 1071 compounds. Over 63% of the reactions have Arabidopsis genes/enzymes assigned and 1759 unique genes are assigned to the pathways. A metabolic pathway is a set of one or more enzymatic transformations, involved in processes such as biosynthesis, degradation, conversion,

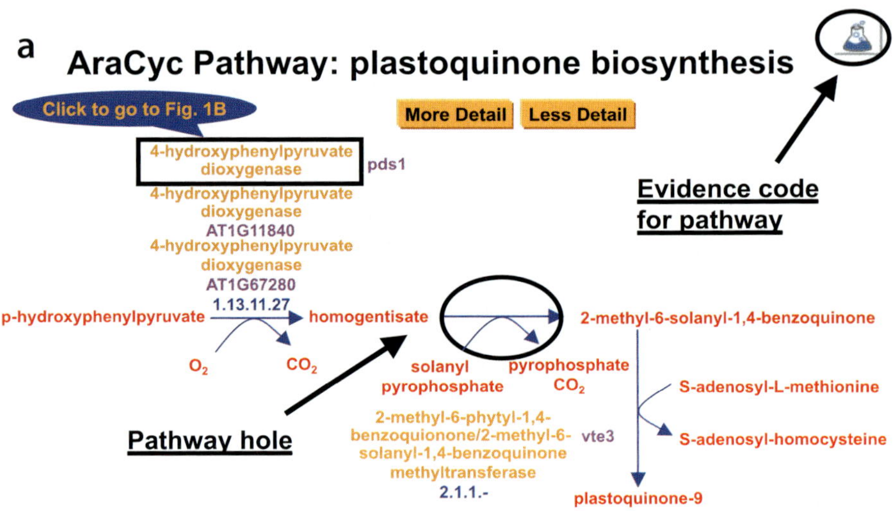

Fig. 1. An example of an AraCyc pathway: **a** pathway evidence, which could be either computational (*indicated by a computer icon*) or experimental (*indicated by a flask icon*), provides assertion of the existence of the pathway in Arabidopsis; **b** (see next page) similarly, evidence attached to an enzyme provides assertion of its catalytic activity involved in a specific reaction. Each piece of evidence is associated with a citation where the source of the evidence can be found (*inset*). Pathway can be zoomed to show various levels of details. Compounds, reactions, enzymes and genes on a pathway detail page are clickable to retrieve more information. (Reprinted with permission from Plant Physiology)

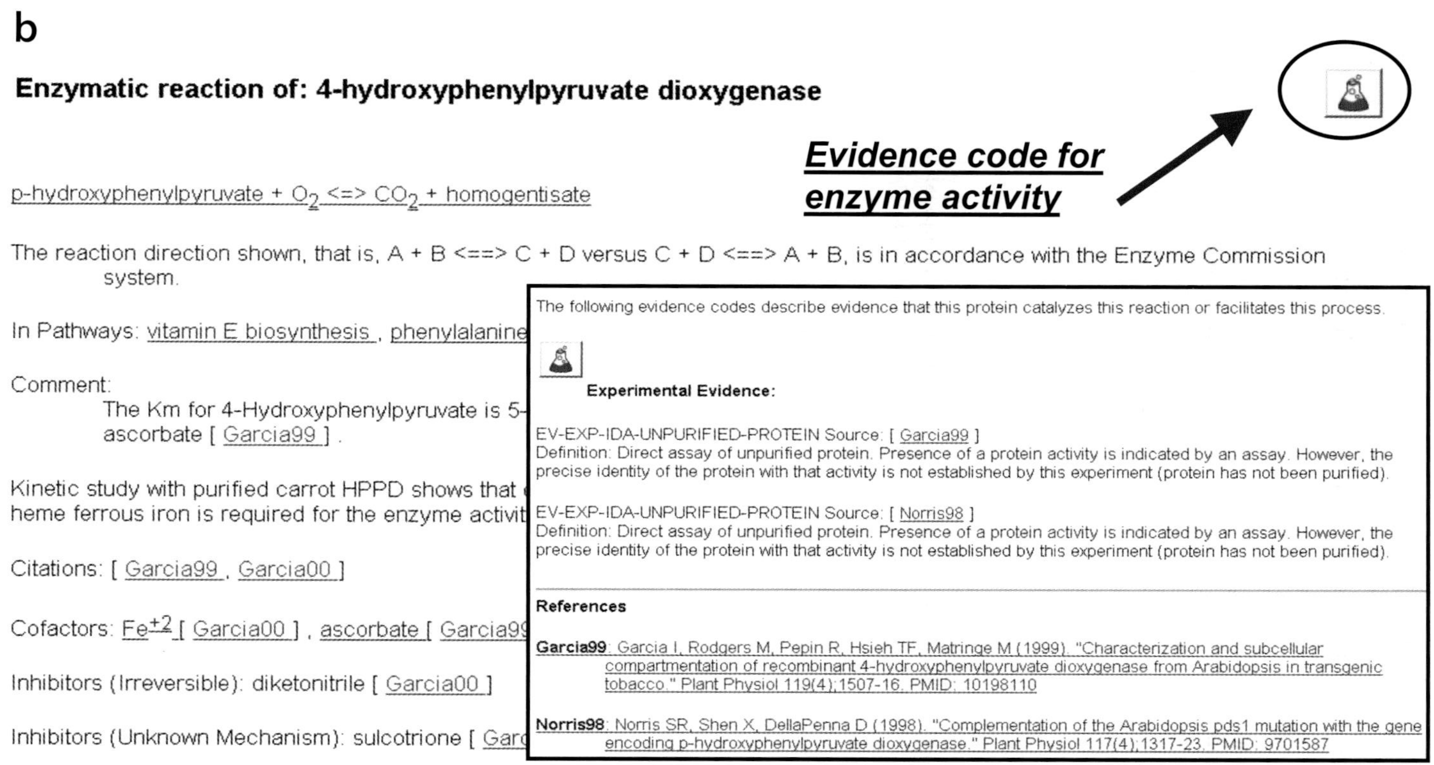

Fig. 1. (continued)

or utilization, as it occurs in a particular organism (Krieger et al. 2004). In addition to the 197 individual pathways, AraCyc has 15 super-pathways. A super-pathway is an aggregation of two or more individual pathways that are related in some way (Krieger et al. 2004). The reactions in AraCyc have EC numbers (Enzyme Commission Nomenclature, http://www.chem.qmul.ac.uk/iubmb/enzyme/) assigned, when available (Fig. 1a). Chemical structures are annotated to the compounds. The assignments of the enzymes to the reactions are based on the characterization of the enzymes or the functional annotations of the genes, which could be either experimentally determined or computationally derived. For example, cellulose synthases CesA1 and At3g02230 are both assigned to EC 2.4.1.12 on the cellulose biosynthesis pathway. However, the cellulose synthase activity is supported by functional studies of the enzyme only for CesA1, and the annotation of 'cellulose synthase' for At3g02230 comes from computational prediction based on sequence similarity. To distinguish the different levels of annotation qualities, an evidence code is provided along with the assignment of an enzyme to a reaction. These evidences can be easily recognized through the use of intuitive evidence icons (computer screen for computational, flask for experimental), which label each enzyme detail page (Fig. 1b). Citations are provided along with the evidence so that users can obtain more details about the source of the annotation.

The 197 pathways are classified into three main categories: "Biosynthesis", "Degradation/Utilization/Assimilation", and "Generation of Precursor Metabolites and Energy" (Table 1). Biosynthesis of all 20 protein amino acids, all DNA/RNA purine and pyrimidine nucleosides and nucleotides, commonly occurring sugars and polysaccharides, major fatty acid and lipid classes (including triacylglycerol, phospho- and glyco-lipids), cofactors, prosthetic groups and electron carriers, and six known major plant hormone classes are represented. In addition, biosynthesis of the major molecules found in plant primary and secondary cell wall and epidermal structures, including cellulose, homogalacturonan (a component of pectin), lignin, suberin, wax and cutin are included. Pathways for central energy metabolism are well represented. It is not easy to assess the comprehensiveness of pathways under "Degradation/Utilization/Assimilation" as there is much less information available for catabolism than for biosynthesis in plants. Secondary metabolism is not yet covered comprehensively but this situation will change in the near future (see Sect. 6).

Data objects in AraCyc, such as pathways, reactions and compounds, as well as subcellular compartments (for annotating enzyme locations), and evidence types are structured in hierarchical ontologies (Gruber 1993; Karp 2000; Karp et al. 2004). Each ontology describes concepts (terms) and relationships between them. Terms are organized into classes according to the primary 'is-a' relationship. The 'is-a' relationship classifies what type of a concept a term is. The broader concepts, or parent terms, appear on higher levels of the hierarchy tree. The more specific concepts, or children terms are grouped under the broader

Table 1. Summary of AraCyc database content

Total pathways excluding super-pathways	210 (197)[a]
Biosynthesis	137 (127)[a]
Amino acids	36
Cell structure	8
Cofactors, prosthetic groups, electron donors	24
Fatty acids and lipids	15
Plant hormones	14
Nucleosides and nucleotides	4
Secondary metabolism	17
Sugars and polysaccharides	9
Others	10
Degradation	58 (57)[b]
Amino acids	21
Fatty acids and lipids	7
Inorganic nutrients	5
Sugar derivatives	2
Sugars and polysaccharides	11
Others	12
Generation of precursor metabolites and energy	15[b]
Total unique reactions of pathways	979
Total unique compounds of pathways	1071
Total unique genes of pathways	1759

[a] Some pathways are classified to more than one pathway class. The numbers in parenthesis are unique number counts

[b] Two pathways are classified under both 'Degradation' and 'Generation of precursor metabolites and energy'

concepts. For example, within the pathway ontology, alanine biosynthesis is classified to Pathways→Biosynthesis→Amino acids→Individual amino acids (http://www.arabidopsis.org:1555/ARA/NEW-IMAGE?type=PATHWAY{\&} object=ALANINE-SYN2-PWY). Unlike terms in a simple list, the organization of terms into a hierarchical ontology allows more robust queries and makes retrieval of related information easier.

3 Search, Browse, and Analyze Functionalities

AraCyc can be freely accessed through the web (http://arabidopsis.org/tools/aracyc) using a common web browser, or downloaded as text files. The database is available to download with an 'open source' license (http://arabidopsis.org/aracyc/form.html). Installing AraCyc database and desktop version of the software on a local computer has a few advantages such as allowing

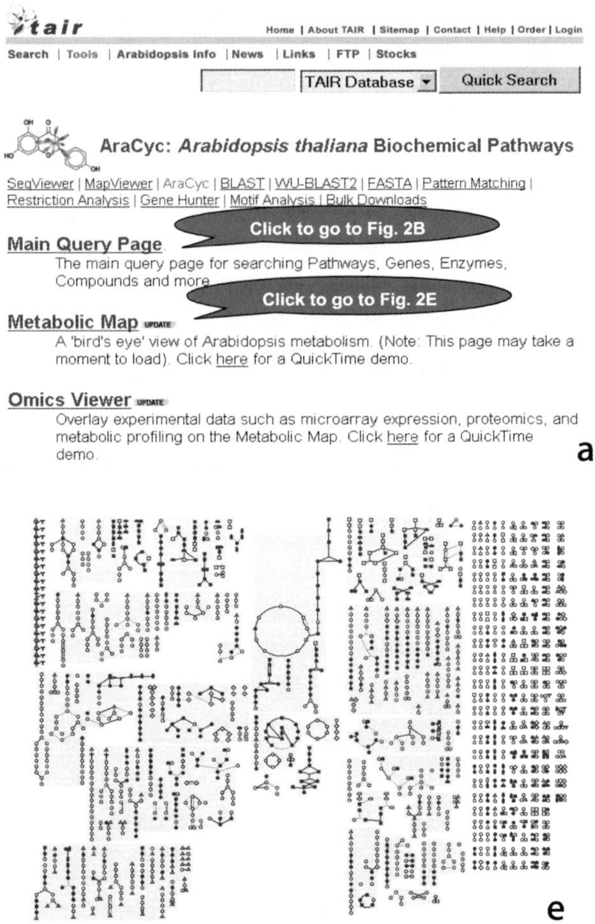

Fig. 2. Accessing the metabolic data: **a,b** from the AraCyc main page, data can be queried by names and browsed from alphabetic lists or hierarchy ontologies; **c** query results are displayed grouped by data types–different data types such as compounds, reactions, pathways, enzymes and genes are interlinked from individual data detail pages; **d** an example of a compound detail page; **e** the Metabolic Map depicts all of the pathways in one diagram. Clicking on a pathway glyph will open up the pathway detail page. Experimental data such as those resulting from gene expression and metabolic profiling experiments can be painted onto the Metabolic Map using the Omics Viewer tool

the user to update the database with proprietary data and to perform more advanced queries. Using either the web or the desktop version, a user can browse, query, and visualize the data (Fig. 2a). One can navigate through all of the pathways, for example, from an alphabetic list or from the hierarchy ontology browser (http://www.arabidopsis.org:1555/ARA/class-instances?object=Pathways). Substrings, or partial words, for example, 'gibber' in 'gibberellin', can be queried against names of a specific data type such as com-

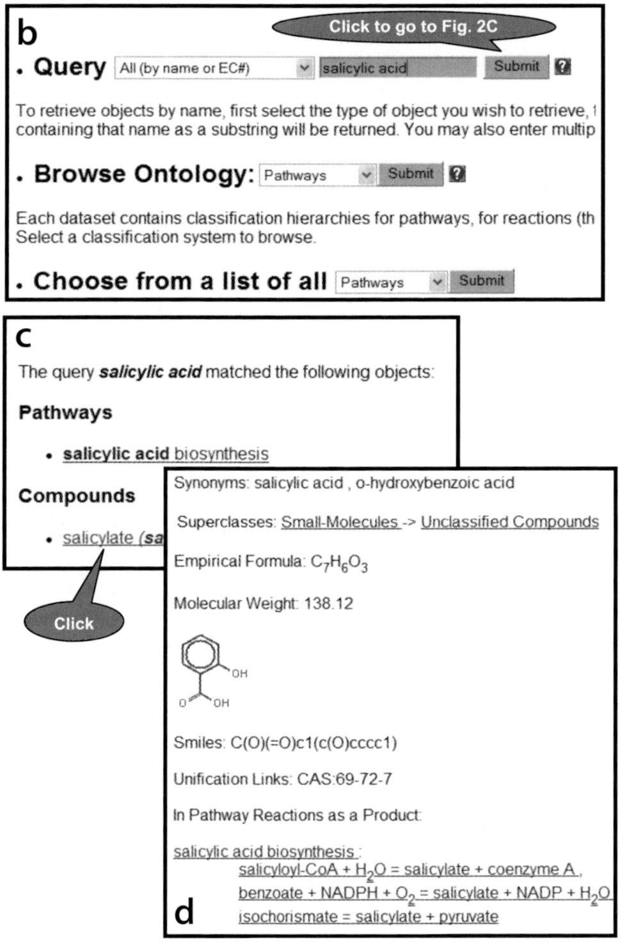

Fig. 2. (continued)

pounds only, or all of the data types (Fig. 2b). In the latter case, results are grouped according to the different data types (Fig. 2c). Each result is linked to its corresponding detail page. In the example shown in Fig. 2, clicking on the compound 'salicylate' from the query result page opens up the compound detail page for salicylate (Fig. 2d). Many data are interconnected by hyperlinks. Pathways and reactions shown on the compound salicylate detail page are linked to the corresponding pathway detail pages and reaction detail pages, and vice versa. The 'Metabolic Map' tool shows a bird's eye view of all of the pathways grouped by the pathway classes (Fig. 2e). In addition to the searching and browsing options using a web browser or the desktop application, datasets are also provided as downloadable text files, (ftp://ftp.arabidopsis.org/home/tair/Pathways/), such as a pathway dump file that lists all of the pathways and the genes and enzymes assigned to each pathway. A user who has a list of genes of interest can use this file to quickly sort out what pathways the genes are involved in. The files are updated with each AraCyc release.

An important component of the Pathway Tools software package (Karp et al. 2002) is the Omics Viewer (http://aracyc.stanford.edu:1555/expression.html). It allows the analysis of changes in the levels of transcripts, proteins, and

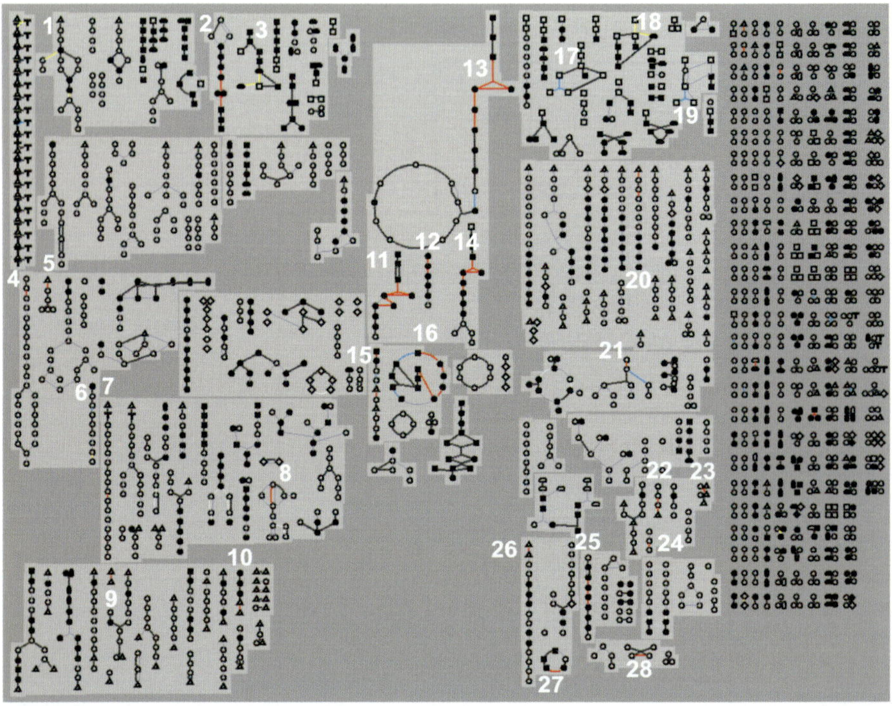

Fig. 3. Proteomics data from Weckwerth et al. (2004) overlaid on AraCyc Omics Viewer. Each glyph represents a pathway in which reactions are represented by *lines* and metabolites are represented by *triangles* (amino acids), *squares* (carbohydrates), or other metabolites (*circles*). Individual reactions that are not placed within a metabolic pathway are *listed on the right side of the diagram*. The 281 proteins that were detected in the leaves of Columbia and C24 accessions were overlaid onto the metabolic map. 104 proteins were found to carry out reactions in 28 pathways on the map, which are highlighted. *Red lines*, proteins assigned to reactions, which were detected only in Columbia; *blue lines*, proteins that were detected in both ecotypes, and *yellow lines*, proteins that were detected only in C24. Labeled pathways are: *1*, lignin biosynthesis; *2*, gluconeogenesis; *3*, sucrose biosynthesis; *4*, brassinosteroid biosynthesis; *5*, ethylene biosynthesis; *6*, jasmonic acid biosynthesis; *7*, chlorophyll biosynthesis; *8*, formylTHF biosynthesis; *9*, glutamine biosynthesis I; *10*, serine biosynthesis; *11*, glycolysis IV; *12*, glyceraldehyde 3-phosphate catabolism; *13*, glycolysis; *14*, glycolysis; *15*, phosphorespiration; *16*, Calvin cycle; *17*, starch degradation; *18*, sucrose degradation; *19*, glycogen catabolism; *20*, methionine degradation I; *21*, lipoxygenase pathway; *22*, nitrate assimilation; *23*, ammonium assimilation; *24*, cyanate catabolism; *25*, aerobic glycerol catabolism; *26*, serine isocitrate lyase pathway; *27*, xylulose monophosphate cycle; *28*, removal of superoxide radicals

metabolites by overlaying results of genome-wide gene expression, proteomics, or metabolite profiling data onto the metabolism overview diagram (Fig. 3). Each reaction (represented as a line connecting the compounds) can be color-coded according to the expression level of the gene or protein that catalyzes the reaction. Metabolite levels can be depicted by color-coding the symbols for compounds (represented as squares or triangles connected by the reaction lines). Note that only those genes and compounds that are included in AraCyc can be displayed on the metabolic map. However, it is possible to extrapolate from the Omics Viewer to identify additional components of a pathway. For example, if a set of genes from an expression array appeared to be all involved in the same pathway and showed similar changes in expression values, one could cluster the original dataset to identify other genes having a similar expression profile. These genes, in turn, may represent components of the pathway that are missing from AraCyc. Specific usage examples of this tool are described in the next section.

4 Applications of AraCyc

4.1 Putting Functional Genomics Data into a Metabolic Network Framework

The Omics Viewer is a convenient way of quickly assessing the metabolic changes from a large-scale experiment such as gene expression profiling or metabolite profiling under different environmental or genotypic conditions, either to test a specific hypothesis or explore the trends in a large-scale data set. Alternatively the viewer can be used to annotate a set of genes grouped by certain criteria such as gene families or co-expressed genes. For example, Arabidopsis accessions Columbia (Col-0) and C24 are known to have a number of polymorphisms that lead to differences in a variety of phenotypic traits (Rohde et al. 2004). For example, Columbia can acclimate to cold and tolerate freezing much better than C24. However, the exact molecular, biochemical, and physiological differences that result in this phenotypic trait are not known. One hypothesis is that their metabolic state is different. Weckwerth and colleagues compared the protein and metabolite content of Columbia and C24 leaves (Weckwerth et al. 2004). They detected 297 proteins, of which 153 were detected in both accessions and 144 were detected in only one of the ecotypes. An equivalent number of proteins was found in each accession (30% specific in Columbia, 30% in C24, and 40% detected in both). Overlaying 281 of the differentially expressed proteins on the AraCyc Omics viewer shows that about 37% of the proteins (104/281) were placed within the metabolic framework, specifically onto 28 pathways (Fig. 3). The mapping shows that Columbia is much more active metabolically than C24, with 19 out of 28 pathways that were mapped with proteins specific to Columbia. It is possible that Columbia's more active metabolic state may be relevant for its ability to acclimate to cold better than C24. Pathways that were more prominent in C24 than in Columbia include lignin biosynthesis, sucrose biosynthesis, and sucrose degradation. While it is premature to derive definitive conclusions from this exercise, it demonstrates the ability of this tool to explore large-scale datasets quickly and efficiently.

Similarly, the Omics Viewer can be used to address hypotheses using large-scale data such as gene expression microarrays. For example, it is well-established that CBF (CRT/DREB Binding Factor) transcription factors play an important role in cold acclimation (Thomashow 1999; Cook et al. 2004; Gilmour et al. 2004). Also, increases in sugars such as sucrose, glucose, fructose, and raffinose, and in the amino acid proline, are correlated with the ability to tolerate freezing, perhaps because these compounds act as osmoprotectants (compatible solutes) (Strand et al. 1999; Taji et al. 2002; Shinozaki et al. 2003; Uemura et al. 2003; Zuther et al. 2004). It is, however, not known whether the cold acclimation process via the CBF pathway directly affects the increase in the production of these metabolites (Stitt and Hurry 2002). Recently Vogel and colleagues asked which genes are regulated by the CBF pathway in the cold acclimation process by examining the global gene expression profiles of cold-treated wild type plants and lines overexpressing the *CBF2* gene (Vogel et al. 2005). They found that 93 genes were affected in both of these lines as compared to wild type and considered these genes to be involved in the CBF cold-response pathway. By overlaying the expression profiles of these genes on the Omics Viewer, we can quickly assess which metabolic pathways are affected via the CBF cold-response pathway (Fig. 4). The results show that 31 out of 93 genes were placed within the metabolic context. Six pathways are induced by CBF overexpression and cold treatment and one pathway, glucosinolate biosynthesis, is reduced by CBF overexpression and cold treatment. Pathways in which genes encoding enzymes carrying out reactions are induced include sucrose biosynthesis, flavonoid biosynthesis, flavonol biosynthesis, anaerobic glycolysis, homogalacturonan degradation, and sucrose degradation. This result suggests that sucrose biosynthesis may be directly affected by the CBF pathway.

4.2 Application of AraCyc Data to Other Software Environments

In addition to addressing specific questions using the data and tools in AraCyc as exemplified above, AraCyc data can be used in other software environments to explore metabolism data in conjunction with other data such as gene expression profiling, metabolite profiling, and proteomics data. Examples of third-party software that have used data from AraCyc include MapMan (Thimm et al. 2004), MetNetDB (Wurtele et al. 2003), FCModeler (Wurtele et al. 2003), and 3D virtual reality visualization environment (Dickerson et al. 2003). MapMan is a user-driven software environment that allows the display and analysis of large-scale datasets in the context of functional categories in-

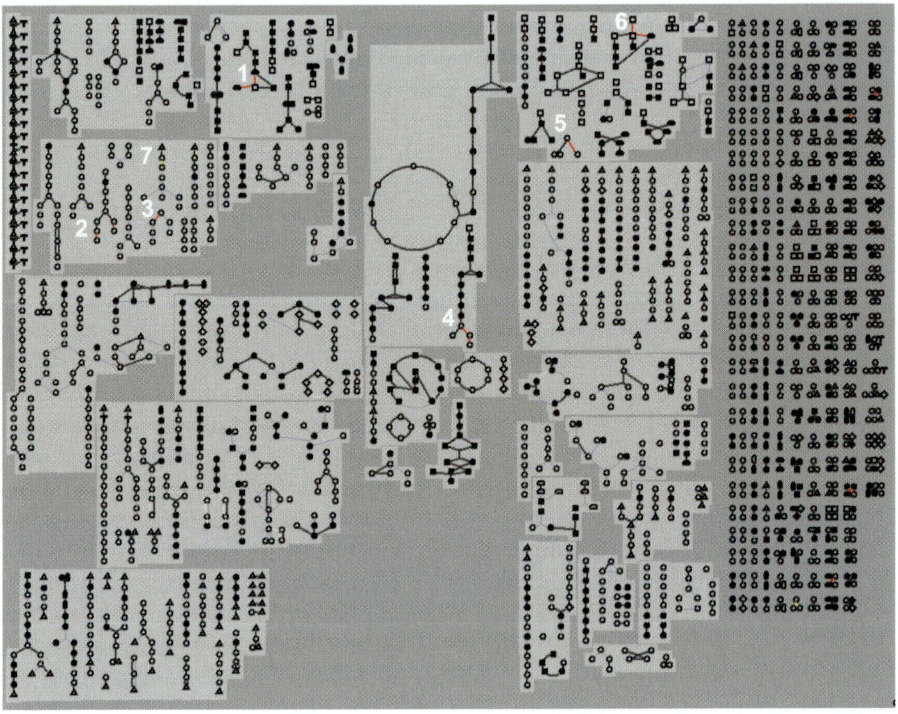

Fig. 4. Genes that belong to the CBF regulon in Arabidopsis from Vogel et al. (2005) overlaid on AraCyc Omics Viewer. Ninety three genes were used as input and 31 genes were placed on the metabolic framework into seven pathways. *Red* (induced) and *yellow* (reduced) *lines* indicate reactions that are performed by the proteins encoded by genes whose transcripts are induced or reduced by the CBF transcription factor overexpression as well as treatment with cold, respectively. Six pathways are induced by CBF overexpression and cold treatment and one pathway, glucosinolate biosynthesis, is reduced by CBF overexpression and cold treatment. Labeled pathways are: *1*, sucrose biosynthesis; *2*, flavonoid biosynthesis; *3*, flavonol biosynthesis; *4*, anaerobic glycolysis; *5*, homogalacturonan degradation; *6*, sucrose degradation; *7*, glucosinolate biosynthesis

cluding metabolic pathways. It is composed of two modules, SCAVENGER and IMAGEANNOTATOR. SCAVENGER designates measured values of an experiment onto a set of metabolic pathways and other processes that are organized into bins and IMAGEANNOTATOR allows users to generate a custom view of the annotated bins and the measured parameters according to their specifications and needs. MetNetDB is an Arabidopsis interactions database that is used as a basis for FCModeler software. FCModeler software uses fuzzy cognitive maps to allow biologists to generate models of regulatory and metabolic pathways from data in MetNetDB and large-scale datasets such as those resulting from genome-wide gene expression profiling experiments. The 3D virtual reality environment uses the data in MetNetDB and allows visualization of the network data in 3D in a virtual reality cave such that users can 'get inside' the pathways and explore the data from particular areas within the network. All of the applications mentioned here have imported the AraCyc data to be visualized and analyzed in a number of flexible and creative ways.

5 Current Issues and Future Directions

About 86% (170 pathways) of the pathways have been manually validated, meaning that the pathway diagrams have been validated and corrected according to the latest literature information. The remaining 27 pathways were predicted to exist in Arabidopsis but no experimental support was found in the literature irrevocably confirming their existence in plants. Pathways of secondary metabolism are under-represented in AraCyc. Curation of new secondary metabolic pathways is an ongoing task. In addition, we plan to curate and integrate transporters into their relevant pathways. Users are encouraged to contact us (curator@arabidopsis.org) for data submissions, including updating or correcting an existing pathway, or submitting a new pathway.

Starting in 2005, updates to AraCyc are released on a quarterly basis. Each release includes manual updates, corrections of the existing pathway data, and manual curation of new pathway data. A major release at the end of each year is planned to take advantage of the progress in the functional annotations of the Arabidopsis genome (Berardini et al. 2004; Zhang et al. 2005). A gene whose function was previously unknown may now have an annotated function and thus may be assigned to a corresponding AraCyc pathway.

Many enhancements to the data visualization capabilities provided by the Pathway Tools software are planned, such as the ability to display pathways in the context of subcellular location information.

6 Conclusions

Currently we are experiencing a rapidly increasing rate of production of large-scale data such as genome sequences, genome-wide gene expression profiles, proteomics and metabolomics data. The necessity to organize all of these data into a biological framework has been, in part, the motivation for the work described in this review. While we have created a comprehensive database that describes the metabolic network of a model plant species, *Arabidopsis thaliana*, the database is far from being either complete or error-free. Many of the pathways are in need of manual curation using the current literature and many more pathways, particularly those for secondary metabolism and those that include transport reactions, need to be brought into the database. As with any other database project, the content of the AraCyc database is dynamic and will continue to undergo enhancement, additions, and modifications to make it more useful.

Acknowledgements. We are grateful to Tanya Berardini and Leonore Reiser for their careful reading of the manuscript. Work described in this project is funded in part by grant #1-R01-GM65466-01 from the NIH National Institute of General Medical Sciences and grant # 0417062 from NSF Biological Database and Informatics division.

References

AGI (2000) Analysis of the genome sequence of the flowering plant *Arabidopsis thaliana*. Nature 408:796–815

Berardini TZ, Mundodi S, Reiser L, Huala E, Garcia-Hernandez M, Zhang P, Mueller LA, Yoon J, Doyle A, Lander G, Moseyko N, Yoo D, Xu I, Zoeckler B, Montoya M, Miller N, Weems D, Rhee SY (2004) Functional annotation of the Arabidopsis genome using controlled vocabularies. Plant Physiol 135:745–755

Cook D, Fowler S, Fiehn O, Thomashow MF (2004) A prominent role for the CBF cold response pathway in configuring the low-temperature metabolome of Arabidopsis. Proc Natl Acad Sci USA 101:15243–15248

Dickerson JA, Yang Y, Blom K, Reinot A, Lie J, Cruz-Neira C, Wurtele ES (2003) Using virtual reality to understand complex metabolic networks. Atlantic symposium on computational biology and genomic information systems and technology, september, pp 950–953

Gilmour SJ, Fowler SG, Thomashow MF (2004) Arabidopsis transcriptional activators CBF1, CBF2, and CBF3 have matching functional activities. Plant Mol Biol 54:767–781

Goff SA, Ricke D, Lan TH, Presting G, Wang R, Dunn M, Glazebrook J, Sessions A, Oeller P, Varma H, Hadley D, Hutchison D, Martin C, Katagiri F, Lange BM, Moughamer T, Xia Y, Budworth P, Zhong J, Miguel T, Paszkowski U, Zhang S, Colbert M, Sun WL, Chen L, Cooper B, Park S, Wood TC, Mao L, Quail P, Wing R, Dean R, Yu Y, Zharkikh A, Shen R, Sahasrabudhe S, Thomas A, Cannings R, Gutin A, Pruss D, Reid J, Tavtigian S, Mitchell J, Eldredge G, Scholl T, Miller RM, Bhatnagar S, Adey N, Rubano T, Tusneem N, Robinson R, Feldhaus J, Macalma T, Oliphant A, Briggs S (2002) A draft sequence of the rice genome (*Oryza sativa* L. ssp. japonica). Science 296:92–100

Gruber TR (1993) A translation approach to portable ontology specifications. Knowledge Acquisition 5:199–220

Kanehisa M (2002) The KEGG database. Novartis Found Symp 247:91–101; discussion 101–103, 119–128, 244–152

Kanehisa M, Goto S, Kawashima S, Okuno Y, Hattori M (2004) The KEGG resource for deciphering the genome. Nucleic Acids Res 32:D277–D280

Karp PD (2000) An ontology for biological function based on molecular interactions. Bioinformatics 16:269–285

Karp PD, Paley S, Romero P (2002) The pathway tools software. Bioinformatics 18 [Suppl 1]:S225–S232

Karp PD, Paley S, Krieger CJ, Zhang P (2004) An evidence ontology for use in pathway/genome databases. Pacific Symp Biocomput 9:190–201

Krieger CJ, Zhang P, Mueller LA, Wang A, Paley S, Arnaud M, Pick J, Rhee SY, Karp PD (2004) MetaCyc: a multiorganism database of metabolic pathways and enzymes. Nucleic Acids Res 32:D438-D442

Maranas CD, Burgard AP (2001) Review of EcoCyc and MetaCyc databases. Metab Eng 3:98–99

Mueller LA, Zhang P, Rhee SY (2003) AraCyc: a biochemical pathway database for Arabidopsis. Plant Physiol 132:453–460

Ogata H, Goto S, Sato K, Fujibuchi W, Bono H, Kanehisa M (1999) KEGG: Kyoto encyclopedia of genes and genomes. Nucleic Acids Res 27:29–34

Paley SM, Karp PD (2002) Evaluation of computational metabolic-pathway predictions for *Helicobacter pylori*. Bioinformatics 18:715–724

Rohde P, Hincha DK, Heyer AG (2004) Heterosis in the freezing tolerance of crosses between two *Arabidopsis thaliana* accessions (Columbia-0 and C24) that show differences in non-acclimated and acclimated freezing tolerance. Plant J 38:790–799

Selkov E, Basmanova S, Gaasterland T, Goryanin I, Gretchkin Y, Maltsev N, Nenashev V, Overbeek R, Panyushkina E, Pronevitch L, Selkov E Jr, Yunus I (1996) The metabolic pathway collection from EMP: the enzymes and metabolic pathways database. Nucleic Acids Res 24:26–28

Shinozaki K, Yamaguchi-Shinozaki K, Seki M (2003) Regulatory network of gene expression in the drought and cold stress responses. Curr Opin Plant Biol 6:410–417

Stitt M, Hurry V (2002) A plant for all seasons: alterations in photosynthetic carbon metabolism during cold acclimation in Arabidopsis. Curr Opin Plant Biol 5:199–206

Strand A, Hurry V, Henkes S, Huner N, Gustafsson P, Gardestrom P, Stitt M (1999) Acclimation of Arabidopsis leaves developing at low temperatures. Increasing cytoplasmic volume accompanies increased activities of enzymes in the Calvin cycle and in the sucrose-biosynthesis pathway. Plant Physiol 119:1387–1398

Taji T, Ohsumi C, Iuchi S, Seki M, Kasuga M, Kobayashi M, Yamaguchi-Shinozaki K, Shinozaki K (2002) Important roles of drought- and cold-inducible genes for galactinol synthase in stress tolerance in *Arabidopsis thaliana*. Plant J 29:417–426

Thimm O, Blasing O, Gibon Y, Nagel A, Meyer S, Kruger P, Selbig J, Muller LA, Rhee SY, Stitt M (2004) MAPMAN: a user-driven tool to display genomics data sets onto diagrams of metabolic pathways and other biological processes. Plant J 37:914–939

Thomashow MF (1999) Plant cold acclimation: freezing tolerance genes and regulatory mechanisms. Annu Rev Plant Physiol Plant Mol Biol 50:571–599

Uemura M, Warren G, Steponkus PL (2003) Freezing sensitivity in the sfr4 mutant of Arabidopsis is due to low sugar content and is manifested by loss of osmotic responsiveness. Plant Physiol 131:1800–1807

Vogel JT, Zarka DG, Van Buskirk HA, Fowler SG, Thomashow MF (2005) Roles of the CBF2 and ZAT12 transcription factors in configuring the low temperature transcriptome of Arabidopsis. Plant J 41:195–211

Weckwerth W, Wenzel K, Fiehn O (2004) Process for the integrated extraction, identification and quantification of metabolites, proteins and RNA to reveal their co-regulation in biochemical networks. Proteomics 4:78–83

Wurtele ES, Li J, Diao L, Zhang H, Foster CM, Fatland B, Dickerson JA, Brown A, Cox Z, Cook D, Lee E-K, Hofmann H (2003) MetNet: Software to build and model the biogenetic lattice of Arabidopsis. Comp Funct Genom 4:239–245

Yu J, Hu S, Wang J, Wong GK, Li S, Liu B, Deng Y, Dai L, Zhou Y, Zhang X, Cao M, Liu J, Sun J, Tang J, Chen Y, Huang X, Lin W, Ye C, Tong W, Cong L, Geng J, Han Y, Li L, Li W, Hu G, Li J, Liu Z, Qi Q, Li T, Wang X, Lu H, Wu T, Zhu M, Ni P, Han H, Dong W, Ren X, Feng X, Cui P, Li X, Wang H, Xu X, Zhai W, Xu Z, Zhang J, He S, Xu J, Zhang K, Zheng X, Dong J, Zeng W, Tao L, Ye J, Tan J, Chen X, He J, Liu D, Tian W, Tian C, Xia H, Bao Q, Li G, Gao H, Cao T, Zhao W, Li P, Chen W, Zhang Y, Hu J, Liu S, Yang J, Zhang G, Xiong Y, Li Z, Mao L, Zhou C, Zhu Z, Chen R, Hao B, Zheng W, Chen S, Guo W, Tao M, Zhu L, Yuan L, Yang H (2002) A draft sequence of the rice genome (*Oryza sativa* L. ssp. indica). Science 296:79–92

Zhang P, Foerster H, Tissier CP, Mueller LA, Paley S, Karp PD, Rhee SY (2005) MetaCyc and AraCyc: metabolic pathway databases for plant research. Plant Physiol 138:27–37

Zuther E, Buchel K, Hundertmark M, Stitt M, Hincha DK, Heyer AG (2004) The role of raffinose in the cold acclimation response of *Arabidopsis thaliana*. FEBS Lett 576:169–173

II.5 KaPPA-View: A Tool for Integrating Transcriptomic and Metabolomic Data on Plant Metabolic Pathway Maps

T. Tokimatsu[1,2], N. Sakurai[1], H. Suzuki[1], and D. Shibata[1]

1 Introduction

Recent advances in DNA array technology (Aharoni and Vorst 2002; Donson et al. 2002) and compound separation techniques coupled to mass spectrometry (MS), such as gas chromatography (GC)-MS, liquid chromatography (LC)-MS and capillary electrophoresis (CE)-MS of metabolites (Sumner et al. 2003; Sato et al. 2004) have produced large amounts of transcriptomic and metabolomic ('omic') quantitative data. The interpretation of omic data is one of the major challenges for researchers identifying gene functions. Multigene families are considerably more prevalent among plant genomes than among animal genomes (Arabidopsis Genome Initiative 2000). Consequently, a relatively common characteristic of plants is that several homologous gene products are often assigned to a single enzymatic reaction, which complicates an understanding of the individual contributions of gene functions in plant metabolism. Metabolic pathway databases and tools are therefore crucial to the interpretation of behaviors of individual genes from omic data and to understanding their functions.

An analytical tool called KaPPA-View was developed to facilitate the display user transcript and/or metabolite data on a set of comprehensive plant metabolic pathway maps (Tokimatsu et al. 2005). Using an Internet browser, users can access the tool at http://kpv.kazusa.or.jp/kappa-view/. Here, we present a concise introduction to the functions of the tool and discuss the limitations of the present version and possible improvements that could be made thereto.

2 General Features of the KaPPA-View Tool

The architecture of the web-based visualization tool is shown in Fig. 1. The application engine of the system processes a user's comma separated value (CSV)-formatted dataset of transcripts and/or metabolites over the Internet and returns the Scalable Vector Graphics (SVG)-formatted files to the user's

[1] Kazusa DNA Research Institute, 2-6-7 Kazusa-kamatari, Kisarazu, Chiba 292-0818, Japan, e-mail: shibata@kazusa.or.jp

[2] *Present address:* Graduate School of Frontier Sciences, The University of Tokyo, 5-1-5 Kashiwanoha, Kashiwa, Chiba, 277-8561, Japan

Biotechnology in Agriculture and Forestry, Vol. 57
Plant Metabolomics (ed. by K. Saito, R.A. Dixon, and L. Willmitzer)

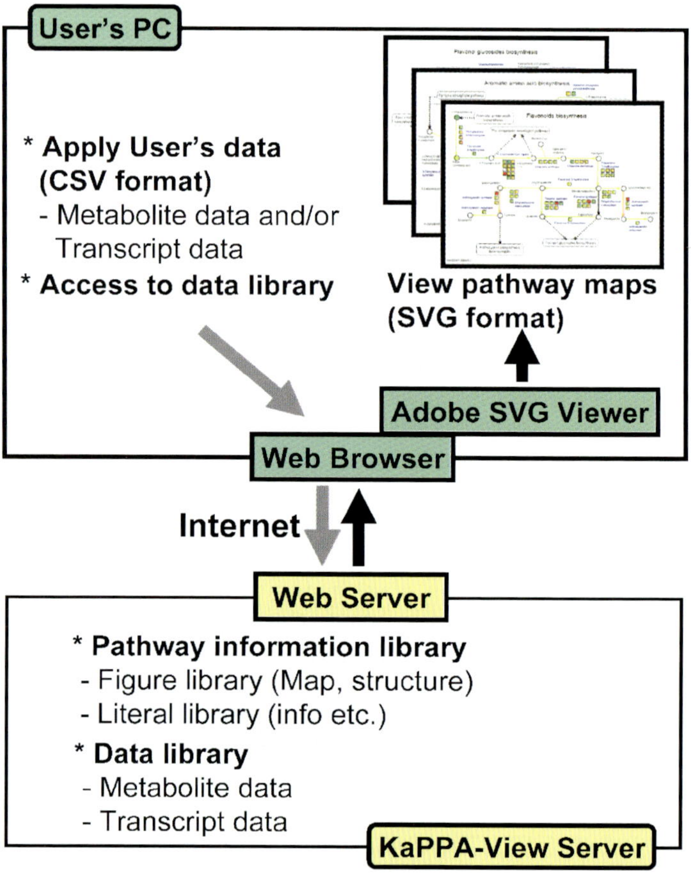

Fig. 1. System architecture of the KaPPA-View tool.

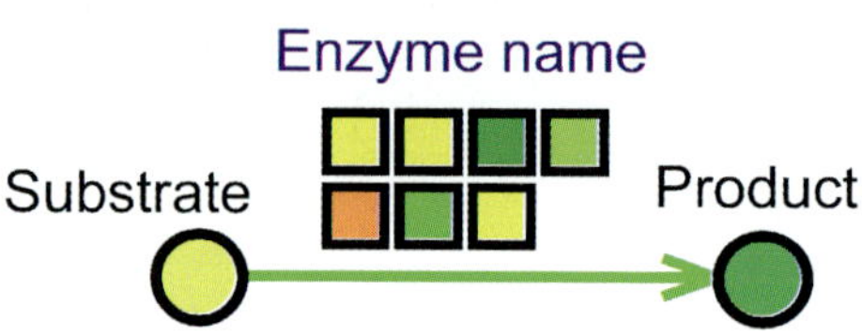

Fig. 2. Diagrammatic representation of metabolic pathway maps. *Circles* depict substrates and reaction products, *arrows* represent reactions and *squares* depict the various transcripts involved in, or putatively assigned to, the reaction being examined

personal computer (PC) through the Adobe SVG Viewer installed on the PC. Users can also access the transcript and metabolite data library on the KaPPA-View server.

The metabolic pathway maps of KaPPA-View were designed so that users would be able to picture the quantitative changes associated with individual

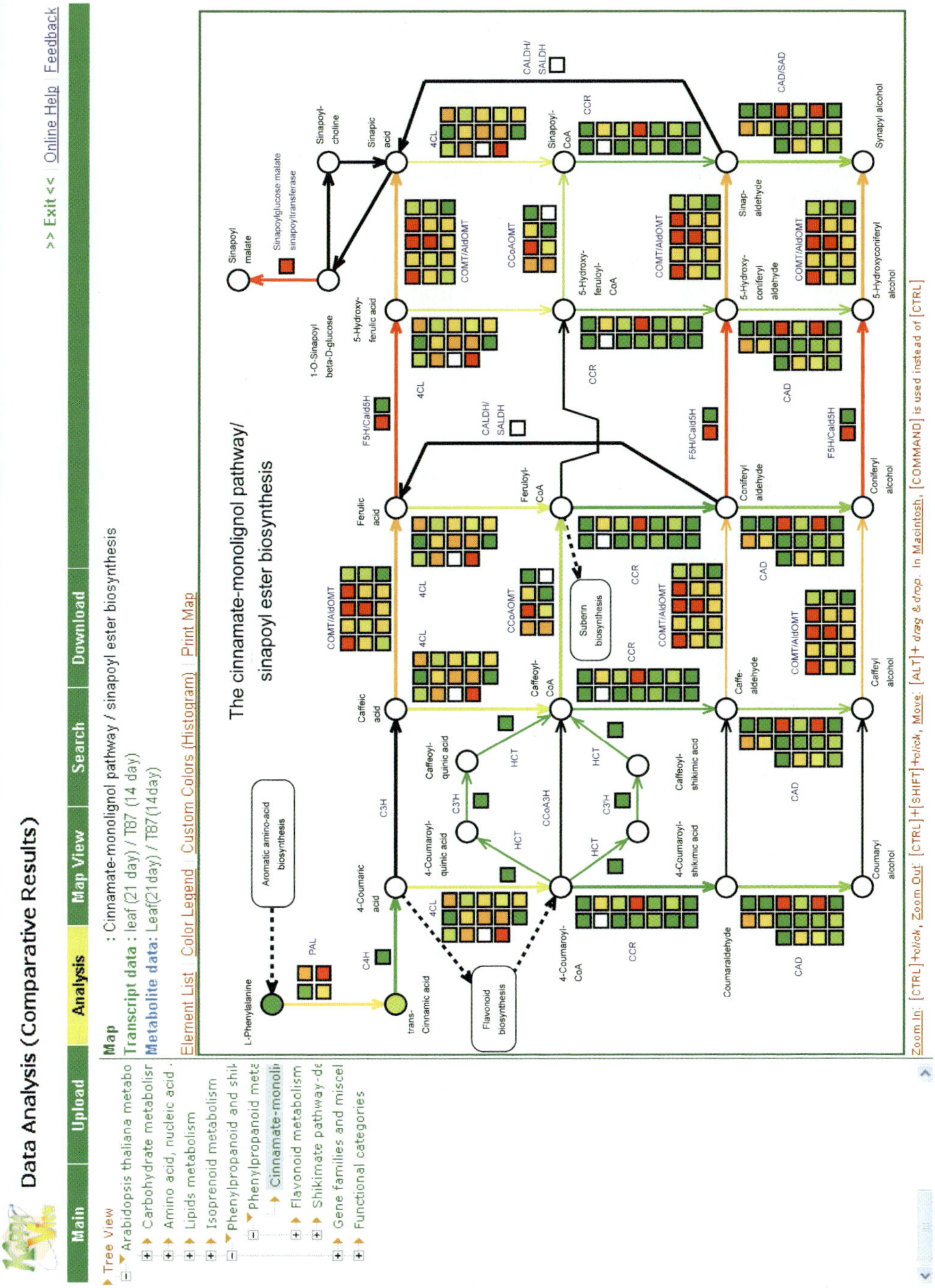

Fig. 3. Representation of an SVG-formatted metabolic pathway map for cinnamate-monolignol pathway/sinapoyl ester biosynthesis with quantitative transcript data

transcripts and metabolites involved in enzymatic reactions displayed on metabolic pathway maps. The metabolites, enzymatic reactions and transcripts involved in, or putatively assigned to a given reaction are represented as circles, arrows and squares, respectively (Fig. 2). Quantitative values for transcripts and/or metabolites submitted by the user as CSV-formatted text over the Internet are represented using symbols of different colors, as exemplified in the cinnamate-monolignol pathway/sinapoyl ester biosynthesis map shown in Fig. 3.

All of the information for an individual metabolite and reactions on a map being displayed on a user's browser can be retrieved from the screen. Compound and enzyme names are given along with their symbols on the metabolic pathway maps. Information on the metabolites, enzymes and transcripts associated with individual pathways can be found in the reference pages, which can be retrieved for each map using popup windows activated by clicking on symbols (Microsoft Windows users) or from the elements list for each map (all users). Gene and compound identifiers in the list are linked to the relevant gene and compound pages of The *Arabidopsis* Information Resource (TAIR) database (http://www.arabidopsis.org/) and the Kyoto Encyclopedia of Genes and Genomes (KEGG/PATHWAY) database (http://www.genome.ad.jp/kegg/pathway.html) (Goto et al. 2002), respectively. Furthermore, the names of pathways immediately up- or downstream are indicated on each map, and related pathway maps can be displayed in popup windows at the user's request.

To facilitate ease of use, SVG format was employed to generate maps and achieve a dynamic graphical representation of the quantitative changes in transcripts and/or metabolites on a user's browser. The maps and the associated quantitative data on the browser can be downloaded to the user's PC and modified for presentation. Given the variety of chemical and genetic nomenclature currently in use, in the likely event that users wish to alter information such as the names of metabolites and genes on the maps for presentation purposes, they can edit the SVG source text files using a text editor. The commercial editor, Adobe Illustrator is well suited for this as it can edit the SVG files.

Given that KaPPA-View is a web-based application, it is platform independent and can be used on a variety of popular operating systems that have the SVG Viewer plug-in supplied by Adobe. However, users are recommended to use Windows 2000/XP and Internet Explorer 6.0 or higher, which permit users to access the full range of KaPPA-View functions.

3 Plant Metabolic Pathway Maps

The initial version of the application contained a set of comprehensive metabolic pathway maps for the model plant, *Arabidopsis thaliana*, for which annotated genome information is available for identifying, or putatively identifying, metabolism-related genes. The metabolic pathways are classified as being one of 25 subclasses that are further subdivided into seven metabolic categories. In the current version, these categories contained 1263 enzymatic reactions (release 1.0). The metabolic pathways that are classified as belonging to functional categories, such as "plant hormones" and "secondary metabolism" in KEGG/PATHWAY (Kanehisa et al. 2002; Goto et al. 2002) and AraCyc (Mueller et al. 2003; Rhee et al. 2006; Zhang et al. 2005), are positioned at the branches of the reactions representing metabolic flows. However, in addition to the standard classification of such reactions, several genes are also classified as being involved in cofactor metabolism, vitamin metabolism or plant hormone metabolism and are therefore positioned in the "Functional categories" for the user's convenience.

As a tool for representing the full extent of the quantitative changes associated with transcripts and metabolites, the dimensions of the metabolic pathway maps generated were considered to be suitable for display on monitors routinely used by desktop users. This meant that related metabolic reactions were integrated into single maps, but care was taken to avoid too much integration.

4 Integration of Transcriptomic and Metabolomic Data on Pathway Maps

Transcriptomic and metabolomic data were used to complement each other and facilitate the identification of gene function in metabolism. While in excess of 5000 metabolites are thought to exist in a single plant species, only a limited number have been identified using compound separation techniques coupled to mass spectrometry (Aharoni et al. 2002). In general, given that metabolic intermediates exist at very low levels in plant tissue, their detection, even with state-of-the-art mass spectrometry, is considerably difficult and metabolomic analyses tend to rely on pooled metabolites and final products. Recent versions of Arabidopsis DNA arrays cover all of the genes annotated in the genome. Most of the transcripts listed on the metabolic pathway maps in KaPPA-View are detected, as can be seen when users analyze the transcriptomic data for leaves and suspension-cultured cells provided with the tool. Consequently, the detection of transcripts involved in reactions involving intermediate metabolites works well for generating hypotheses of gene function in the metabolic pathway being examined.

5 Comparison with Other Databases and Tools

Given that KaPPA-View displays quantitative data for various individual transcripts in different reactions and for metabolites in the various metabolic pathway maps, it is complementary in function to other omic-data tools. The well-cited pathway database of KEGG/PATHWAY (http://www.genome.ad.jp/kegg/pathway.html) has metabolic pathways for 218 organisms, including Arabidopsis and rice, linked to metadata (Goto et al. 2002). The Arabidopsis pathway database, AraCyc, (http://www.arabidopsis.org/tools/aracyc/) is a comprehensive metabolic database for Arabidopsis (Rhee et al. 2006). They also provide tool function for attaching transcript data onto the metabolic overview diagram. However, individual transcript data are not shown on individual metabolic pathway maps. The user-driven tool MAPMAN is designed to present the quantitative data obtained for all known Arabidopsis transcripts that have been categorized on the basis of the functionality of their products using diagrams of various biological processes (Thimm et al. 2004). Although users may include own diagrams, a limited numbers of metabolic pathway diagrams are provided in the present versions of MAPMAN. A new tool for integrating Arabidopsis transcriptomic and metabolomic data, BioPathAt, was released recently (Lange and Ghassemian 2005). However, it only operates as a visual interface for an expensive commercial software package, GeneSpring. The KaPPA-View tool differs conceptually from other omic-data tools, including MetNet (Wurtele et al. 2003), PathMAPA (Pan et al. 2003), Pathway Processor (Grosu et al. 2002) and GiGA (Breitling et al. 2004), all of which are well suited for representing statistical data of metabolism or generating metabolic networks but not for representing specific transcript data on metabolic pathway maps.

6 Limitations and Future Improvements

Information of the cellular location of metabolites is not explicitly given on the metabolic pathway maps of the KaPPA-View tool. In cases where the same metabolites are localized in distinct subcellular compartments, such as lipid metabolism in the cytoplasm and plastids, both pathways were shown using distinct metabolic pathway maps. Information on the destination of gene products (proteins), which can be estimated using the neural network-based TargetP (Emanuelsson et al. 2000), will be incorporated into each map in future versions. This might be achieved by simply inserting single letters (C for cytosol, P for plastid and M for mitochondrion) in squares that symbolize individual transcripts. Such a medication would improve our understanding of the metabolism within organelles.

► **Fig. 4.** Manual-drawn gene expression network for starch and maltose degradation: **a** genes involved in the starch and maltose degradation pathway and that have correlation coefficients exceeding 0.6 are used to construct a gene expression network, calculated using the Arabidopsis thaliana trans-factor and cis-element prediction database (ATTED-II, http://www.atted.bio.titech.ac.jp/); **b** the network shown in **a** is then overlaid on the KaPPA-View map and appear as *red lines*. Note that alpha-amylase genes occur twice in the degradation pathway

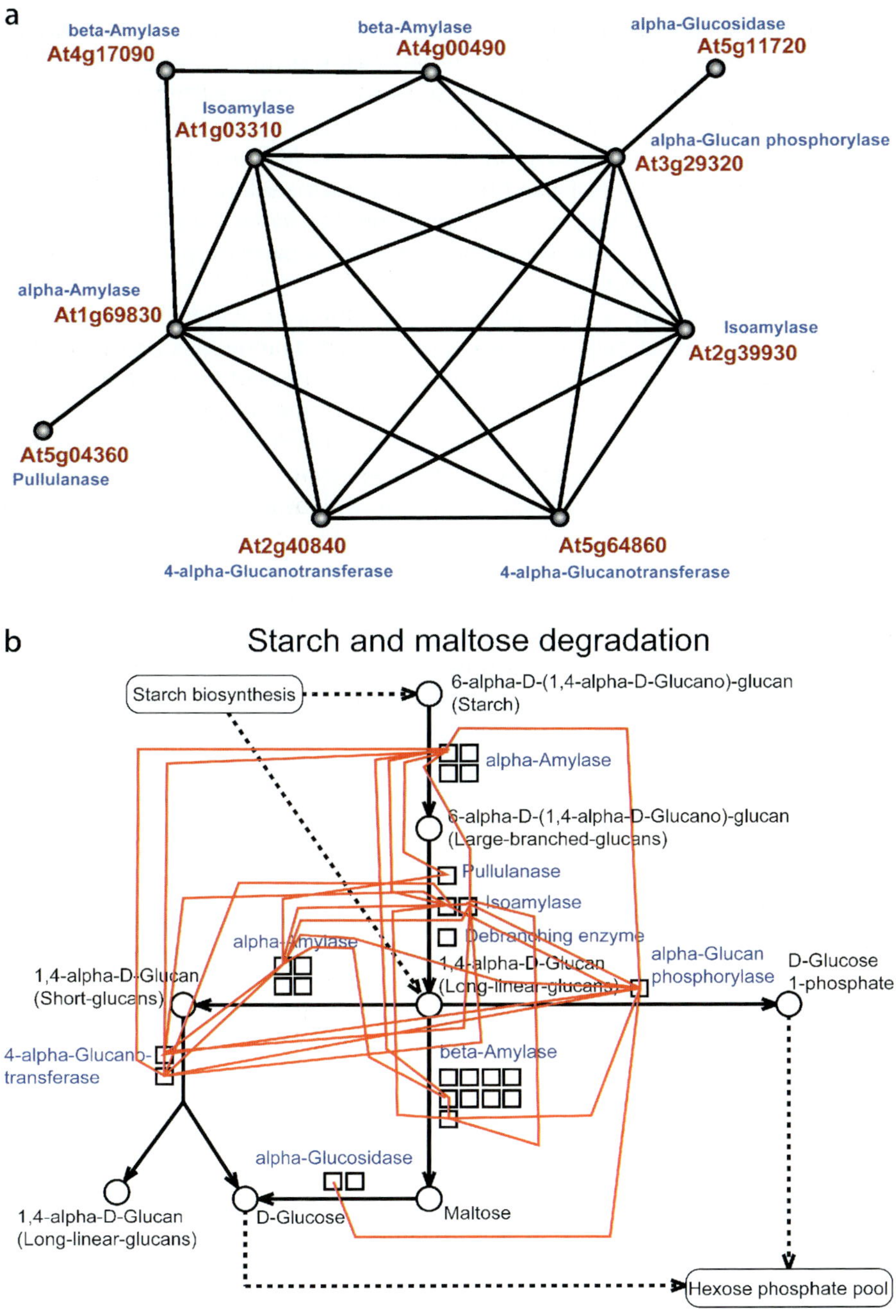
a
beta-Amylase
At4g17090
beta-Amylase
At4g00490
alpha-Glucosidase
At5g11720
Isoamylase
At1g03310
alpha-Glucan phosphorylase
At3g29320
alpha-Amylase
At1g69830
Isoamylase
At2g39930
At5g04360
Pullulanase
At2g40840
4-alpha-Glucanotransferase
At5g64860
4-alpha-Glucanotransferase
b
Starch and maltose degradation
Starch biosynthesis
6-alpha-D-(1,4-alpha-D-Glucano)-glucan
(Starch)
alpha-Amylase
6-alpha-D-(1,4-alpha-D-Glucano)-glucan
(Large-branched-glucans)
Pullulanase
Isoamylase
Debranching enzyme
alpha-Amylase
1,4-alpha-D-Glucan
(Short-glucans)
1,4-alpha-D-Glucan
(Long-linear-glucans)
alpha-Glucan
phosphorylase
D-Glucose
1-phosphate
4-alpha-Glucano-
transferase
beta-Amylase
alpha-Glucosidase
1,4-alpha-D-Glucan
(Long-linear-glucans)
D-Glucose
Maltose
Hexose phosphate pool

The present version of KaPPA-View only provides the user with a way to compare the transcripts/metabolites of two data sets and does not permit the presentation of multiple data sets such as those derived from time course experiments. As an alternative, multiple SVG-formatted metabolic maps with data values can be pasted into applications such as Microsoft Power Point for a slide show presentation or used for making animated gif files. An example of a gif animation for representing time course changes for transcripts in nine metabolic maps can be viewed at http://kpv.kazusa.or.jp/kappa/images/KPV-animation0.2sec.gif.

The present version of the visualization tool has no option for generating images of networks of gene co-expression. Figure 4a depicts a co-expression network of genes involved in starch and maltose degradation generated using correlation coefficient values calculated from the Arabidopsis transcriptomic data derived from various sources (http://www.atted.bio.titech.ac.jp/). The network is overlaid on the degradation pathway in KaPPA-View (Fig 4b). The overlaid diagram can be used to infer how the individual genes in starch and maltose degradation are regulated. Automation of such drawing tasks will decrease the time it takes for the user to gain an appreciation of the various transcripts and expression networks represented in the KaPPA-View metabolic pathway maps.

The present version using SVG-formatted maps is currently only available for Arabidopsis. Consequently, if users wish to use the maps for other plant species, they need to prepare a correspondence table between the AGI numbers for the Arabidopsis genes and gene identification numbers for the plant species of interest based on sequence homology. However, as the number of genes assigned to a reaction is greater than that assigned to the reaction on the Arabidopsis map, the number of genes selected by a user must correspond to the number assigned to Arabidopsis genes. If square symbols are used to represent flexible individual transcripts and ordered according to the numbers assigned to the reaction, such an improvement would increase the applicability of the tool to other plant species. Several maps of species-specific metabolic pathways such as those of isoflavonoids and alkaloids that are not found in Arabidopsis also are needed when applied to other plant species.

7 Conclusions

Simultaneous presentation of transcripts and metabolites on metabolic pathway maps allows users to quickly isolate differences in samples being examined. The KaPPA-View tool is user friendly and intuitive and simplifies the potentially tedious task of reviewing and comparing the complicated data associations typically associated with metabolic analyses. The visualization tool is useful for generating hypotheses rather than for making definitive conclusions regarding the roles of specific genes.

Acknowledgements. This work was supported by New Energy and Industrial Technology Development (NEDO) (as part of the project called, "Development of Fundamental Technologies for Controlling the Process of Material Production of Plants").

References

Aharoni A, Vorst O (2002) DNA microarrays for functional plant genomics. Plant Mol Biol 48:99–118

Aharoni A, Ric de Vos CH, Verhoeven HA, Mariepaard CA, Kruppa G, Bino R, Goddenowe DB (2002) Nontargeted metabolome analysis by use of Fourier Transform Ion Cyclotron Mass Spectrometry. OMICS 6:217–234

Arabidopsis Genome Initiative (2000) Analysis of the genome sequence of the flowering plant *Arabidopsis thaliana*. Nature 408:796–815

Breitling R, Amtmann A, Herzyk P (2004) Graph-based iterative group analysis enhances microarray interpretation. BMC Bioinformatics 5:100

Donson J, Fang Y, Espiritu-Santo G, Xing W, Salazar A, Miyamoto S, Armendarez V, Volkmuth W (2002) Comprehensive gene expression analysis by transcript profiling. Plant Mol Biol 48:75–97

Emanuelsson O, Nielsen H, Brunak S, von Heijne G (2000) Predicting subcellular localization of proteins based on their N-terminal amino acid sequence. J Mol Biol 300:1005–1016

Goto S, Okuno Y, Hattori M, Nishioka T, Kanehisa M (2002) LIGAND: database of chemical compounds and reactions in biological pathways. Nucleic Acids Res 30:402–404

Grosu P, Townsend JP, Hartl DL, Cavalieri D (2002) Pathway processor: a tool for integrating whole-genome expression results into metabolic networks. Genome Res 12:1121–1126

Kanehisa M, Goto S, Kawashima S, Nakaya A (2002) The KEGG database at GenomeNet. Nucleic Acids Res 30:42–46

Lange BM, Ghassemian M (2005) Comprehensive post-genomic data analysis approaches integrating biochemical pathway maps. Phytochemistry 66:413–451

Mueller LA, Zhang P, Rhee SY (2003) AraCyc: a biochemical pathway database for arabidopsis. Plant Physiol 132:453–460

Pan D, Sun N, Cheung KH, Guan Z, Ma L, Holford M, Deng X, Zhao H (2003) PathMAPA: a tool for displaying gene expression and performing statistical tests on metabolic pathways at multiple levels for *Arabidopsis*. BMC Bioinformatics 4:56

Rhee SY, Zhang P, Foerster H (2006) AraCyc: overview of an Arabidopsis metabolism database and its applications for plant research. In: Saito K, Dixon R, Willmitzer L (eds) Plant metabolomics (Biotechnology in Agriculture and Forestry, Vol.57). Springer, Berlin Heidelberg New York, Chap. II.4

Sato S, Soga T, Nishioka T, Tomita M (2004) Simultaneous determination of the main metabolites in rice leaves using capillary electrophoresis mass spectrometry and capillary electrophoresis diode array detection. Plant J 40:151–163

Sumner LW, Mendes P, Dixon RA (2003) Plant metabolomics: large-scale phytochemistry in the functional genomic era. Phytochemistry 62:817–836

Thimm O, Bläsing O, Gibon Y, Nagel A, Meyer S, Krüger P, Selbig J, Müller LA, Rhee SY, Stitt M (2004) MAPMAN: a user-driven tool to display genomics data sets onto diagrams of metabolic pathways and other biological processes. Plant J 37:914–939

Tokimatsu T, Sakurai N, Suzuki H, Ohta H, Nishitani K, Koyama T, Umezawa T, Misawa N, Saito K, Shibata D (2005) KaPPA-view: a web-based analysis tool for integration of transcript and metabolite data on plant metabolic pathway maps. Plant Physiol 138:1289–1300

Wurtele ES, Li J, Diao L, Zhang H, Foster CM, Fatland B, Dickerson J, Brown A, Cox Z, Cook D, Lee EK, Hofmann H (2003) MetNet: software to build and model the biogenetic lattice of *Arabidopsis*. Comp Funct Genom 4:239–245

Zhang P, Foerster H, Tissier CP, Mueller L, Paley S, Karp PD, Rhee SY (2005) MetaCyc and AraCyc. Metabolic pathway databases for plant research. Plant Physiol 138:27–37

II.6 KNApSAcK: A Comprehensive Species-Metabolite Relationship Database

Y. Shinbo[1], Y. Nakamura[2,3], M. Altaf-Ul-Amin[2], H. Asahi[2], K. Kurokawa[2], M. Arita[4], K. Saito[5], D. Ohta[6], D. Shibata[7], and S. Kanaya[2]

1 Introduction

Determination of gene functions on a large scale is one of the major challenges in biology today (Boyes et al. 2001). Given the completion of genome sequencing of *Arabidopsis thaliana* (Arabidopsis Genome Initiative 2000) and rice (Feng et al. 2002; Goff et al. 2002; Sasaki et al. 2002; Yu et al. 2002), transcriptomic approaches have attracted much attention for identifying gene functions as post-genomics research. Metabolomics, by which the whole sets of metabolites of organisms are studied holistically in conjunction with other 'omics' approaches, is also an emerging area in plant sciences. As large numbers of plant metabolites have been used as food, medicines and industrial materials, metabolomics could have a strong impact on applied technology as well as basic sciences for understanding biological systems. Analytical technologies for plant metabolites have been well reviewed in recent articles (Oliver et al. 1998; Tweeddale et al. 1998; Bailey et al. 2000; Fraser et al. 2000; Roberts 2000; Arlt et al. 2001; Bligny and Douce 2001; Ratcliffe and Shachar-Hill 2001; Huhman and Sumner 2002; Soga et al. 2002). Fourier transform ion cyclotron mass spectrometry (FT/ICR-MS) is a powerful technology for comprehensive analysis of metabolites because it measures masses of >2000 compounds with such high accuracy that only one or very few molecular formulae are associated to a single mass in the spectra (Aharoni et al. 2002; Barrow et al. 2004). Information on numerous metabolites originated from various organisms has been collected in several databases, most of which place emphasis on biological pathways (Bairoch 2000; Goto et al. 2002; Kanehisa et al. 2002). However, the

[1] New Energy and Industrial Technology Development Organization, Toshima, Tokyo 170–6028, Japan
[2] Department of Bioinformatics and Genomics, Graduate School of Information Science, Nara Institute of Science and Technology (NAIST), Takayama-cho 8916–5, Ikoma, Nara 630–0101, Japan, e-mail: skanaya@gtc.naist.jp
[3] Ehime Women's College, Baba 421, Ibuki-cho, Uwazima, Ehime 798–0025, Japan
[4] Department of Computational Biology, Graduate School of Frontier Sciences, The University of Tokyo, Kashiwanoha 5–1–5, Kashiwa Chiba 277–8561, Japan
[5] Metabolomics Research Group, RIKEN Plant Science Center, Suehiro-cho 1–7–22, Tsurumi-ku, Yokohama, Kanagawa, 230–0045, Japan
[6] Department of Plant Genes and Physiology, Graduate School of Agriculture and Biological Sciences, Osaka Prefecture University, Gakuen-cho 1–1, Sakai, Osaka 599–8531, Japan
[7] Kazusa DNA Research Institute, Kazusakamatari 2–6–7, Kisarazu, Chiba 292–0818, Japan

Biotechnology in Agriculture and Forestry, Vol. 57
Plant Metabolomics (ed. by K. Saito, R.A. Dixon, and L. Willmitzer)

relationships between metabolites and their biological origins have not been addressed systematically in the previous databases.

We collected information on 18,210 metabolite-species pairs encompassing 7462 metabolites and 6324 species, and designed a database, called KNApSAcK, which allows users to search various aspects of metabolite-species relationships and to retrieve some detailed information about the metabolites. As easy access to metabolite information obtained from analytical techniques and stored in the database is supportive for interpretation of mass spectral data, we provide a tool for retrieving a list of candidate metabolites that correspond to a particular molecular weight on a mass spectrum and from which information on individual metabolites can be obtained. By thorough investigation of the metabolite-species relationships in the database, we observed power-law distribution for frequencies of the number of metabolites in taxonomic units such as species and genera. Also we classified organisms based on metabolite-organism relationship using a graph clustering algorithm.

2 Search Options of the KNApSAcK Database

Information on metabolites in the database can be searched for by names of metabolite or organism, molecular weight and molecular formula (Fig. 1). A list of metabolites that belong to a taxonomic class can be obtained by search with the taxonomic name (Fig. 2), from which information of individual metabolites can be retrieved.

2.1 Search by Metabolite or Organism Name

A list of metabolites or organisms that hit to the query word is displayed by the search function. For example, Fig. 2a,b shows the results of metabolite search with the query "chrysosplenol" and organism search with the query "Citrus", respectively. From a list that is generated by a search, users may select a metabolite or species for detailed information on the metabolite, the molecular weight, formula, structure and biological functions. The display of the molecular structure can be enlarged with the zooming function.

2.2 Search by Molecular Weight

Molecular weight search allows users to enter desired molecular weight and a margin value (Fig. 2c). As approximate masses of compounds are obtained from most MS analyses, the search with a margin value is useful to find candidate metabolites in the database.

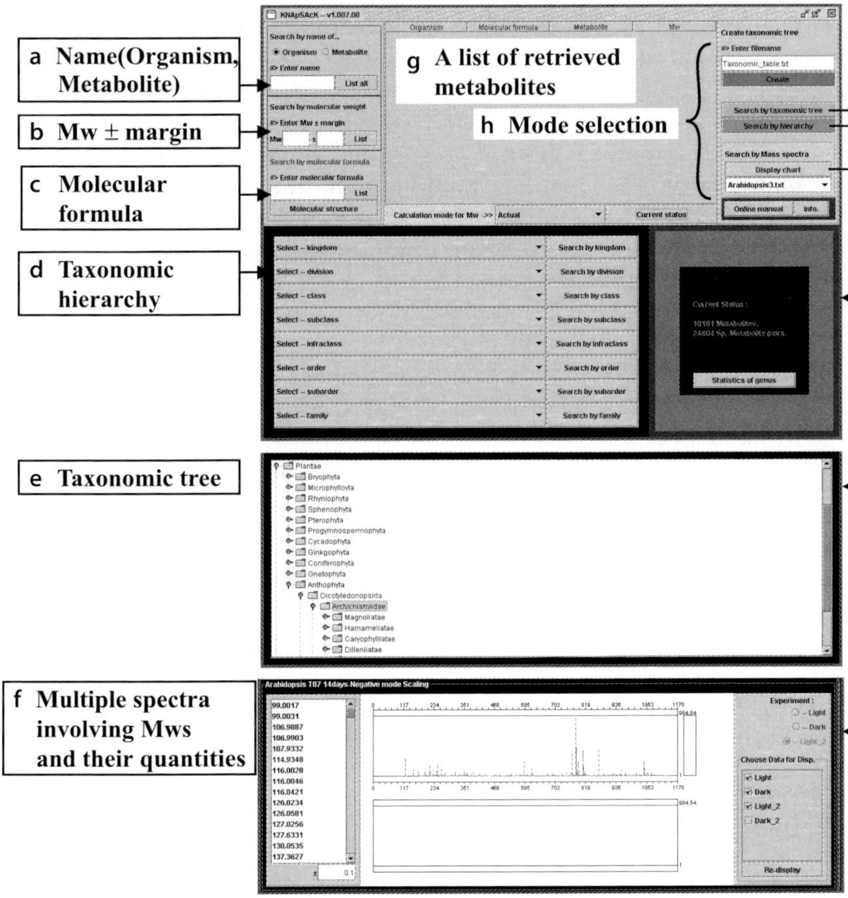

Fig. 1. Search window and major panels of the proposed database system. Information on metabolites can be searched by names of metabolites or organism, molecular weight with ± margin, molecular formula, taxonomic hierarchy, taxonomic tree, and multiple mass spectra which are shown from (**a**) to (**f**), respectively. The search results are displayed in (**g**). To conduct search by taxonomic hierarchy, taxonomic tree, or multiple mass spectra, one can use three corresponding panels (**d**–**f**). One can switch to any of these three panels by using the mode selection region (**h**) of the search window

2.3 Search by Molecular Formula

Metabolite names and origins of the metabolites are listed by molecular formula search (Fig. 2d). The molecular structures of all the metabolites listed are shown in a separate window (Fig. 2e).

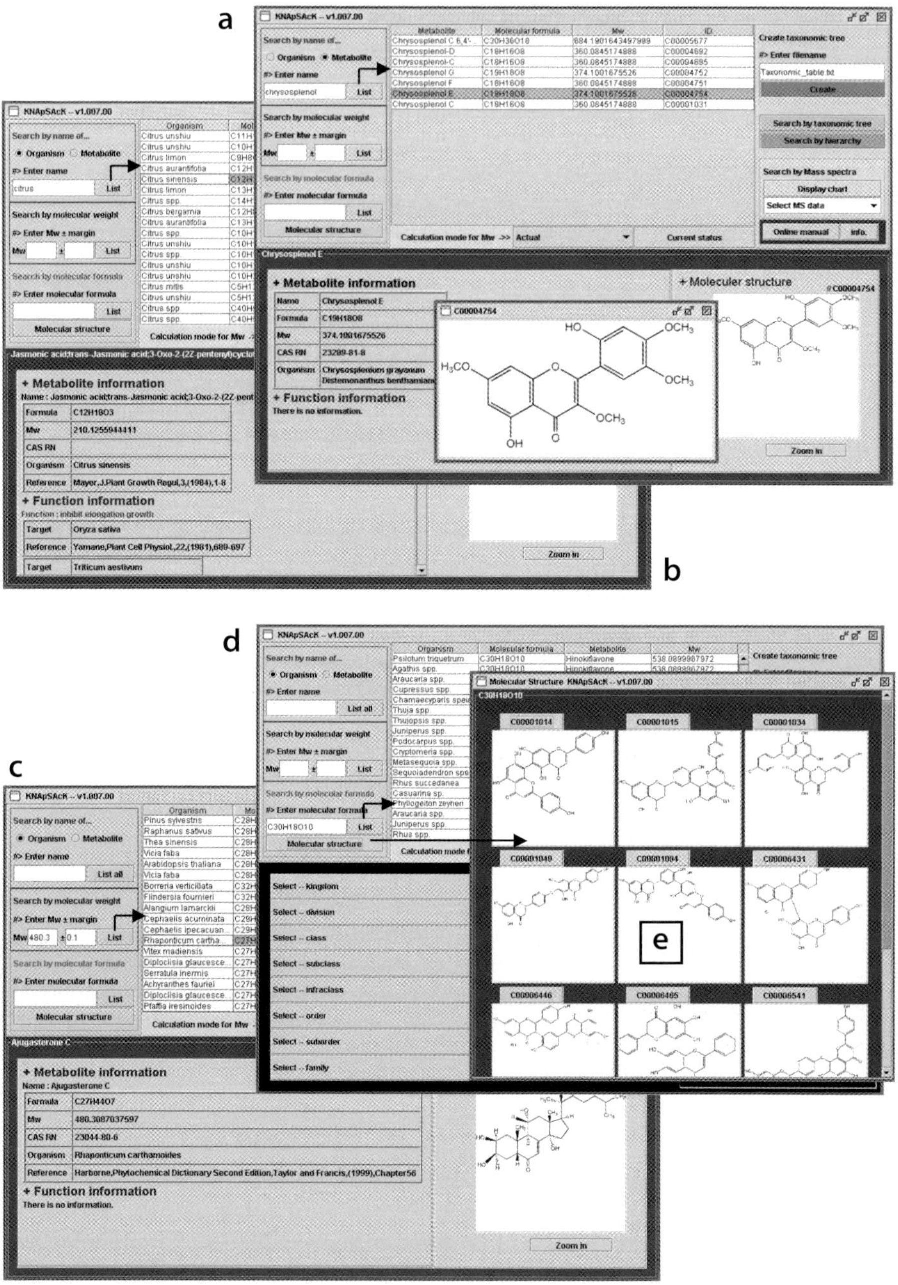
a
b
c
d
e
KNApSAcK -- v1.007.00
Search by name of...
Organism
Metabolite
Enter name
chrysosplenol
List
Search by molecular weight
Enter Mw ± margin
Search by molecular formula
Enter molecular formula
Molecular structure
Metabolite
Molecular formula
Mw
ID
Create taxonomic tree
Enter filename
Create
Search by taxonomic tree
Search by hierarchy
Search by Mass spectra
Display chart
Select MS data
Online manual
info.
Calculation mode for Mw ->>
Actual
Current status
Chrysosplenol E
+ Metabolite information
Name
Chrysosplenol E
Formula
C19H18O8
Mw
374.1001675526
CAS RN
23289-81-8
Organism
Chrysosplenium grayanum
Distemonanthus benthamianus
+ Function information
There is no information.
+ Moleculer structure
C00004754
Zoom in
citrus
Jasmonic acid;trans-Jasmonic acid;3-Oxo-2-(2Z-pentenyl)cyclo
Formula
C12H18O3
Mw
210.1255944411
Organism
Citrus sinensis
Reference
Mayer,J.Plant Growth Regul,3,(1984),1-8
Function : inhibit elongation growth
Target
Oryza sativa
Reference
Yamane,Plant Cell Physiol.,22,(1981),689-697
Target
Triticum aestivum
Molecular Structure KNApSAcK -- v1.007.00
C30H18O10
C00001014
C00001015
C00001034
C00001049
C00001094
C00006431
C00006446
C00006465
C00006541
Select -- kingdom
Select -- division
Select -- class
Select -- subclass
Select -- infraclass
Select -- order
Select -- suborder
Select -- family
Ajugasterone C
Name : Ajugasterone C
Formula
C27H44O7
Mw
480.3087037597
CAS RN
23044-80-6
Organism
Rhaponticum carthamoides
Reference
Harborne,Phytochemical Dictionary Second Edition,Taylor and Francis,(1999),Chapter56
There is no information.

◀ **Fig. 2.** Search methods by entering search items: **a** search by metabolite name; **b** search by organism name; **c** search by molecular weight of metabolite; **d** search by molecular formula; **e** molecular structures of entered molecular formulae. First, the radiobutton with name-tag "Metabolite" or "Organism" (**a** and **b**) is selected in searching by metabolite name or organism name, respectively, and then some metabolite or organism name is entered. To search the database with some accurate molecular weight, one enters it and a margin value (for example, 480.3 and 0.1) and clicks "List" button. Then species-metabolite information is listed (**c**). To search the database with some molecular formula, one enters it and clicks the "List" button, then the result is listed in a similar way (**d**). By clicking the "Molecular structure" button, the molecular structures of all the corresponding metabolites can be displayed in a separate window (**e**)

2.4 Search by Taxonomic Tree and Hierarchy

The molecular formula, molecular weight and origin of the metabolites that belong to a taxonomic class are listed by search with a taxonomic name of family, suborder, order, infraclass, subclass, class, division or kingdom (Fig. 3) and by browsing up to the species level using the taxonomic tree (Fig. 1e).

2.5 Search of Compounds in Mass Spectra

The KNApSAcK package installed in the user's computer provides tools for analyzing his/her own datasets of mass spectra in the files that are prepared according to the instruction of the program. A file may contain a number of mass spectral data. Out of these, up to three spectra can be displayed and analyzed simultaneously by the proposed system. By selecting a data file (Fig. 4a), the spectra selected are overlaid with different colors and shown in the middle panel (Fig. 4c). Any spectrum can be brought to the front by spectrum selection (Fig. 4b). The spectrum data for display are selectable (Fig. 4b). Any region of the spectra can be enlarged by stretching the cursor horizontally as shown in the lower panel (Fig. 4c). All masses in the files are displayed on the left side of the panel (Fig. 4d). By selecting a mass from the list, the black vertical line pointer moves to the position of the peak of the mass on the spectra (Fig. 4c,h), and simultaneously possible metabolites with the molecular mass or close masses in the database are shown in the upper panel (Fig. 4e). The margins of mass values are changeable. As it is helpful to show the mass value with the value of an additive ion such as H^+ and K^+ depending on the solvent used for sample preparation, the species of additive ions are selectable (Fig. 4f,g). This tool is applicable to any data that contain masses and the corresponding intensities, and is especially useful for analyzing the datasets of Fourier transform ion cyclotron resonance mass spectrometry (FT/ICR-MS), as the mass spectrometry determines more than 2000 compounds simultaneously with such high accuracy that an accurate mass corresponds to only one or very few molecular formula.

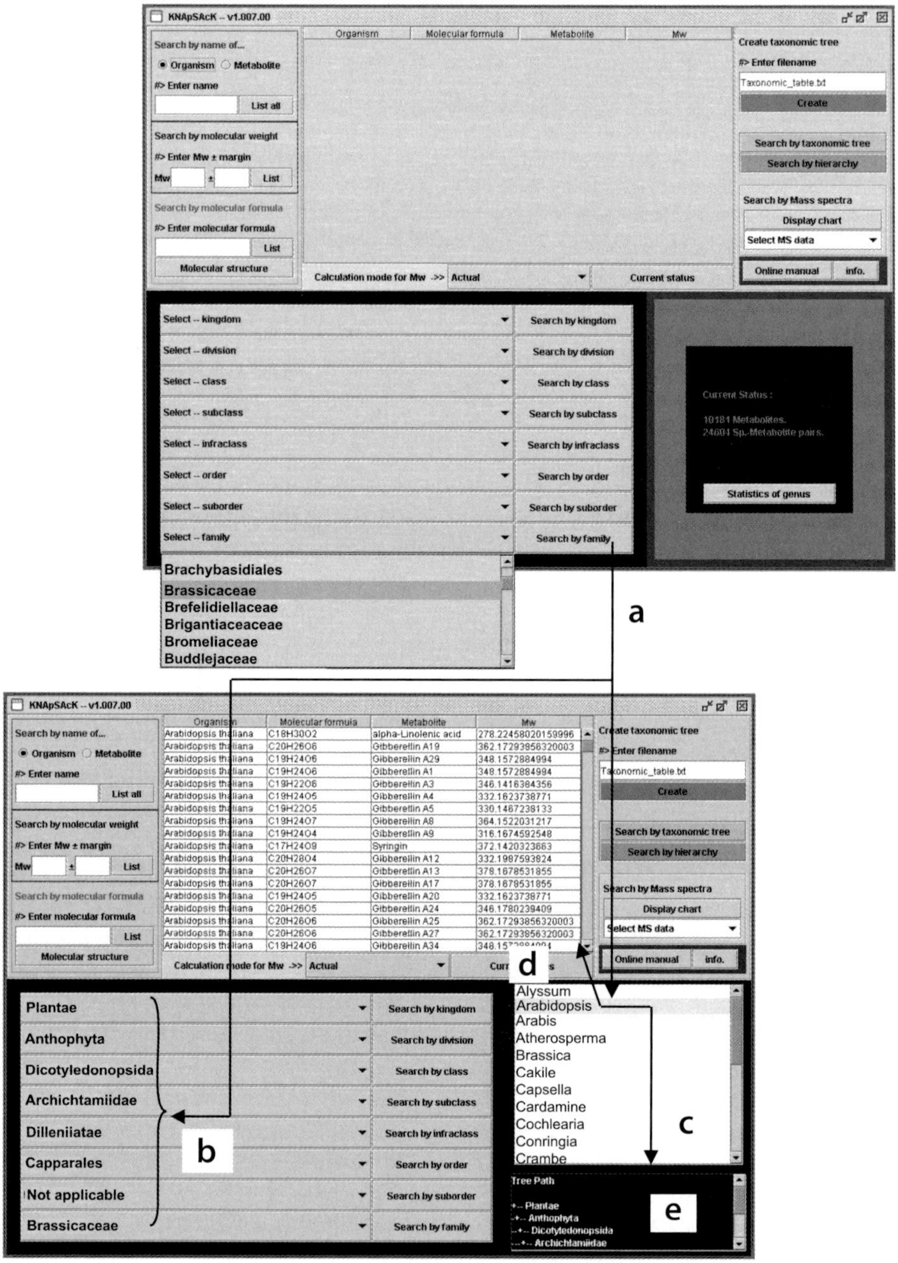

Fig. 3. Search methods by taxonomic hierarchy. When one selects some particular name associated to some hierarchy, say "Brassicaceae" associated to "family", and clicks the "Search by family" button (**a**), taxonomical names associated with "Brassicaceae" for levels higher than family are automatically determined and displayed (**b**). The genera included in the selected item (here "Brassicaceae") are listed on the right side of the window (**c**). By selecting any genus, say "Arabidopsis", the information about related metabolites can be displayed (**d**), and the hierarchical tree path of the selected genus is displayed (**e**)

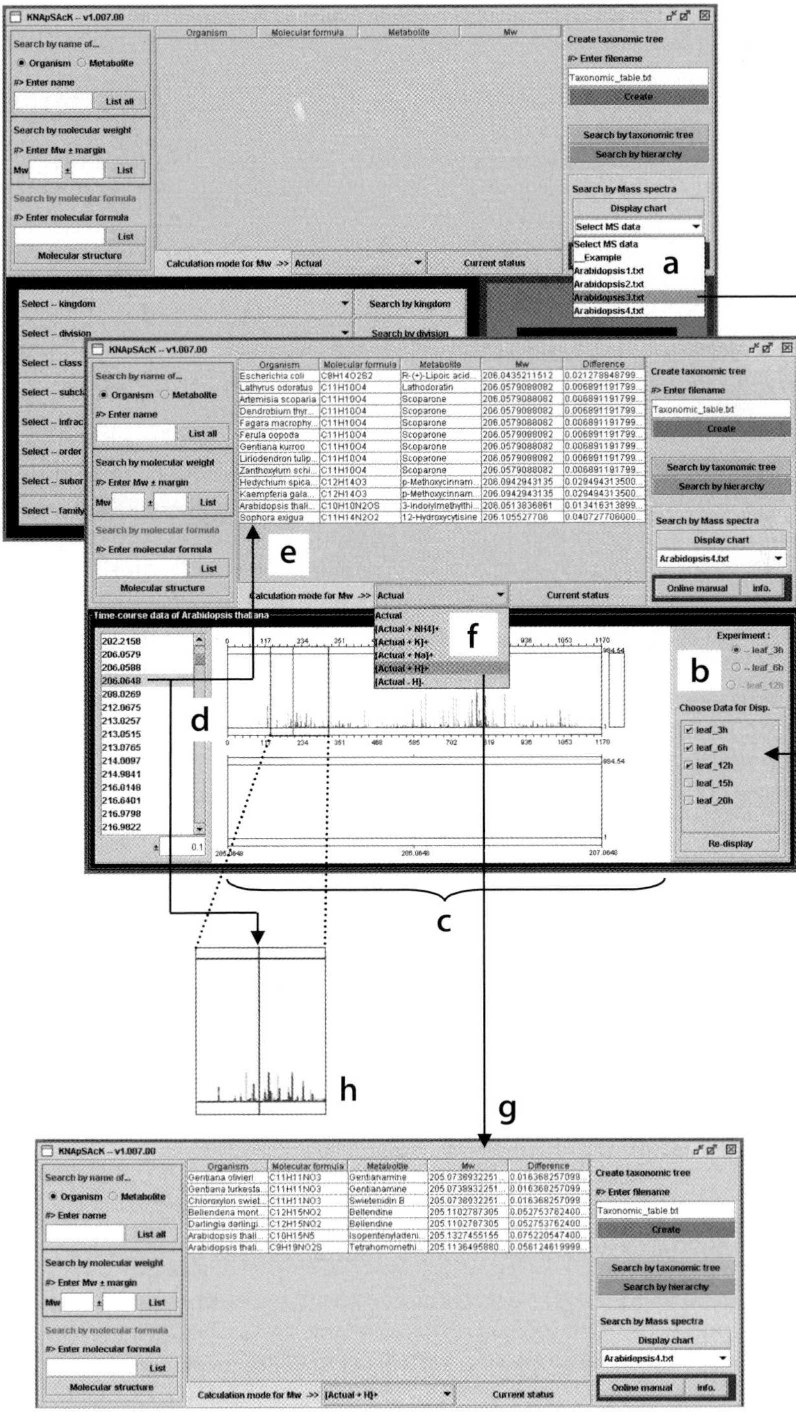

Fig. 4. Search method by molecular weight of mass spectra

3 Statistics of the Database

The latest release of the database (July 7th, 2005) contains 18,210 species-metabolite pairs involving 7462 metabolites and 6324 species. The total number of secondary metabolites for which molecular structures have been elucidated is estimated as 50,000 (De Luca and St Pierre 2000). So, around 15% of the metabolites have been compiled in the present database. The summary of the data contained in the database is shown in Table 1. Presently, the species of Fabaceae, Asteraceae, Brassicaceae, and Rutaceae provide the highest number of metabolites. Species belonging to these families have highly diverse metabolic pathways and they also play important roles as food, medicines and industrial materials.

We observed power-law distribution for the number of species and the associated number of metabolites (Fig. 5a). Such distribution was also observed at genus level (Fig. 5b). The correlations decrease gradually to higher taxonomical levels (Fig. 5c,d). At class level, the power-law is almost not preserved indicating that some random process is involved (Fig. 5e). Although power-law study was initiated in the word usages of text documents (Zipf 1949), more recently power-law has been found in various biological networks or systems; protein-protein interaction networks (Park et al. 2001), the usage of short nucleotide sequences (Mantegna et al. 1994), periodic arrangements of nucleotides in genomes (Fukushima et al. 2002), populations of gene families in genomes (Gerstein 1997; Huynen and van Nimwegen 1998), protein superfamilies and folds in genomes (Qian et al. 2001), and metabolic networks (Jeong et al. 2000). Furthermore, power-law distribution is not a special characteristic of natural scale based on the relationship between power-law distribution and lognormal distribution (Arita 2005).

Though the number of identified metabolites in plants is 50,000 (de Luca and St Pierre 2000), the number of species-metabolite relations collected in the present study (18,210 pairs involving 7462 metabolites) is good enough for statistical analysis to estimate the property of the relation. Therefore, the observed power-law distribution is a significant character of the dataset and is roughly explained by a stationary property of multiplicative process (Gabaix 1999). Power law observed in the degree distribution of a simple network is explained as follows: the network expands continuously by the addition of new vertices and new vertices attach preferentially to sites that are already well connected (Barabási and Albert 1999). The species metabolite relationship database can be represented as a bipartite network where one of the set of vertices are the species and the other set of vertices are the metabolites. Thorough research has been conducted on a few species resulting in discovery of many of their metabolites and hence these well studied species are connected to a good number of metabolites. On the other hand, many other novel species have been checked only to find rare plant metabolites and the therefore majority of the species are connected to one or a few metabolites. In species metabolite database, information on species-metabolite pairs is continuously accumu-

Table 1. Summary of database (July 7th, 2005)

Order	Family	NG[a]	NM[a]
Anthocerotales	****	3	1
Apiales	Apiaceae	46	153
	Araliaceae	7	26
	Garryaceae	1	5
	Nyssaceae	1	2
Arales	Araceae	10	23
	Lemnaceae	2	12
Araucariales	****	2	31
Arecales	Arecaceae	7	32
Aristolochiales	Aristolochiaceae	4	23
Asterales	Asteraceae	112	832
Bromeliales	Bromeliaceae	2	13
Callitrichales	Hippuridaceae	1	1
Campanulales	Campanulaceae	6	18
	Goodeniaceae	1	1
Capparales	Brassicaceae	23	595
	Capparaceae	3	15
	Resedaceae	1	5
	Tovariaceae	1	4
Caryophyllales	Aizoaceae	6	8
	Amaranthaceae	5	17
	Cactaceae	13	30
	Caryophyllaceae	7	66
	Chenopodiaceae	9	98
	Didiereaceae	1	1
	Molluginaceae	2	13
	Nyctaginaceae	3	10
	Phytolaccanceae	1	15
	Portulacaceae	1	10
Casuarinales	Casuarinaceae	1	7
Catales	Cataceae	1	4
Celastrales	Aquifoliaceae	1	6
	Celastraceae	9	26
	Icacinaceae	3	4
Commelinales	Commelinaceae	3	6
	Xyridaceae	1	1
Cornales	Alangiaceae	1	17
	Cornaceae	2	15
Cycadales	****	5	15
Cyperales	Cyperaceae	3	14
	Poaceae	31	224
Dicranales	****	2	5
Dilleniales	Ancistrocladaceae	1	1
Dilleniales	Dilleniaceae	1	6
	Paeoniaceae	1	11
Dipsacales	Caprifoliaceae	4	28
	Dipsacaceae	2	5
	Valerianaceae	1	14
Ebenales	Ebenaceae	3	20
	Sapotaceae	4	4
	Styracaceae	1	3
	Symplocaceae	1	9
Ephedrales	****	1	11
Equisetum	****	1	43
Ericales	Empetraceae	1	9
	Ericaceae	10	85
	Monotropaceae	1	1
	Myrsinaceae	3	3
	Pyrolaceae	2	7
Eriocaulales	Eriocaulaceae	1	2
Eubryales	****	4	28
Eucommiales	Eucommiaceae	1	14
Euphorbiales	Buxaceae	2	9
	Dichapetalaceae	1	1
	Euphorbiaceae	25	132
	Pandaceae	1	1
Fabales	Caesalpiniaceae	5	79
	Fabaceae	133	943
Fabales	Mimosaceae	4	73
Fagales	Betulaceae	4	56
	Fagaceae	5	46
	Juglandaceae	3	12
	Nothofagaceae	1	4
Filicales	Adiantaceae	2	36
	Aspleniaceae	3	37
	Athyriaceae	1	3
	Blechnaceae	3	12
	Cyatheaceae	1	15
	Davalliaceae	1	1
	Dennstaedtilaceae	1	6
	Dicksoniaceae	2	14
	Dryopteridaceae	5	11
	Hymenophyllaceae	1	3
	Loxsomaceae	1	2
	Osmundaceae	1	1
	Parkeriaceae	3	84
	Polypodiaceae	3	8
	Pteridaceae	1	9
	Schizaeaceae	2	8
	Woodsiaceae	4	10

Table 1. (continued)

Order	Family	NG[a]	NM[a]
Funariales	****	1	1
Gentianales	Apocynaceae	41	152
	Asclepiadaceae	7	25
	Gentianaceae	5	73
	Loganiaceae	3	33
Geraniales	Erythroxylaceae	1	19
	Geraniaceae	3	27
	Limnanthaceae	1	13
	Oxalidaceae	1	4
	Tropaeolaceae	1	7
Ginkgoales	****	1	22
Gnetales	****	1	7
Grimmiales	****	1	2
Haloragales	Haloragaceae	1	1
Hamamelidales	Cercidiphyllacea	1	3
	Davidsoniaceae	1	5
	Hamamelidaceae	4	14
	Platanaceae	1	17
Hydrocharitales	Hydrocharitaceae	1	3
Hyponobryales	****	2	8
Illiciales	Illiciaceae	1	8
	Schisandraceae	2	18
	Schizandraceae	1	1
Isobryales	****	1	6
Isoetales	****	1	1
Juncales	Juncaceae	2	4
Jungermanniales	****	9	32
Laminales	Boraginaceae	23	44
	Lamiaceae	29	322
	Phrymaceae	1	3
	Verbenaceae	7	67
Laurales	Calycanthaceae	2	4
	Hernandiaceae	3	9
	Lauraceae	17	115
	Monimiaceae	7	12
Lecythidales	Lecythidaceae	2	4
Liliales	Agavaceae	1	4
	Aloeaceae	2	8
	Dioscoreaceae	2	25
	Haemodoraceae	1	2
	Iridaceae	10	49
	Liliaceae	49	205
	Philydraceae	1	5
	Pontederiaceae	1	2
Liliales	Smilacaceae	2	18
	Stemonaceae	1	1
	Taccaceae	1	2
	Velloziaceae	1	27
	Xanthorrhoeaceae	1	5
Linales	Linaceae	1	33
Lycopodiales	****	1	37
Magnoliales	Annonaceae	12	41
	Austrobaileyaceae	1	1
	Eupomatiaceae	1	1
	Himantandraceae	1	3
	Magnoliaceae	3	60
	Myristicaceae	5	34
	Trimeniaceae	1	1
	Winteraceae	3	3
Malvales	Bombacaceae	2	2
	Elaeocarpaceae	3	10
	Gyrostemonaceae	1	1
	Malvaceae	7	100
	Sterculiaceae	6	22
	Tiliaceae	2	8
Marattiales	****	1	2
Marchantiales	****	4	23
Metzgeriales	****	3	14
Myricales	Myricaceae	1	12
Myrtales	Combretaceae	5	21
	Lythraceae	6	9
	Melastomataceae	3	4
	Myrtaceae	11	87
	Onagraceae	3	21
	Punicaceae	1	7
	Sonneratiaceae	1	2
	Thymelaeaceae	4	43
	Trapaceae	1	1
Najadales	Juncaginaceae	1	1
	Potamogetonaceae	2	3
Najadales	Zosteraceae	2	5
Nepenthales	Droseraceae	2	4
	Nepenthaceae	1	1
	Sarraceniaceae	1	2
Nymphaeales	Nelumbonaceae	1	3
	Nymphaeaceae	3	10
Orchidales	Orchidaceae	27	41
Pandanales	Pandanaceae	1	1
Papaverales	Capparidaceae	1	4

Table 1. (continued)

Order	Family	NG[a]	NM[a]
Papaverales	Papaveraceae	14	77
Piperales	Chloranthaceae	1	1
	Piperaceae	2	49
	Saururaceae	2	4
Plantaginales	Plantaginaceae	1	13
Plumbaginales	Plumbaginaceae	4	35
Polygalales	Malpighiaceae	2	4
	Polygalaceae	2	23
Polygonales	Polygonaceae	6	94
Polytrichales	****	1	1
Primulales	Myrisinaceae	2	14
Primulales	Primulaceae	5	70
Proteales	Elaeagnaceae	1	9
	Proteaceae	10	16
Psilotales	****	1	6
Rafflesiales	Rafflesiaceae	1	1
Ranunculales	Berberidaceae	7	73
	Coriariaceae	1	6
	Lardizabalacea	1	2
	Menispermaceae	16	25
	Podophyllaceae	1	15
	Ranunculaceae	18	134
	Sabiaceae	1	1
Restionales	Restionaceae	1	1
Rhamnales	Rhamnaceae	10	62
Rhamnales	Vitaceae	3	53
Rhizophorales	Rhizophoraceae	4	9
Ricciales	****	2	3
Rosales	Chrysobalanaceae	2	1
	Connaraceae	1	1
	Crassulaceae	5	91
	Grossulariaceae	3	17
	Hydrangeaceae	2	7
	Pittosporaceae	2	2
	Rosaceae	24	213
	Saxifragaceae	7	62
Rubiales	Rubiaceae	35	110
Salicales	Salicacea	2	64
Salviniales	Azollaceae	1	1
Santalales	Balanophoraceae	1	3
	Loranthaceae	2	12
	Santalaceae	1	4
Sapindales	Aceraceae	1	13
	Anacardiaceae	9	66
Sapindales	Balsaminaceae	1	8
	Zygophyllaceae	6	51
	Burseraceae	2	8
	Hippocastanaceae	2	18
	Meliaceae	6	23
	Rutaceae	68	410
	Sapindaceae	5	18
	Simaroubaceae	4	16
Scrophulariales	Acanthaceae	7	39
	Bignoniaceae	14	42
	Buddlejaceae	1	6
	Gesneriaceae	3	8
	Globulariaceae	2	10
	Lentibulariaceae	1	1
	Myoporaceae	2	9
	Oleaceae	8	52
	Orobanchaceae	1	2
	Pedaliaceae	2	13
	Scrophulariaceae	17	102
Selaginellales	****	1	15
Solanales	Convolvulaceae	9	67
	Cuscutaceae	1	5
	Hydrophyllaceae	3	17
	Polemoniaceae	3	9
	Solanaceae	22	246
Sphagnales	****	1	1
Takakiales	****	1	7
Taxales	****	3	35
Taxodiales	Cupressaceae	7	79
	Pinaceae	9	152
	Podocarpaceae	3	80
	Taxodiaceae	6	47
Theales	Actinidiaceae	1	19
	Clusiaceae	14	73
	Dipterocarpa-ceae	5	7
	Ochnaceae	1	11
	Theaceae	4	51
Trochodendrales	Tetracentraceae	1	1
	Trochodendra-ceae	1	2
Typhales	Sparganiaceae	1	2
	Typhaceae	1	10
Urticales	Cannabaceae	2	43
	Moraceae	9	113
Urticales	Ulmaceae	4	14
	Urticaceae	4	13

Table 1. (continued)

Order	Family	NG[a]	NM[a]
Violales	Begoniaceae	1	16
	Bixaceae	1	1
	Caricaceae	2	10
	Cistaceae	2	41
	Cucurbitaceae	9	72
	Flacourtiaceae	7	7
	Passifloraceae	5	41
	Stachyuraceae	1	4
	Tamaricaceae	1	19
	Violaceae	1	7
Zingiberales	Cannaceae	1	1
	Marantaceae	1	2
	Musaceae	1	12
	Zingiberaceae	7	41

a NG and NM represent number of genera and metabolites, respectively.
**** family names are unknown.

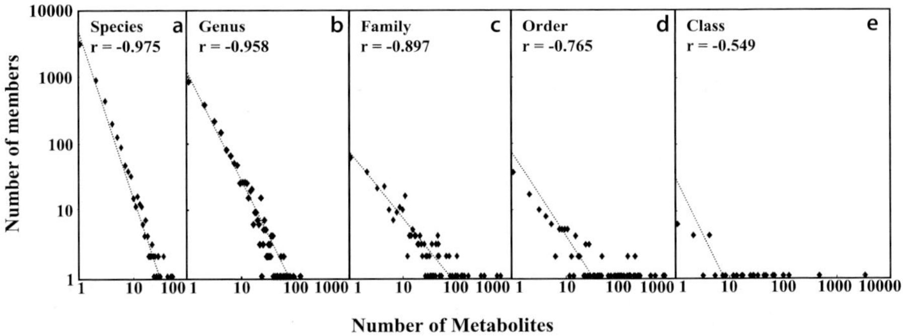

Fig. 5. Power-law distribution in the metabolite-taxonomic category relationships. The number of taxonomic units: **a** species; **b** genera; **c** families; **d** orders; **e** classes is plotted against the number of associated metabolites

lated. Therefore when a novel metabolite is added it might be connected to a novel species. However, it is likely that some of the novel metabolites might also be connected to one or more well studied species that are already connected to a number of metabolites. As a consequence, a power-law is revealed when the frequencies of species are plotted against the number of associated metabolites. Interestingly, power-law is also seen at the genus or higher level (Fig. 6). This is also likely due to the fact that research activity seeks novel metabolites in novel plant categories. Such hierarchical power-law structure is the characteristic of the bipartite network that represents the database. It is noteworthy that the metabolic pathways of individual organisms also follow power law,

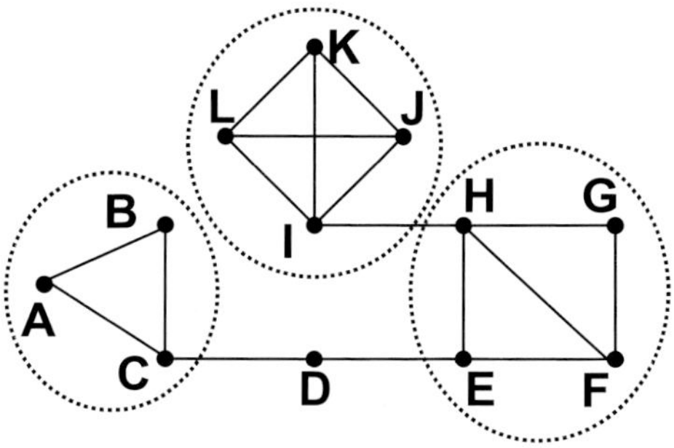

Fig. 6. The concept of graph clustering algorithm

i. e., the probability $P(k)$ that a metabolite interacts with k other metabolites in the metabolic pathway decays as a power-law, following $P(k) \sim k^{-r}$, where r is a constant (Ravasz et al. 2002). The power-law distribution obtained in the present study can be interpreted as human research activity for searching small k vertices (rare metabolites) of the metabolic pathway networks.

4 Classification Based on Common Metabolites

We represented the genera included in the database as a network based on the frequency of occurrences of common metabolites between pairs of genera and classified them into clusters using a graph clustering algorithm (Table 2; Amin and Kanaya 2004). Figure 6 illustrates the basic concept of the graph clustering algorithm. For the purpose of classifying the genera, we first represent them as a simple, undirected graph $G = (N, E)$. In this graph, the set of nodes N represents plant genera and E represents the set of edges. We consider an edge between any two genera if there are at least four common metabolites between them. A typical such graph consisting of 12 genera is shown in this figure. By intuition, we can identify three cohesive groups ({A, B, C}, {E, F, G, H}, {I, J, K, L}), on the basis of the fact that the densities of these groups are 1.00, 0.83 and 1.00, respectively, which are much higher than that of the whole graph (0.26). Here, the density d of a graph/subgraph is the ratio of the number of edges present $[n(E)]$ and the maximum possible number of edges $[n(E)_{\max}]$. Mathematically $d = n(E)/n(E)_{\max} = 2{*}n(E)/\{n(N){*}(n(N)-1)\}$ where $n(N)$ and n(E) are the numbers of nodes and edges, respectively, in the graph/subgraph. The density of a graph/subgraph is a real number ranging from 0 to 1. The classification of the nodes in this figure is possible by visualization because it is very small in size. In the present analysis, we handled a graph of 2092 genera and it is too big to classify its nodes by visualization. The threshold for determining clusters is set to $d = 0.70$.

Table 2. Classification of genera based on common metabolites

Kingdom	Order	Family	Number of Genus	Number of Species	Cluster ID 1	2	3	4	5	6	7	8	9	10	11	12	13	14	15	16	17	18	19	20	21	22	23	24	25	26	27
Plantae	Apiales	Apiaceae	46	110	8	6																				3					
	Sapindales	Anacardiaceae	11	23						2																					
		Meliaceae	6	14			1		1																						
		Rutaceae	84	230	4		8	7	2						2																
		Sapindaceae	5	7					1																						
		Simaroubaceae	4	12					1																						
	Fabales	Fabaceae	146	398						7	7	6	5							1											
	Rosales	Rosaceae	25	68	1																										
	Geraniales	Erythroxylaceae	1	3																								1			
	Urticales	Moraceae	9	39	1																										
	Fagales	Betulaceae	4	17														2													
	Salicales	Salicaceae	3	19	1																										
	Theales	Clusiaceae	14	51					1																						
	Caryophyllales	Chenopodiaceae	8	20								1																			
		Phytolaccanceae	1	6										1																	
		Portulacaceae	1	2										1																	
		Aizoaceae	6	10										3																	
	Ranunculales	Berberidaceae	8	21											3																
		Menispermaceae	18	28											1																
		Ranunculaceae	19	93											2																
	Papaverales	Papaveraceae	15	68											3	6															
	Laurales	Lauraceae	18	51													4														
		Monimiaceae	7	18													1														
	Magnoliales	Magnoliaceae	3	21													1														
		Myristicaceae	1	2													1														
	Illiciales	Illiciaceae	1	3													1														
	Piperales	Piperaceae	1	16	1																										
	Asterales	Asteraceae	123	550	2													9	9	5	5	5									
	Rubiales	Rubiaceae	39	77																			7								
	Gentianales	Apocynaceae	44	132																				7	5		1				
	Laminales	Lamiaceae	32	136													1									7	5				
		Verbenaceae	6	17																			1								
	Solanales	Hydrophyllaceae	3	8														1													
		Solanaceae	23	107	1																							6			
	Scrophulariales	Myoporaceae	2	7													1														
	Liliales	Iridaceae	10	20																			1								
		Liliaceae	48	121																									9		
	Orchidales	Orchidaceae	28	51																										6	
	Taxodiales	Cupressaceae	7	34													1														
		Pinaceae	9	48	1																										
Fungi	Lecanorales	Cladoniaceae	2	3																											1
		Lecanoraceae	1	2																											1
		Parmeliaceae	3	3																											3
		Ramalinaceae	1	1																											1

Numbers of genera are written in boldface type when total number of genera of identical order in a cluster is 5 or more.
Numbers of genera and species represent those in the database.

We dealt with 2092 genera and found 108 clusters each containing 5 or more genera. We extracted 27 out of 108 clusters such that each of them has at least 5 or more genera belonging to an identical order (Table 2). Out of the 27 clusters, 16 clusters (Cluster 2, 3, 4, 5, 7, 9, 10, 12, 15, 17, 18, 20, 21, 25, 26 and 27) consist of genera of the same orders, suggesting that the graph clustering tends to put together the genera that are related taxonomically. Members of Clusters 3 and 5 belong to different groups of the Sapindales order, consistent with the previously-determined taxonomic relationship on the basis of *rbcL* and *atpB* sequence variations, in which Meliaceae, Rutaceae, Sapindaceae and Simaroubaceae are related as sister groups (Chase et al. 1999). The genera of Cluster 10 belong to only the Caryophyllales order, which is consistent with the finding that Phytolaccaceae (one genus), Portulaceae (one genus) and Aizoaceae (three genera) are closely related on the basis of the existence of a common intron sequence in the chloroplast gene *rpoC1* (Wallace and Cota 1996).

As individual species in the plant kingdom synthesize both common and intrinsic metabolites (Wink 1988, 2003), integration of species-metabolite relationships with the aid of graph clustering might help taxonomic classification on the basis of common metabolites as suggested in this study. Further accumulation of information on species-metabolite relations will facilitate our understanding of metabolite diversity in species and classification of organisms in a global context.

5 Conclusion and Remarks

We prepared a database, KNApSAcK for accumulation and search of metabolite-species relationships. The power-law distribution observed in the present study is likely to be associated with research activity for finding novel metabolites from nature. In addition, it seems to be derived from searching rare metabolites from the organisms originally exhibiting power-law in the degree distribution of their metabolic networks. This suggests that the database contains chemical diversity of metabolites which occurred through evolution of species. Graph clustering is shown to be useful to extract taxonomic relationships on the basis of common metabolites. As we are continuously accumulating metabolite-species pairs in the database, we continue to advance our understanding of species-metabolite relations in taxonomic hierarchy. Furthermore, we plan to add an option for searching metabolite structures by entering partial structures, which will be helpful for metabolite research.

6 Access to KNApSAcK

KNApSAcK web version is available at http://kanaya.naist.jp/KNApSAcK/KNApSAcK.php. If and when a user wants to customize KNApSAcK for some purpose, Java j2sdk-1.4.2 is required to be installed in the user's computer. First, the compressed file KNApSAcK_database_forWIN.zip or KNApSAcK_database_forMAC.zip must be downloaded from http://kanaya.naist.jp/KNApSAcK/. By uncompressing the file, users can find the instruction file ReadMe (KNApSAcK) in the KNApSAcK_database_forWIN or KNApSAcK_database_forMAC folder to find how to use the database.

References

Arabidopsis Genome Initiative (2000) Analysis of the genome sequence of the flowering plant *Arabidopsis thaliana*. Nature 408:796–815

Aharoni A, Ric de Vos CH, Verhoeven HA, Maliepaard CA, Kruppa G, Bino R, Goodenowe D (2002) Non-targeted metabolomic profiling using Fourier transform ion cyclotron mass spectrometry (FTMS). OMICS J Integr Biol 6:217–234

Amin MA, Kanaya S (2004) Detection of protein complexes in large interaction networks. Proc the 8th world multi-conference on systemics, cybernetics and informatics, vol VII, pp 119–123

Arita M (2005) Scale-freeness and biological networks. J Biochem 138 (in press)

Arlt K, Brandt S, Kehr J (2001) Amino acid analysis in five pooled single plant cell samples using capillary electrophoresis coupled to laser-induced fluorescence detection. J Chromatogr A 926:319–325

Bailey NJ, Stanley PD, Hadfield ST, Lindon JC, Nicholson JK (2000) Mass spectrometrically detected directly coupled high performance liquid chromatography/nuclear magnetic resonance spectroscopy/mass spectrometry for the identification of xenobiotic metabolites in maize plants. Rapid Commun Mass Spectrom 14:679–684

Bairoch A (2000) The ENZYME database in 2000. Nucleic Acids Res 28:304–305

Barabási AL, Albert R (1999) Emergence of scaling in random networks. Science 286:509–512

Barrow MP, Headley JV, Perub KM, Derrick PJ (2004) Fourier transform ion cyclotron resonance mass spectrometry of principal components in oils ands naphthenic acids. J Chromatogr A 1058:51–59

Bligny R, Douce R (2001) NMR and plant metabolism. Curr Opin Plant Biol 4:191–196

Boyes DC, Zayed AM, Ascenzi R, McCaskill AJ, Hoffman NE, Davis KR, Görlach J (2001) Growth stage-based phenotypic analysis of Arabidopsis: A model for high throughput functional genomics in plants. Plant Cell 13:1499–1510

Chase MW, Morton CM, Kallunki JA (1999) Phylogenetic relationships of Rutaceae: A cladistic analysis of the subfamilies using evidence from rbcL and atpB sequence variation. Am J Bot 86:1191–1199

De Luca V, St Pierre B (2000) The cell and developmental biology of alkaloid biosynthesis. Trends Plant Sci 5:168–173

Feng Q, Zhang Y, Hao P, Wang S, Fu G, Huang Y, Li Y, Zhu J, Liu Y, Hu X et al. (2002) Sequence and analysis of rice chromosome 4. Nature 420:316–320

Fraser PD, Pinto ME, Holloway DE, Bramley PM (2000) Application of high-performance liquid chromatography with photodiode array detection to the metabolic profiling of plant isoprenoids. Plant J 24:551–558

Fukushima A, Ikemura A, Kinouchi M, Oshima T, Kudo Y, Mori H, Kanaya S (2002) Periodicity in prokaryotic and eukaryotic genomes identified by power spectrum analysis. Gene 300:203–211

Gabaix X (1999) Zipf's law for cities: An explanation. Q J Econ 114:739–767

Gerstein M (1997) A structural census of genomes: comparing bacterial, eukaryotic, and archaeal genomes in terms of protein structure. J Mol Biol 274:562–576

Goff SA, Ricke D, Lan TH, Presting G, Wang R, Dunn M, Glazebrook J, Sessions A, Oeller P, Varma H et al. (2002) A draft sequence of the rice genome (*Oryza sativa* L. ssp. *japonica*). Science 296:92–100

Goto S, Okuno Y, Hattori M, Nishioka T, Kanehisa M (2002) LIGAND: Database of chemical compounds and reactions in biological pathways. Nucleic Acids Res 30:402–404

Huhman DV, Sumner LW (2002) Metabolic profiling of saponins in *Medicago sativa* and *Medicago truncatula* using HPLC coupled to an electrospray ion-trap mass spectometer. Phytochemistry 59:347–360

Huynen MA, van Nimwegen E (1998) The frequency distribution of gene family sizes in complete genomes. Mol Biol Evol 15:583–589

Jeong H, Tombor B, Albert R, Oltvai ZN, Barabasi AL (2000) The large-scale organization of metabolic networks. Nature 407:651–654

Kanehisa M, Goto S, Kawashima S, Nakaya A (2002) The KEGG databases at GenomeNet. Nucleic Acids Res 30:42–46

Mantegna RN, Buldyrev SV, Goldberger AL, Havlin S, Peng C, Simons M, Stanley HE (1994) Linguistic features of noncoding DNA sequences. Phys Rev Lett 73:3169–3172

Oliver SG, Winson MK, Kell DB, Baganz F (1998) Systematic functional analysis of the yeast genome. Trends Biotech 16:373–378

Park J, Lappe M, Teichmann SA (2001) Mapping protein family interactions; intramolecular and inter molecular protein family interaction repertories in the PDB and yeast. J Mol Biol 307:929–938

Qian J, Luscombe NM, Gerstein M (2001) Protein family and fold occurance in genomes: Power-law behaviour and evolutionary model. J Mol Biol 313:673–681

Ratcliffe RG, Shachar-Hill Y (2001) Probing plant metabolism with NMR. Annu Rev Plant Physiol Plant Mol Biol 52:499–526

Ravasz E, Somera AL, Mongru DA, Oltvai ZN, Barabasi AL (2002) Hierarchical organization of modularity in metabolic networks. Science 297:1551–1555

Roberts JKM (2000) NMR adventures in the metabolic labyrinth within plants. Trends Plant Sci 5:30–34

Sasaki T, Matsumoto T, Yamamoto K, Sakata K, Baba T, Katayose Y, Wu J, Niimura Y, Cheng Z, Nagamura Y et al (2002) The genome sequence and structure of rice chromosome 1. Nature 420:312–316

Soga Y, Ueno Y, Naraoka H, Ohashi Y, Tomita M, Nishioka T (2002) Simultaneous determination of anionic intermediates for *Bacillus subtilis* metabolic pathways by capillary electrophoresis electrospray ionization mass spectrometry. Anal Chem 74:2233–2239

Tweeddale H, Notley-McRobb L, Ferenci T (1998) Effect of slow growth on metabolism of *Escherichia coli*, as revealed by global metabolite pool ("metabolome") analysis. J Bacteriol 180:5109–5116

Wallace RS, Cota JH (1996) An intron loss in the chloroplast gene rpoC1 supports a monophyletic origin for the subfamily Cactoideae of the Cactaceae. Curr Genet 29:275–281

Wink M (1988) Plant breeding: importance of plant secondary metabolites for protection against pathogens and herbivores. Theor Appl Genet 75:225–233

Wink M (2003) Evolution of secondary metabolites from an ecological and molecular phylogenetic perspective. Phytochemistry 64:3–11

Yu J, Hu S, Wang J, Wong GK, Li S, Liu B, Deng Y, Dai L, Zhou Y, Zhang X et al (2002) A draft sequence of the rice genome (*Oryza sativa* L. ssp. *indica*). Science 296:79–92

Zipf GK (1949) Human behavior and the principle of least effort: an introduction to human ecology. Addison-Wesley, Cambridge, MA

Section III Applications

III.1 Systems Biology: A Renaissance of the Top-down Approach for Plant Analysis

F. Carrari[1,2], N. Schauer[1], L. Willmitzer[1], and A.R. Fernie[1]

"Everything touches everything"

Jorge Luis Borges – La Biblioteca de Babel

1 Introduction

The previous chapters have essentially been concerned with technical issues and biological applications of metabolite profiling in isolation. The concept of metabolite profiling has been around for decades because of the fundamental importance of metabolites as constituents of metabolic pathways, the importance of certain metabolites in the human diet and their use as diagnostic markers for a wide range of biological conditions and response to chemical treatment (Brindle et al. 2002; Fernie et al. 2004; Oksman-Caldentey and Saito 2005). Following the trend set by other post-genomic approaches, recent years have seen an explosion both in the interest afforded to metabolite profiling and in the range of metabolites that can be measured using these techniques (Fiehn et al. 2000; Sumner et al. 2003). We can, however, currently only measure a fraction of the metabolic diversity inherent in plants, which has been estimated to be in excess of 200,000 compounds (de Luca and St Pierre 2000). That said, metabolite profiling has found great utility as a technology platform for diagnostics (Fernie et al. 2004), and is beginning to be used in combination with other technologies for gene functional analysis and systems biology.

This chapter will concentrate on the growing role of metabolite profiling in plant systems biology. It will begin with a historical evaluation of holistic approaches to biochemistry. Thereafter its major focus will be on the integration of broad range metabolite profiling into genomics approaches (Sweetlove et al. 2003; Fernie and Sweetlove 2004). Furthermore, we argue the case for the inclusion of both steady-state and dynamic metabolite profiling in such strategies and present examples of the value of them both, in the evaluation of the regulation of metabolic networks. We will also briefly touch on the use of metabolic profiling in gene functional identity elucidation; however, this

[1] Max Planck Institute of Molecular Plant Physiology, Am Mühlenberg 1, 14476, Golm, Germany, e-mail: fernie@mpimp-golm.mpg.de
[2] Istituto de Biotecnología, CICV-INTA, Las Cabañas y Los Reseros, BWAA 1712, Buenos Aires, Argentina

Biotechnology in Agriculture and Forestry, Vol. 57
Plant Metabolomics (ed. by K. Saito, R.A. Dixon, and L. Willmitzer)

is detailed comprehensively in later chapters of this book (see Chaps. III.2 and III.5). Finally, we give a perspective for the future development of such combinatorial approaches.

2 Re-emergence of Top-down Thinking

To grasp fully the shift in perspective that has recently been afforded biologists by recent technological advances (which allow the acquisition of large and diverse data sets) it is important to understand the historical context in which this has occurred. For most of the last century, biochemistry was tackled in a reductionist manner whereby we reduce a cell to its component parts and examine these in isolation. The assumption inherent is that the parts will function in isolation as they do in the cell. Although a large amount of biological knowledge was accrued via such approaches, there are several problems with the reductionist philosophy. The most apparent of these are metabolic complexity and the fact that biological processes are rarely controlled by a single molecular entity and that control is generally shared across the different entities rather than being exerted at a single point (ter Kuile and Westerhoff 2001). The analysis of metabolic control in the early 1970s formalised a mathematical description of system properties of enzymes and metabolites (Heinrich and Rapoport 1973; Kacser and Burns 1974). Their method – metabolic control analysis – allows the effect of small changes in enzymes or metabolites on the whole system to be quantified. These studies were a radical departure from the prevailing ideas of the time and laid the foundation for network analysis of metabolism. Unfortunately, though the experimental determination of the parameters needed to calculate metabolic control has been quite laborious and although our understanding of the regulation of some pathways in plants has been greatly enhanced (Geigenberger et al. 2004), it probably never fulfilled its true potential (Fernie and Sweetlove 2004). The emergence of high-throughput genomic tools that allow the cataloguing of changes in metabolites, transcripts and proteins in parallel with significant advances in our ability to quantify metabolic flux (Mack 2004; Stephanopoulos et al. 2004; Westerhoff and Palsson 2004; Fernie et al. 2005; Ratcliffe and Shacher-Hill 2005) offer fresh hope for successful application of the holistic approach.

3 Systems Biology in Non-plant Systems

Despite the fact that the plant metabolomics community was highly prominent in the development of this research field (Fiehn et al. 2000; Roessner et al. 2001; Sumner et al. 2003), the microbial and mammalian research fields lead the way in terms of integrative genomics analysis (Yang et al. 2002; Even et al. 2003;

Sauer 2004). For that reason in this section we outline important studies that have been carried out in these fields. Before we begin though it is important that we define what we mean by systems biology. Although historically the term systems biology was applied exclusively to mathematical-modelling strategies (see for example Edwards and Cobb 1999), it is now more widely applied particularly in genomics. For the purposes of this article we will stick to the definition of experimental systems biology given by Sweetlove et al. (2003) – the comprehensive multidimensional representation of all the major biosynthetic reactions of the cell. Given that there are already many excellent reviews covering mathematical and conceptual aspects of systems biology (Ideker et al. 2001a; Oltvai and Barabasi 2002; Kitano 2004; Somerville et al. 2004), we will concentrate here on the integrative experimental approaches taken to date. Perhaps the most significant early experiments in this vein were the combined studies of the transcriptome and a subset of the proteome that were carried out in yeast (Futcher 2000; Gygi et al. 2000; Ideker et al. 2001b). These studies all revealed that the relationship between the levels of transcripts and the proteins that they encode is generally relatively low and as such were in keeping with the control analysis study of ter Kuile and Westerhoff (2001). Despite the suggestion of low connectivity between transcription and metabolic regulation, Askenazi et al. (2003) were able to utilize transcript profiling in combination with a limited metabolite profiling to aid in the metabolic engineering of medicinally important polyketides in *Aspergillus terreus*. Another study that merits discussion is the observation that analysis of the intracellular concentrations of metabolites in wild type and mutant yeast strains revealed phenotypes for proteins active in metabolic regulation (Raamsdoonk et al. 2001). In this study the quantification of the change of several metabolite concentrations relative to the concentration change of a selected metabolite was demonstrated to reveal the site of action, in the metabolic network, of silent genes, demonstrating the utility of metabolite profiling in gene functional analysis. Whilst the above-mentioned studies allowed the identification of highly correlated systems elements they did not in their own right address where systemic control generally resided within the systems under study. For this purpose a higher order integration of data of multiple molecular entities will likely be required. It has been postulated that genomic correlations of transcripts, proteins and metabolites in mammalian disease models versus controls could be used as early biomarkers of disease (Mack 2004); however, it is worth noting that the levels of metabolites alone have been demonstrated utility for medical diagnostics (Brindle et al. 2002). A second interesting recent example of systems biology was carried out using the cytoscape software platform which allows the integration of transcriptional data through global biomolecular interaction databases to identify active biomolecular interactions (Shannon et al. 2003). Ideker et al. (2001a) utilized this platform to demonstrate that differential transcript data placed into the context of biomolecular networks can yield valuable insight into pathways. To illustrate this they showed – using galactose-pathway gene perturbations in yeast – that putatively interacting genes were more likely to be synchronously differential active. Figure 1 illustrates a specific example in which differential transcription data reveal the pathways activated and deactivated upon the removal of the GAL80 repressor protein. It is notable that, although the primary active pathway includes GAL80 and the surrounding genes, not all active pathways are directly connected to GAL80. This example thus highlights the somewhat noisy and incomplete nature of biomolecular networks. However, this fact aside, it illustrates the possibility of identifying relationships that become active after intervention rational gene targets for disruption become apparent (Stephanopoulos et al. 2004). It is furthermore likely that the incorporation of data describing further molecular entities and associations will improve both the coverage and fidelity of such biomolecular networks.

4 Systems Biology in Plant Systems

In plants the development of systems biology has followed a similar trajectory and, like in non-plant systems, is dominated by gene expression studies (Harmer et al. 2000; Espinosa-Soto et al. 2004; Thimm et al. 2004). That said, several interesting observations have been made on the correlative behaviour of metabolites, both in isolation and in combination with information concerning other molecular entities. The measurement of many metabolites in parallel gives insights into the complex regulatory circuits that underpin metabolism. Initial systems-based approaches that used data from metabolite profiling involved comprehensive correlation analyses between all metabolites profiled in potato tubers. These studies showed that most metabolites had little correlation, although some were tightly regulated and others were non-linearly regulated (Roessner et al. 2001). We initially believed that the identification of non-linearly regulated metabolite pairs was highly exciting since two of those identified in our initial studies had previously been described to be linked by enzymes that were subject to feedforward and feedback mechanisms of regulation (Roessner et al. 2001). However, the low number of metabolites displaying this behaviour in subsequent studies in our laboratories (A.R. Fernie and L. Willmitzer, unpublished results), has somewhat tempered our early optimism that this would provide a rapid way to elucidate novel mechanism of allosteric regulation within metabolite pathways. Since this initial study, more sophisticated metabolic-network analysis have been undertaken by plotting networks in which the metabolites are represented by nodes that are interconnected by lines representing the correlative behaviour that intertwines them (Weckwerth et al. 2004). Such studies allow the actions of a small theoretical network on the strength of correlations between the metabolites that constitute these networks. However, published data is only available for a severely limited number of conditions and caution must be exercised here since the field is still in its infancy, especially given that what is known of network structure in plants falls someway short of that for microbial systems.

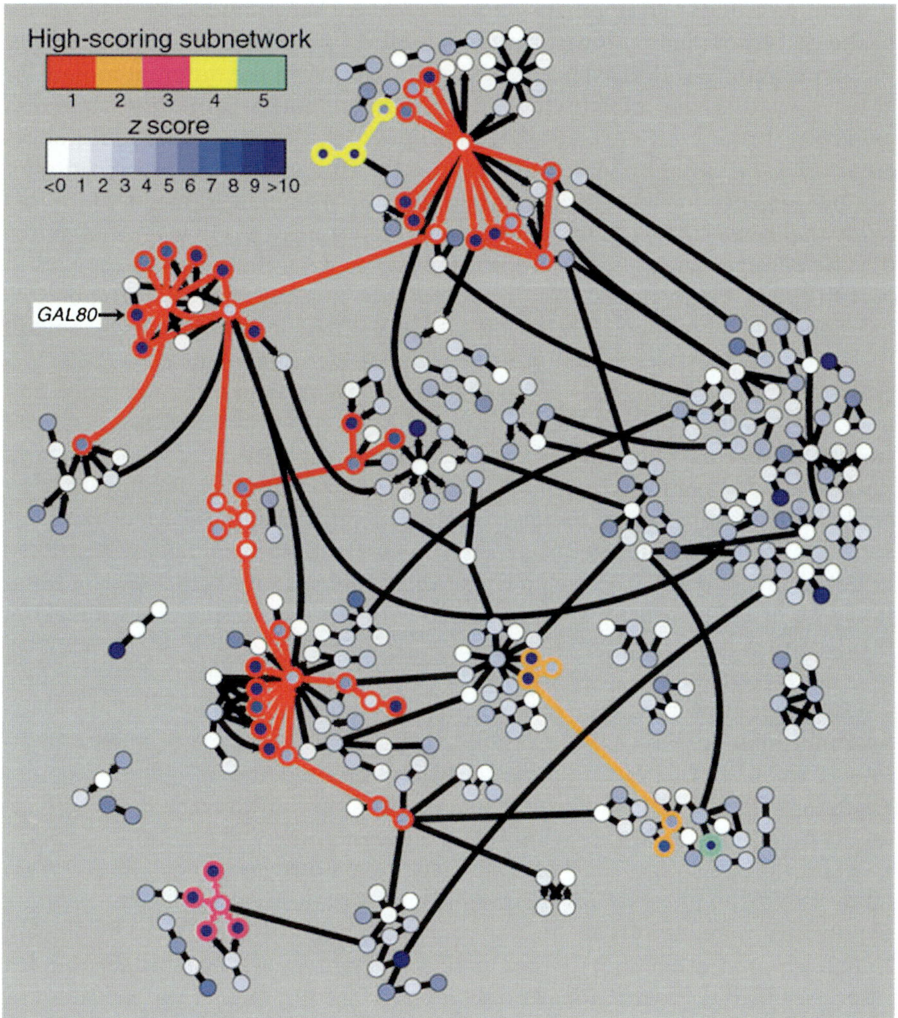

Fig. 1. Determining active pathways after removing a transcription factor repressor. Identification of active pathways helps define gene targets. In this experiment the GAL4 repressor gene GAL80 was deleted. Using microarray data superimposed on a predetermined set of protein: protein and protein: DNA interactions for yeast, the differential gene expression after gene deletion reveals the corresponding activated subnetwork illustrated above. Even without galactose present, removal of the GAL80 triggers cellular galactose-processing pathways by eliminating the repression of the GAL4 transcription factor. *Node colour* indicates differential expression statistical significance for the particular gene, whereas *node outline colour and interaction edges between nodes* indicates activated subnetworks. Significance of differential expression does not distinguish between upregulation and downregulation states; thus both GAL80 (eliminated) and GAL1 (highly upregulated) will possess high z scores indicating high differential activity. As is evident from this figure, a single modification of a gene can have a cascade effect throughout the biomolecular interaction network. Reprinted with permission from Stephanopoulos et al. (2004)

Perhaps a more comprehensive approach is that provided by integration of the data provided by measuring metabolites, proteins and/or mRNA from the same sample and to assess the connectivity across the different molecular entities. In this vein metabolite-transcript correlations from large data sets collected throughout development in wild type lines and transgenic tubers engineered to have enhanced sucrose metabolism allows the identification of candidate genes for biotechnology (Urbanczyk-Wochniak et al. 2003). In this study the transcript levels of approximately 280 transcripts that showed reproducible changes with respect to control samples were compared to changes in metabolite levels in paired samples. A total of 517 out of the 26,616 possible pairs showed significant correlation (at the $P < 0.01$ level). Although some of these correlations were already known, most were new and contained several strong correlations between transcript levels and nutritionally important metabolites. The use of metabolomics to assign gene function has also been carried in several other studies with profiling being used in conjunction with knock-out mutagenesis to identify the specific function of PAL1 and PAL2 genes of phenylpropanoid metabolism (Rohde et al. 2004), the Myb-like transcription factor PAP1 (Tohge et al. 2005); however these examples are discussed in detail in Chap. III.5. Targeted metabolite analysis of volatiles in combination with broad transcript analysis has allowed the elucidation of several genes involved in volatile synthesis (Fridman et al. 2005), and secondary metabolites in general (Goosens et al. 2003).

Gain of function analysis is emerging as a particularly powerful approach for functional genomics. For example, the analysis of a gene of known function, a member of the threonine aldolase family, that was introduced into *A. thaliana*, both confirmed the expected function and revealed new effects on the metabolic network, including the upregulation of the methionine pathway and the downregulation of the isoleucine pathway (Fernie et al. 2004). In many cases, the effects of overexpression or mutation of a gene can be quite pleiotropic: for example, in the case of the *dgd*1 mutant there were significant changes in half of the metabolites analysed (Fiehn et al. 2000). Such complex responses involve interactions among metabolic components and interactions between the metabolic network and the mechanisms of gene and protein expression. However, it is important that these network behaviours are well understood if functional genomics is to maximise its potential for metabolite engineering.

Of all genomics tools, metabolite profiling offers arguably the best combinations of practical performance and cost per sample. The expression of almost every gene from the yeast and *E. coli* genomes in *A. thaliana* and subsequent metabolite profiling with GC-MS and LC-MS/MS has recently been achieved (Fernie et al. 2004; Fig. 2). This approach was deliberately non-biased with respect to both the choice of gene and metabolites measured, because of twin objectives – to explore gene function at the level of protein activity and to explore the consequences of introducing new proteins into the metabolic network of *Arabidopsis*. The approach of focussing on individual genes can be

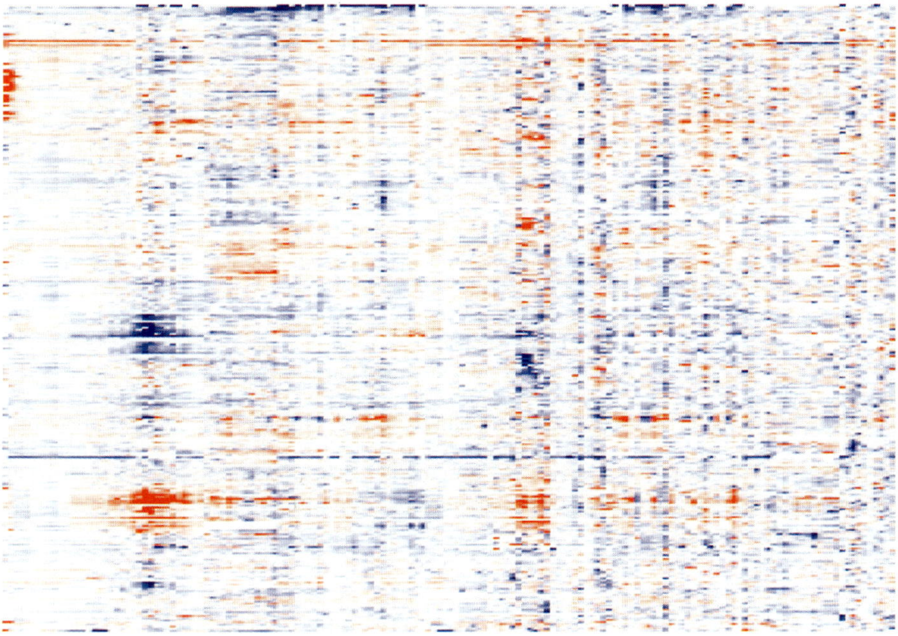

Fig. 2. Overexpression and metabolic profiling at the transgenomic level. An example of a heat map of the metabolite profiles of the leaves of around 19,000 mature plants including plant lines that each overexpress essentially every gene of the yeast genome (R.N.T and A.J.K., unpublished results). Most of this map is *white*, which reflects the fact that overexpression does not result in a change in metabolite content compared with control plants in most cases. Regions of *red* and *blue* indicate that the metabolite content is either increased or decreased, respectively, following overexpression. The colour scale is nonlinear and the maximum increases and decreases detected are around 100-fold. A total of 158 metabolites that have been derived from GC-MS and LC-MS analyses are shown; the chemical identity is known for around 60% of them. The chemical classes covered include amino acids, organic acids, sugars, sugar alcohols, vitamins and pigments. Although the individual metabolite columns can be visually distinguished, the pixel resolution of the image is not sufficient to distinguish the rows (which represent the plants and plant lines). The software that is used to generate the images uses smoothening software algorithms to circumvent this limitation. Such datasets provide a rich resource for the identification of novel gene-function relationships and provide a foundation for systems-biology approaches. Reprinted with permission from Fernie et al. (2004)

easily extended to exploring the phenotypic relevance of genome regions. Recently, GC-MS profiling of breeding populations of tomato, wherein genomic segments of the wild species tomato *Solanum lycopersicum* have been introgressed into the elite cultivated species *Solanum pennellii*, has been initiated. As a first step in this project we profiled the metabolite contents of leaf and mature fruit samples from five wild species tomato that can be readily bred with *S. lycopersicum* (Schauer et al. 2005). Changes in metabolite content were identified in these species that are potential important with respect both to stress response and nutritional importance (Fig. 3), suggesting the incorporation of

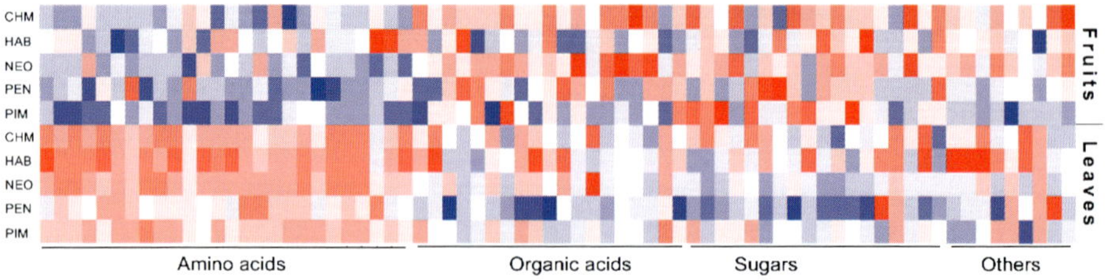

Fig. 3. Survey of metabolite levels in leaves and mature fruits of the *Solanum lycopersicum* complex. Data of tomato fruit and leaf tissue of five wild species tomato (*S. pennellii (PEN)*, *S. habrochaites (HAB)*, *S. neorickii (NEO)*, *S. pimpinellifolium (PIM)*, *S. chmielewskii (CHM)*) is represented as a heat map of relative changes with respect to *S. lycopersicum cv.* M82. The data has been normalised. The colour scale is nonlinear with maximum increases (*red*) and decreases (*blue*) being around 100-fold. *White areas* reflect no changes in metabolite content with respect to *S. lycopersicum cv.* M82 or not being detected. A total of 58 metabolites of different compound classes have been measured in a GC-MS based survey. The identified metabolic changes are of potential importance for breeding stress tolerant and nutritional beneficial traits into elite cultivars. Thus showing the use of natural diversity for crop amelioration by conventional breeding techniques. Experimental data taken from Schauer et al. (2005)

genetic material from wild species could represent an attractive alternative to transgenic approaches for crop improvement. Stress responses in plants are in their own right also starting to be evaluated by multiple genomics tools with recent studies revealing important information on responses to sulphur starvation (Hirai et al. 2004; Nikiforova et al. 2004), and nitrogen (Scheible et al. 2004) and low temperature stresses (Cook et al. 2004; Kaplan et al. 2004). To recapitulate the integration of metabolite profiling with other genomics tools is starting to prove very effective in gene functional annotation and the emerging field of systems biology, despite the large challenges presented in attempting such ambitious projects in multi-cellular organisms. A further example of the complexity of plants lies in their cellular spatial complexity that is brought about by extensive compartmentation and even microcompartmentation of metabolism (Jorgensen et al. 2005). It is unlikely that steady-state metabolite profiling will offer much insight into these processes; however, as is argued in the next section, dynamic metabolite profiling may provide insight into these phenomena.

5 Dynamic Profiling in Plant Cells

The measurement of steady-state metabolite levels provides clues about the metabolic activity of a tissue and about the metabolic response of an organism to an environmental or genetic perturbation (Fernie et al. 2005). The majority of metabolites determined by current metabolite profiling techniques are, however, not end-products but intermediates of metabolic pathways; these

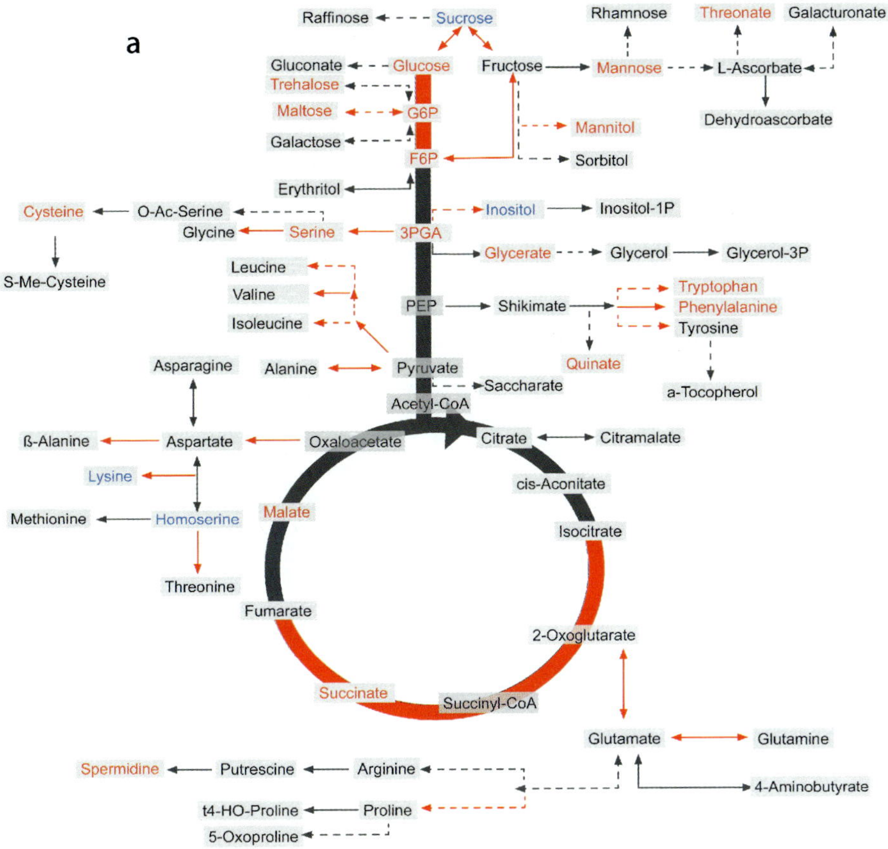

Fig. 4. Pathway based representation of changes in steady-state metabolite levels and determined intermolecular fluxes following incubation of potato tuber slices in ^{13}C labelled glucose. Two well-characterised transgenic situations are represented here – tubers expressing either: **a** yeast invertase (Sonnewald et al. 1997) or; **b** (*on next page*) bacterial sucrose phosphorylase (Trethewey et al. 1999). Metabolites that are set in *red* type are significantly increased in the transgenics with respect to wild type whilst those set in *blue* type are significantly decreased in the transgenics with respect to the wild type (those set in *pale grey* type were not detected using the experimental platform used). Similarly, significant changes in the net relative flux are denoted by the colour of the reaction arrow (*red* indicating increased fluxes, in this example no fluxes were decreased). Experimental data taken from Roessner et al. (2001) and Roessner-Tunali et al. (2002)

techniques are generally incapable of resolving issues of molecular dynamics (Fernie et al. 2005). This is illustrated in Fig. 4, wherein data from steady-state and dynamic metabolite profiling are overlaid on one another. In this example, although in the case of the invertase expressing potato lines the dynamic and steady-state measurements are in keeping with one another, several discrepancies exist between the two types of measurement for the sucrose phosphorylase expressing lines. Thus, this example highlights that in certain

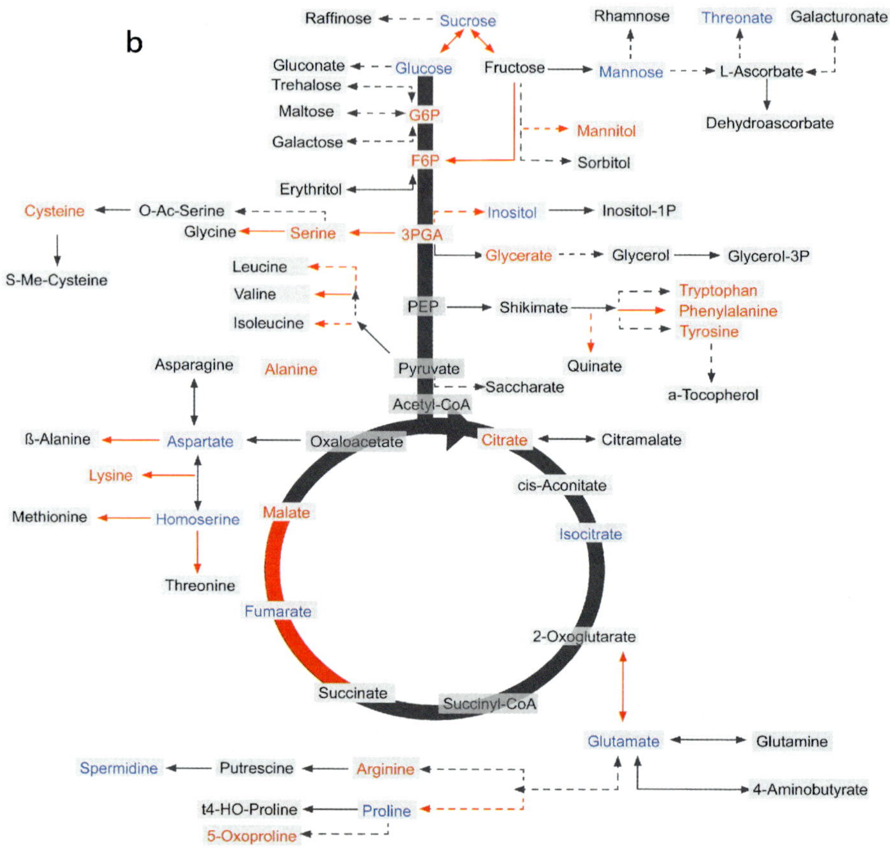

Fig. 4. (continued)

situations the dynamic alterations in a system may be under represented or even misinterpreted on the basis of steady-state measurements alone. Whilst flux measurements are admittedly technically difficult (Fernie et al. 2005), recent technological advances have allowed a broadening of the basis of the information accessible via high throughput flux analysis (Roessner-Tunali et al. 2004; Sauer 2004; Sriram et al. 2004). As mentioned above, flux measurements have been used in systems biology in the framework of metabolic control analysis for many years. Furthermore, by analogy to experimentation in the microbial field (Hellerstein and Nesse 1999), network analysis has also been carried out in this context in plant systems. Historically, isotope labelling studies have played a cardinal role in the definition of metabolic pathways, perhaps the most important recent example of this being the demonstrated of the role of ribulose 1,5-bisphosphate carboxlase/oxygenase (Rubisco) in a previously undefined metabolic context (Schwender et al. 2004). Similar experiments on the central metabolic pathways of tomato fruit metabolism (Rontein et al. 2002), and on the compartmentation of carbohydrate oxidation within *B. napus* embryos (Schwender et al. 2003), have also highlighted the utility of this approach in functional genomics. Even greater spatial resolution was attained by a combination of proteomics, cell biology, traditional enzymology and stable isotope feeding experiments which cumulatively demonstrated the functional association of glycolysis with the mitochondria in *Arabidopsis* (Giege et al. 2003). Given the prominence of molecular associations between enzymes (Jorgensen et al. 2005), it is anticipated that the use of dynamic profiling is likely to identify further microcompartmented pathways.

6 Conclusions and Future Perspectives

The past years have revealed a great application for metabolite profiling as a diagnostic tool, and its growing importance in gene functional analysis is currently apparent. In contrast, attempts to use metabolite profiling as a tool in systems biology are in their infancy. It is likely that the development of systems biology depends to a large extent on technological improvements to improve our coverage of the metabolome. The integrative genomics approaches taken to date have given rich descriptive data networks. Whilst the challenge remains to elucidate the mechanisms underlying behaviour in these networks, the fact that the phenotype of any biological system is largely determined by its metabolite composition provides ample reason to develop further on the foundation studies described in this chapter.

References

Askenazi M, Driggers EM, Holtzman DA, Norman TC, Iverson S, Zimmer DP, Boers ME, Blomquist PR, Martinez EJ, Monreal AW, Feibelman TP, Mayorga ME, Maxon ME, Sykes K, Tobin JV, Cordero E, Salama SR, Trueheart J, Royer JC, Madden KT (2003) Integrating transcriptional and metabolite profiles to direct the engineering of lovastatin-producing fungal strains. Nat Biotechnol 21:150–156

Brindle JT, Antti H, Holmes E, Tranter G, Nicholson JK, Bethell HW, Clarke S, Schofield PM, McKilligin E, Mosedale DE, Grainger DJ (2002) Rapid and noninvasive diagnosis of the presence and severity of coronary heart disease using 1H-NMR-based metabonomics. Nat Med 8:1439–1444

Cook D, Fowler S, Fiehn O, Thomashow MF (2004) A prominent role for the CBF cold response pathway in configuring the low-temperature metabolome of Arabidopsis. Proc Natl Acad Sci USA 101:15243–15248

De Luca V, St Pierre B (2000) The cell and developmental biology of alkaloid biosynthesis. Trends Plant Sci 5:168–173

Edwards EJ, Cobb AH (1999) The effect of prior storage on the potential of potato tubers (Solanum tuberosum L) to accumulate glycoalkaloids and chlorophylls during light exposure, including artificial neural network modelling. J Sci Food Agr 79:1289–1297

Espinosa-Soto C, Padilla-Longoria P, Alvarez-Buylla ER (2004) A gene regulatory network model for cell-fate determination during Arabidopsis thalianal flower development that is robust and recovers experimental gene expression profiles. Plant Cell 16:2923–2939

Even S, Lindley ND, Cocaign-Bousquet M (2003) Transcriptional, translational and metabolic regulation of glycolysis in Lactococcus lactis subsp. cremoris MG 1363 grown in continuous acidic cultures. Microbiology 149:1935–1944

Fernie AR, Sweetlove LJ (2004) Broad range metabolite analysis: integration into genomics programs. In: Ketner C (ed) Experimental standardization of classification of enzyme characteristics. Logos Verlag, Berlin (Germany)

Fernie AR, Trethewey RN, Krotzky A, Willmitzer L (2004) Metabolite profiling: from diagnostics to systems biology. Nature Rev Mol Cell Biol 5:763–769

Fernie AR, Geigenberger P, Stitt M (2005) Flux an important, but neglected, component of functional genomics. Curr Opin Plant Biol 8:174–182

Fiehn O, Kopka J, Dormann P, Altmann T, Trethewey RN, Willmitzer L (2000) Metabolite profiling for plant functional genomics. Nat Biotechnol 18:1157–1161

Fridman E, Wang J, Iijima Y, Froehlich JE, Gang DR, Ohlrogge J, Pichersky E (2005) Metabolic, genomic, and biochemical analyses of glandular trichomes from the wild tomato species lycopersicon hirsutum identify a key enzyme in the biosynthesis of methylketones. Plant Cell 17:1252–1267

Futcher B (2000) Microarrays and cell cycle transcription in yeast. Curr Opin Cell Biol 12:710–715

Geigenberger P, Stitt M, Fernie AR (2004) Metabolic control analysis and regulation of the conversion of sucrose to starch in growing potato tubers. Plant Cell Environ 27:655–673

Giege P, Heazlewood JL, Roessner-Tunali U, Millar AH, Fernie AR, Leaver CJ, Sweetlove LJ (2003) Enzymes of glycolysis are functionally associated with the mitochondrion in Arabidopsis cells. Plant Cell 15:2140–2151

Goossens A, Hakkinen ST, Laakso I, Seppanen-Laakso T, Biondi S, De Sutter V, Lammertyn F, Nuutila AM, Soderlund H, Zabeau M, Inze D, Oksman-Caldentey KM (2003) A functional genomics approach toward the understanding of secondary metabolism in plant cells. Proc Natl Acad Sci USA 100:8595–8600

Gygi SP, Rist B, Aebersold R (2000) Measuring gene expression by quantitative proteome analysis. Curr Opin Biotechnol 11:396–401

Harmer SL, Hogenesch JB, Straume M, Chang HS, Han B, Zhu T, Wang X, Kreps JA, Kay SA (2000) Orchestrated transcription of key pathways in Arabidopsis by the circadian clock. Science 290:2110–2113

Heinrich R, Rapoport TA (1973) Linear theory of enzymatic chains; its application for the analysis of the crossover theorem and of the glycolysis of human erythrocytes. Acta Biol Med Ger 31:479–494

Hellerstein MK, Neese RA (1999) Mass isotopomer distribution analysis at eight years: theoretical, analytic, and experimental considerations. Am J Physiol 276:E1146–E1170

Hirai MY, Yano M, Goodenowe DB, Kanaya S, Kimura T, Awazuhara M, Arita M, Fujiwara T, Saito K (2004) Integration of transcriptomics and metabolomics for understanding of global responses to nutritional stresses in *Arabidopsis thaliana*. PNAS 101(27):10205–10210

Ideker T, Galitski T, Hood L (2001a) A new approach to decoding life: systems biology. Annu Rev Genom Hum Genet 2:343–372

Ideker T, Thorsson V, Ranish JA, Christmas R, Buhler J, Eng JK, Bumgarner R, Goodlett DR, Aebersold R, Hood L (2001b) Integrated genomic and proteomic analyses of a systematically perturbed metabolic network. Science 292:929–934

Jorgensen K, Rasmussen AV, Morant M, Nielsen AH, Biarnholt N, Zagrobelny M, Bak S, Moller BL (2005) Metabolon formation and metabolic chanelling in the biosynthesis of plant natural products. Curr Opin Plant Biol (in press)

Kacser H, Burns JA (1974) The control of flux. Symp Soc Exp Biol 28:65–104

Kaplan F, Kopka J, Haskell DW, Zhao W, Schiller KC, Gatzke N, Sung DY, Guy CL (2004) Exploring the temperature stress metabolome of Arabidopsis. Plant Physiol 136:4159–4168

Kitano H (2004) Biological Robustness. Nature Rev Gen 5:826–837

Mack GS (2004) Can complexity be commercialised? Nature Biotech 22:1223–1229

Nikiforova VJ, Gakiere B, Kempa S, Adamik M, Willmitzer L, Hesse H, Hoefgen R (2004) Towards dissecting nutrient metabolism in plants: a systems biology case study on sulphur metabolism. J Exp Bot 55:1861–1870

Oksman-Caldentey KM, Saito K (2005) Integrating genomics and metabolomics for engineering plant metabolic pathways. Curr Opin Biotech 8 (in press)
Oltvai ZN, Barabasi AL (2002) Systems biology. Life's complexity pyramid. Science 298:763–764
Raamsdonk LM, Teusink B, Broadhurst D, Zhang N, Hayes A, Walsh MC, Berden JA, Brindle KM, Kell DB, Rowland JJ, Westerhoff HV, van Dam K, Oliver SG (2001) A functional genomics strategy that uses metabolome data to reveal the phenotype of silent mutations. Nat Biotechnol 19:45–50
Ratcliffe RG, Shachar-Hill Y (2005) Revealing metabolic phenotypes in plants: inputs from NMR analysis. Biol Rev Camb Philos Soc 80:27–43
Roessner U, Luedemann A, Brust D, Fiehn O, Linke T, Willmitzer L, Fernie A (2001) Metabolite profiling allows comprehensive phenotyping of genetically or environmentally modified plant systems. Plant Cell 13:11–29
Roessner-Tunalli U, Willmitzer L, Fernie AR (2002) Metabolic profiling and biochemical phenotyping of plant systems. Plant Cell Rep 21:189–196
Roessner-Tunalli U, Liu JL, Leisse A, Balbo I, Melis-Perez A, Willmitzer L, Fernie AR (2004) Kinetic analysis of organic and amino acid metabolism in potato tubers by gas chromatography-mass spectrometry following incubation in 13C labelled isotopes. Plant J 39:668–679
Rohde A, Morreel K, Ralph J, Goeminne G, Hostyn V, de Rycke R, Kushnir S, van Doorsselaere J, Joseleau JP, Vuylsteke M, van Driessche G, van Beeumen J, Messens E, Boerjan W (2004) Molecular phenotyping of the pal1 and pal2 mutants of Arabidopsis thaliana reveals far-reaching consequences on phenylpropanoid, amino acid, and carbohydrate metabolism. Plant Cell 16:2749–2771
Rontein D, Dieuaide-Noubhani M, Dufourc EJ, Raymond P, Rolin D (2002) The metabolic architecture of plant cells. Stability of central metabolism and flexibility of anabolic pathways during the growth cycle of tomato cells. J Biol Chem 277:43948–43960
Sauer U (2004) High-throughput phenomics: experimental methods for mapping fluxomes. Curr Opin Biotech 15:58–63
Schauer N, Zamir D, Fernie AR (2005) Metabolic profiling of leaves and fruit of wild species tomato: a survey of the Solanum lycopersicum complex. J Exp Bot 56:297–307
Scheible WR, Morcuende R, Czechowski T, Fritz C, Osuna D, Palacios-Rojas N, Schindelasch D, Thimm O, Udvardi MK, Stitt M (2004) Genome-wide reprogramming of primary and secondary metabolism, protein synthesis, cellular growth processes, and the regulatory infrastructure of Arabidopsis in response to nitrogen. Plant Physiol 136:2483–2499
Schwender J, Ohlrogge JB, Shachar-Hill Y (2003) A flux model of glycolysis and the oxidative pentosephosphate pathway in developing Brassica napus embryos. J Biol Chem 278:29442–2953
Schwender J, Goffman F, Ohlrogge JB, Shachar-Hill Y (2004) Rubisco without the Calvin cycle improves the carbon efficiency of developing green seeds. Nature 432:779–782
Shannon P, Markiel A, Ozier O, Baliga NS, Wang JT, Ramage D, Amin N, Schwikowski B, Ideker T (2003) Cytoscape: a software environment for integrated models of biomolecular interaction networks. Genome Res 13:2498–2504
Somerville C, Bauer S, Brininstool G, Facette M, Hamann T, Milne J, Osborne E, Paredez A, Persson S, Raab T, Vorwerk S, Youngs H (2004) Toward a systems approach to understanding plant cell walls. Science 306:2206–2211
Sonnewald U, Hajirezaei MR, Kossmann J, Heyer A, Trethewey RN, Willmitzer L (1997) Increased potato tuber size resulting from apoplastic expression of a yeast invertase. Nat Biotechnol 15:794–797
Sriram G, Fulton DB, Iyer VV, Peterson JM, Zhou R, Westgate ME, Spalding MH, Shanks JV (2004) Quantification of compartmented metabolic fluxes in developing soybean embryos by employing biosynthetically directed fractional (13)C labeling, two-dimensional [(13)C, (1)H] nuclear magnetic resonance, and comprehensive isotopomer balancing. Plant Physiol 136:3043–3057
Stephanopoulos G, Alper H, Moxley J (2004) Exploiting biological complexity for strain improvement through systems biology. Nat Biotech 22:1261–1267

Sumner LW, Mendes P, Dixon RA (2003) Plant metabolomics: large scale phytochemistry in the functional genomics era. Phytochemistry 62:817–836

Sweetlove LJ, Last RL, Fernie AR (2003) Predictive metabolic engineering: a goal for systems biology. Plant Physiol 132:420–425

Ter Kuile BH, Westerhoff HV (2001) Transcriptome meets metabolome: hierarchical and metabolic regulation of the glycolytic pathway. FEBS Lett 500:169–171

Thimm O, Blasing O, Gibon Y, Nagel A, Meyer S, Kruger P, Selbig J, Muller LA, Rhee SY, Stitt M (2004) MAPMAN: a user-driven tool to display genomics data sets onto diagrams of metabolic pathways and other biological processes. Plant J 37:914–939

Tohge T, Nishiyama Y, Hirai MY, Yano M, Nakajima J, Awazuhara M, Inou E, Takahishi H, Goodenowe DB, Kitayama M et al (2005) Functional genomics by integrated anaylsis of the metabolome and transcriptome of Arabidopsis thaliana plants overexpressing a MYB transcription factor. Plant J (in press)

Trethewey RN, Riesmeier JW, Willmitzer L, Stitt M, Geigenberger P (1999) Tuber-specific expression of a yeast invertase and a bacterial glucokinase in potato leads to an activation of sucrose phosphate synthase and the creation of a sucrose futile cycle. Planta 208:227–238

Urbanczyk-Wochniak E, Luedemann A, Kopka J, Selbig J, Roessner-Tunali U, Willmitzer L, Fernie AR (2003) Parallel analysis of transcript and metabolic profiles: a new approach in systems biology. EMBO Rep 4:989–993

Weckwerth W, Loureiro ME, Wenzel K, Fiehn O (2004) Differential metabolic networks unravel the effects of silent plant phenotypes. Proc Natl Acad Sci USA 101:7809–7814

Westerhoff HV, Palsson BO (2004) The evolution of molecular biology into systems biology. Nat Biotechnol 22:1249–1252

Yang C, Hua Q, Shimuzu K (2002) Integration of information from gene expression and metabolic fluxes for the analysis of regulatory mechanisms in Synechocystis. Appl Microbiol Biotechnol 58:813–822

III.2 Systems-based Analysis of Plant Metabolism by Integration of Metabolomics with Transcriptomics

M.Y. Hirai[1], T. Tohge[1], and K. Saito[1,2,3]

1 Introduction

Plants produce a wide diversity of compounds used for foods, medicines, flavors and industrial materials. To improve the productivity of plants by modifying the genes involved in the synthesis of useful compounds or by strictly controlling plant growth, it is essential to understand the plant's metabolic processes and their regulatory mechanisms as a whole.

Plants cannot move away from the place in which they live, even if the environmental conditions get worse; hence, plants have evolved a metabolic system which is robust against changes in environmental conditions. Responding to the changes of the external circumstances, metabolite levels are adjusted by modulating gene expression, protein modification and enzymatic activity, leading to a new state of metabolic equilibrium. Such manifold regulations make it hard to understand plant metabolism as a whole solely by 'traditional' biology such as molecular biology, biochemistry, and forward and reverse genetics. In recent years, however, novel technologies for comprehensive analysis of the transcripts, proteins and metabolites open the door for elucidation of metabolic systems as a whole.

2 Understanding Whole Plant Metabolism – Our Aims and Strategy

Our final goal is to elucidate overall plant metabolism as an integrated system. For this purpose, first of all, all genes and metabolites in plant cells should be identified as the components of the system. In the model plant *Arabidopsis thaliana*, approximately 26,000 genes were predicted based on nucleotide sequence information; however, for only half of these genes has a function been annotated based on sequence similarity to known genes, and the functions of

[1] RIKEN Plant Science Center, Suehiro-cho 1-7-22, Tsurumi-ku, Yokohama, Kanagawa 230-0045, Japan

[2] Department of Molecular Biology and Biotechnology, Graduate School of Pharmaceutical Sciences, Chiba University, Yayoi-cho 1-33, Inage-ku, Chiba 263-8522, Japan, e-mail: ksaito@faculty.chiba-u.jp

[3] CREST, Japan Science and Technology Agency, Honcho 4-1-8, Kawaguchi, Saitama 332-0012, Japan

Biotechnology in Agriculture and Forestry, Vol. 57
Plant Metabolomics (ed. by K. Saito, R.A. Dixon, and L. Willmitzer)

only about 11% have been confirmed experimentally. In the case of metabolites, no catalogue of metabolites in the cell is available at this moment. Hence, one of our immediate aims is to identify the functions of unknown genes and to identify the metabolites in the cell. At the same time, we intend to clarify the networks constructed from genes and metabolites, and to obtain an image of overall metabolism with the help of bioinformatics.

With this aim, we have adopted a strategy of integration of metabolomics and transcriptomics. By comprehensive analysis of metabolome and transcriptome, and following multivariate analyses, the networks between pathways, genes and metabolites can be speculated on. Such network analysis enables us to identify the functions of unknown genes. By this strategy, we have successfully identified the functions of the genes involved in sulfur metabolism and in flavonoid accumulation (Hirai et al. 2005; Tohge et al. 2005). In this chapter we introduce these studies and present a novel strategy for functional genomics.

3 Metabolome and Transcriptome Analyses

Metabolome analysis was carried out by combining non-targeted and targeted analyses. Non-targeted metabolome analysis was conducted by Fourier-transform ion cyclotron resonance mass spectrometry (FT-MS) according to Tohge et al. (2005). In brief, extraction of polar and non-polar metabolites was conducted in triplicate from each sample. Extracts were analyzed by two ionization methods, electrospray ionization (ESI) and atmospheric pressure chemical ionization (APCI), in positive and negative ion modes. The mass spectra from each analysis were integrated after calibration, creating a peak list that contained the accurate mass and absolute intensity of each peak. In order to compare and summarize data across different ionization modes and polarities, all detected mass peaks were converted to their corresponding neutral masses assuming hydrogen adduct formation. Approximately 2000–3000 mass peaks were observed in a single sample. For targeted metabolite analyses, anions, organic acids and sugars were measured by capillary electrophoresis, and amino acids by high-performance liquid chromatography (HPLC) (Hirai et al. 2003). Flavonoids were analyzed by HPLC/photodiode array/detection/electrospray ionization mass-spectrometry (HPLC/PDA/ESI-MS; Tohge et al. 2005).

Transcriptome analyses were conducted by using cDNA macroarray, which carried 13,000 non-redundant ESTs corresponding to ca. 9000 *Arabidopsis* genes (Hirai et al. 2003), or DNA microarray which carried ca. 22,000 *Arabidopsis* genes (Agilent *Arabidopsis* 2 oligoarray or Affymetrix Genechip ATH1).

4 Studies on Sulfur Metabolism

4.1 Roles of Sulfur in Plant Cells and the Experimental Design for Understanding the Adaptation Mechanism to Sulfur Deficiency

Sulfur is an essential macronutrient required for plant growth. It is found in amino acids, oligopeptides such as glutathione (GSH) and phytochelatins, vitamins and cofactors, sulfolipids and a variety of secondary products (Saito 2004). Sulfur enters a plant primarily through the roots as inorganic sulfate. After activation to adenosine 5′-phosphosulfate (APS), sulfate is mainly assimilated into sulfide, and then coupled with *O*-acetylserine (OAS) yielding cysteine (Fig. 1). The thiol group of cysteine in proteins plays a role in maintaining protein structure by forming disulfide bonds. GSH plays a role in the various redox reactions in the cell and in detoxification of xenobiotics mediated by GSH *S*-transferase (GST). Besides the major pathway to form cysteine, APS is converted into 3′-phosphoadenosine 5′-phosphosulfate (PAPS), which is used for sulfation of organic compounds such as glucosinolates (GLSs) and flavonoids (Fig. 1). Sulfur-containing secondary metabolites such as GLSs in Cruciferae are defense compounds against herbivores and pathogens. Because of such important roles of sulfur in plant cells, sulfur deficiency (-S) causes stunted growth and chlorosis of plants, and plants finally die of sulfur starvation. However, if the degree of sulfur deficiency is not too severe, plants can grow apparently normally. This means that the plant has a robust metabolic system against -S and manages on limited sulfur supply.

To elucidate how plants can manage on limited sulfur supply and grow normally, we conducted metabolomics and transcriptomics. For elucidation of long-term responses to -S, plants were grown under moderate sulfur-deficient condition in hydroponic culture (Fig. 2a, Experiment 1 condition 2). Sulfur

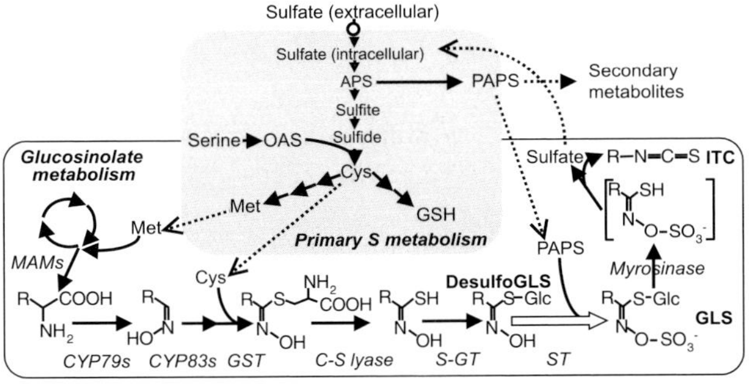

Fig. 1. Sulfur assimilation pathway and glucosinolate metabolism. APS, adenosine-5′-phosphosulfate; PAPS, 3′-phosphoadenosine-5′-phosphosulfate; *MAM*, methylthioalkylmalate synthase; *CYP*, cytochrome P450; S-GT, *S*-glucosyltransferase; ST, sulfotransferase

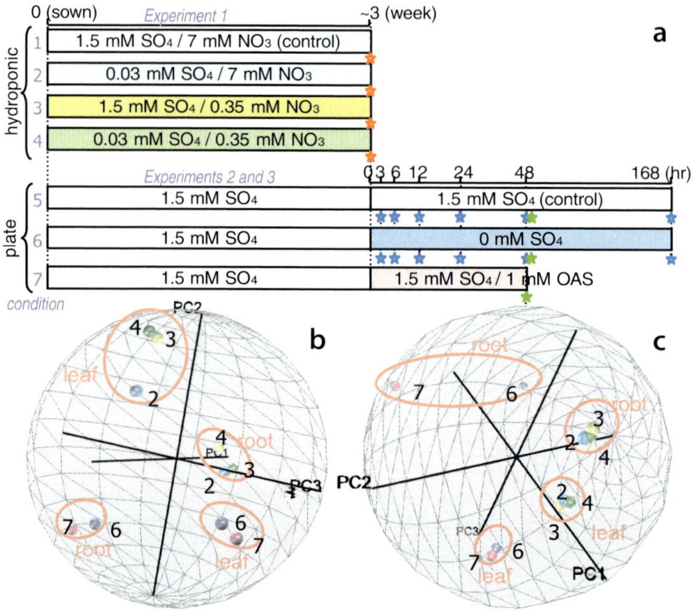

Fig. 2. Plant growth conditions and PCA for elucidation of the changes in metabolome and transcriptome caused by culture conditions: **a** plant growth conditions are described in the text; **b** to classify the samples according to the global change in metabolome, log ratio value of the accumulation level of each metabolite to that in the appropriate control sample was calculated and analyzed by PCA. Each small globe represents the sample. Proportions of the first, second and third components are 27.6%, 18.8% and 15.8%, respectively; **c** to classify the samples according to the global change in transcriptome, log ratio value of the expression level of each gene to that in the appropriate control sample was calculated and analyzed by PCA. Proportions of the first, second and third components are 35.8%, 14.6% and 13.3%, respectively

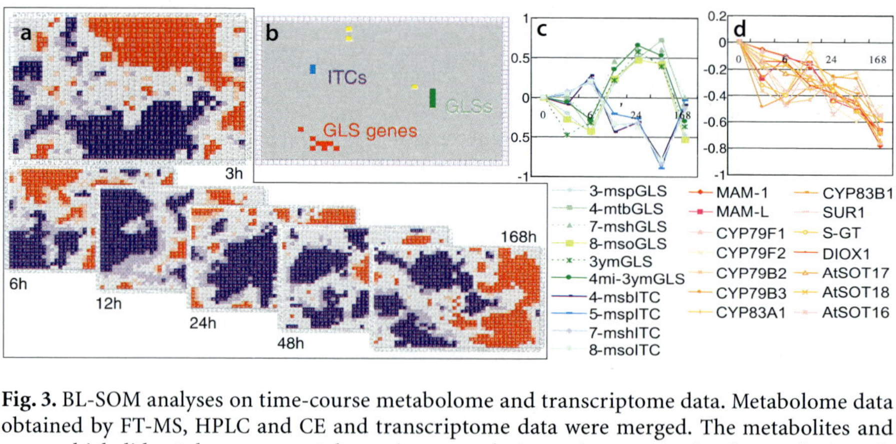

Fig. 3. BL-SOM analyses on time-course metabolome and transcriptome data. Metabolome data obtained by FT-MS, HPLC and CE and transcriptome data were merged. The metabolites and genes which did not show apparent change in accumulation and expression level were eliminated. Approximately 1000 metabolites and 10,000 genes left after elimination were classified by BL-SOM

deficiency is considered as a shortage of sulfur relative to nitrogen nutrition because several responses under -S diminish when nitrogen supply is simultaneously limited. Hence, we also grew *Arabidopsis* in hydroponic culture under reduced nitrate concentration (Fig. 2a, condition 3). Nitrogen-starved plants were expected to exhibit the reverse response to that of sulfur-starved plants. *Arabidopsis* was also grown under sulfur and nitrogen deficient condition, which were expected to cause milder sulfur deficiency (Fig. 2a, condition 4). All these plants were well adapted to the given conditions and grew apparently normally without showing chlorosis throughout their life. Rosette leaves and roots were harvested at ~three weeks after imbibition (shown as orange star in Fig. 2a).

To elucidate short-term response to -S, plants were grown on S-sufficient agar-solidified control medium for three weeks, transferred to sulfur-deprived or control media and harvested at various time points shown in Fig. 2a as blue and green stars (Experiments 2 and 3, conditions 5 and 6). Plants did not show apparent morphological changes until at least one week after transfer. OAS is considered as one of the positive regulators of -S-responsive genes (Saito 2004) and exogenous application of OAS mimics sulfur deficiency. To elucidate the effects of OAS application on the global metabolome and transcriptome, plants grown on control medium for three weeks were transferred to OAS-supplemented medium (Experiment 2, condition 7) and harvested at 48 h after transfer (shown as green star in Fig. 2a). Rosette leaves and roots were subjected to metabolome and transcriptome analyses.

4.2 Multivariate Analyses – Principal Component Analysis and Batch Learning-Self Organizing Mapping

After appropriate normalization of metabolome and transcriptome data, the log ratio of metabolite (mass peak) level and transcript level in the treated sam-

according to time-dependent pattern of change in accumulation and expression: **a** self-organizing map based on the data of the leaf samples. Six maps (corresponding to six time points) are the same except for coloring of the cells. Each cell was colored according to the relative log ratio values of the metabolites and genes therein: when all of the relative log ratio values of the metabolites and genes in the cell were greater or smaller than the average, the cell was colored in *pink* or *pale blue*, respectively. *Red* and *blue* indicated that at least one of the relative log ratio values was greater than the average plus standard deviation or smaller than the average minus standard deviation, respectively; **b** unified map showing clustering pattern of GLSs, ITCs and GLS biosynthesis genes; **c** the change in the content of GLSs (*green lines*) and ITCs (*blue lines*) in leaves. Ordinate scale indicates the relative log ratio value; **d** the change in expression of GLS biosynthesis genes. SUR1, a gene encoding C-S lyase; DIOX1, a gene involved in side chain modification of GLSs. Ordinate scale indicates the relative log ratio value. The following abbreviations indicate the side chain groups of glucosinolates and isothiocyanates: 3-msp, 3-methylsulfinylpropyl; 4-mtb, 4-methylthiolbutyl; 7-msh, 7-methylsulfinylheptyl; 8-mso, 8-methylsulfinyloctyl; i-3ym, indol-3-ylmethyl; 4mi-3ym, 4-methoxyindol-3-ylmethyl; 4-msb, 4-methylsulfinylbutyl; 5-msp, 5-(methysulfinyl)pentyl ple compared to the appropriate control sample was calculated, and subjected to principal component analysis (PCA) and Batch Learning-Self Organizing Mapping (BL-SOM) (Kanaya et al. 2001; Abe et al. 2003). BL-SOM is an improved method of the original SOM (Kohonen 1990; Kohonen et al. 1996) with regard to the fact that the initial weight vectors are set by PCA and the learning process is designed to be independent of the order of input of vectors, and hence the result is reproducible.

4.3 Global Change in Metabolome and Transcriptome in Response to Sulfur Deficiency

4.3.1 Changes Caused by Culture Conditions

To clarify the global change of the metabolome, the data obtained in Experiments 1 and 2 (Fig. 2a, orange and green stars) were subjected to PCA (Fig. 2b). The samples were clustered according to the type of organ (leaves or roots), the method of plant culture (hydroponic or agar-solidified plate) and the period of stress (three weeks or 48 h), indicating several features of global regulation. First, long-term -S, -N and -SN had similar effects on the metabolome. Second, metabolite profiles were quite different between long-term -S and short-term -S. Third, OAS had similar effects as short-term -S, suggesting that OAS is a regulator of global metabolite profiles under short-term -S. A similar clustering pattern was observed in case of the transcriptome (Fig. 2c), indicating that the global transcript profile and metabolite profile were strongly related to each other. As mentioned above, -S is supposed to be caused by shortage of S nutrition relative to N nutrition. From this point of view, it was expected that changes in metabolome and transcriptome under -N might be in the opposite direction to those under -S. However, similar changes occurred under -S and -N, suggesting kinds of general responses to nutritional deficiency in regulation of global metabolome and transcriptome.

The general response to both S and N deficiency and the specific response to either S or N deficiency were observed, for example, in glucosinolate metabolism. GLSs are synthesized from several amino acids such as chain-elongated Met and Trp thorough a number of reactions, and are degraded by a thioglucosidase, myrosinase (Fig. 1). The GLS contents and the expression level of genes encoding GLS biosynthetic and degrading enzymes changed in a treatment-specific manner (Hirai et al. 2004). In the case of roots, for example, the GLS biosynthesis genes were up-regulated both by -S and -N (general response). However, the myrosinase gene was up-regulated by -S and down-regulated by -N, which caused glucosinolate accumulation only in N-starved roots (specific response). One GLS molecule contains two or three S atoms and one N atom; hence GLSs can play a role as S storage source. Both roles of GLSs as defense compounds and as S storage source might determine the metabolic balance of GLSs under S and/or N deficiency.

4.3.2 *Classification of Metabolites and Genes According to Time-dependent Changing Pattern*

To clarify gene-to-gene and metabolite-to-gene networks, metabolome data and transcriptome data of ca. 22,000 genes obtained in Experiment 3 (Fig. 2a, blue stars) were integrated into a single matrix and analyzed by BL-SOM. To classify the metabolites and genes based on the time-dependent changing pattern in response to -S, the metabolites and genes which exhibited an apparent change in accumulation level over 168 h after transfer to -S were selected. For each of ca. 1000 metabolites and ca. 10,000 genes selected, the sum of the square of the six log ratio values at six time points was set equal to one to give relative log ratio values. Ca. 1000 metabolites (or mass peaks) and ca. 10,000 genes were classified by BL-SOM into 40 × 29 (leaves; Fig. 3a,b) and 40 × 24 (roots; data not shown) cells on the map based on the time-dependent pattern of the change in response to -S. A group of metabolites and genes exhibiting similar accumulation pattern were clustered in the same or neighboring cells. In leaves, six GLS molecular species were clustered (Fig. 3b). Interestingly, their degradation products isothiocyanates (ITCs) were also detected by FT-MS and clustered on BL-SOM (Fig. 3b). Accumulation pattern of ITCs was a mirror image of that of GLSs (Fig. 3c). These results suggested that GLS metabolism is coordinately regulated in leaves. In *Arabidopsis,* most of the GLS biosynthesis genes (Fig. 1) were identified. These genes involved in GLS biosynthesis were clustered into the same region on the map (Fig. 3b), supporting the idea of coordinated regulation of GLS metabolism.

4.4 Functional Identification of Novel Glucosinolate Biosynthesis Genes

In GLS biosynthesis desulfoGLSs were known to be subjected to sulfation, but no gene responsible for the sulfation had been identified. On the SOM, 3 out of 18 putative sulfotransferase genes of *Arabidopsis* (AtSOT16, 17 and 18) were clustered with known GLS biosynthesis genes, strongly suggesting their involvement in GLS biosynthesis (Fig. 3b,d). In vitro enzymatic assay using respective gene products proved that these three genes actually encode PAPS:desulfoGLS sulfotransferases (Hirai et al. 2005). In the same way, we could putatively identify the genes encoding C-S lyases, *S*-glucosyltransferase, and GST involved in GLS biosynthesis, some of which were identified late by other groups.

5 Studies on Anthocyanin Metabolism

5.1 Roles of Anthocyanins and the Experimental Design for Elucidation of the Anthocyanin-specific Pathway

Flavonoids, including red-purple anthocyanin pigments, are secondary metabolites which play a role in anti-oxidation protection against strong light, and so on. They are also important for humans owing to their usage as antioxidants and anticancer drugs. However, the chemical structures of flavonoids and their biosynthetic genes in *Arabidopsis* have not yet been completely elucidated. In particular, no genes encoding glycosyltransferases and acyltransferases for the modification of anthocyanin aglycones have been identified yet. For elucidation of anthocyanin structures and their modification enzyme genes in *Arabidopsis*, we focused on *Arabidopsis* lines ectopically expressing the *PAP1* gene which encodes a MYB transcription factor. In a T-DNA activation-tagged line *pap1-D*, the expression of the *PAP1* gene was enhanced by the action of an enhancer sequence in the inserted T-DNA and some phenylpropanoid derivatives such as anthocyanins were over-accumulated (Borevitz et al. 2000). It was shown that several genes involved in anthocyanin biosynthesis were expressed constitutively in the *pap1-D* mutant (Borevitz et al. 2000). However, the transcriptome and metabolome have not been extensively characterized in this mutant. The *PAP1*-overexpressing plants are an ideal model system to elucidate the whole cellular mechanisms at both transcriptome and metabolome levels under the expression of a single transcription factor.

5.2 Metabolome of *PAP1*-overexpressing *Arabidopsis*

Flavonoid accumulation profiles were analyzed by HPLC/PDA/ESI-MS. The metabolites were identified by their UV-visible absorption spectra and mass fragmentation pattern by tandem MS spectroscopy in comparison with the authentic compounds in our laboratory stock, and reported data (Graham 1998; Veit and Pauli 1999; Bloor and Abrahams 2002). In the *PAP1*-overexpressing lines (*pap1-D* mutant and a *PAP1* cDNA-overexpressing transgenic plant), 11 anthocyanin pigments and 3 quercetin glycosides over-accumulated in leaves. Among the 11 anthocyanins, 8 were novel cyanidin derivatives that had never been reported in *Arabidopsis*.

Non-targeted FT-MS metabolome analysis was also conducted on the leaf and root samples of the wild-type plant, *pap1-D* mutant and a *PAP1* cDNA-overexpressing transgenic plant grown on either agar-solidified medium or vermiculite. To elucidate the key determinant factors for the metabolome, PCA was conducted with ca. 1800 peaks detected in non-targeted FT-MS analysis. The results suggested that the major determinant factors for the metabolome were the type of organ (leaves or roots) and the method of plant culture (agar-solidified medium or vermiculite), which was consistent with the result of the sulfur study. This implied that the global metabolome profiles of *PAP1*-overexpressing lines were similar to those of wild-type plants despite the marked difference in total anthocyanin observed. These results suggested that the *PAP1* gene regulates anthocyanin accumulation in a specific manner, causing only a small change in the metabolome.

5.3 Transcriptome of *PAP1*-overexpressing *Arabidopsis*

The transcript levels of ca. 23,000 genes were determined using DNA microarray, and the genes exhibiting reproducible up-regulation in *PAP1*-overexpressing lines were identified. Eight among the 39 up-regulated genes in leaves were annotated as encoding well-known anthocyanin biosynthetic enzymes (*TT3*, *TT4*, *TT5*, *TT7* and *TT19*) or regulatory proteins (*PAP1*, *TTG2* and *TT8*) characterized previously. Combined with the results of metabolite profiling, this result suggested that *PAP1* transcription factor induced specific expression of the genes involved in anthocyanin production.

5.4 Functional Identification of Novel Anthocyanin Biosynthesis Genes by Integration of Metabolomics and Transcriptomics

We could assume that the rest of the 39 genes up-regulated in *PAP1*-overexpressing lines may also be involved in anthocyanin biosynthesis and accumulation. Based on functional annotation and sequence similarity to the previously identified genes, the functions of several genes could be predicted. Among 107 putative UDP-sugar-dependent glycosyltransferase genes in the *Arabidopsis* genome, three genes, At5g54060, At4g14090 and At5g17050, were up-regulated in *PAP1*-overexpressing plants in both leaves and roots, suggesting the involvement of these three gene products in transfer of sugar moieties to anthocyanins. Another putative UDP-sugar-dependent glycosyltransferase gene, At3g21560, was up-regulated only in the leaves of *PAP1*-overexpressing lines. The molecular phylogenetic tree of these genes together with the identified flavonoid glycosyltransferases genes in other plant species (Fig. 4b) showed that At5g17050 and At4g14090 belong to the subfamily of anthocyanin 3-*O*-glucosyltransferase (A3GT) and that of anthocyanin 5-*O*-glucosyltransferase (A5GT), respectively, and that At5g54060 has the highest similarity to Petunia anthocyanin 3-*O*-glucoside-6″-*O*-rhamnosyltransferase (A3G-6″ RT). By flavonoid profiling of *PAP1*-overexpressing lines, it was shown that the most abundant and most extensively modified anthocyanin molecule was cyanidin 3-*O*-[2″ -*O*-(6‴ -*O*-(sinapoyl) xylosyl) 6″ -*O*-(*p*-*O*-(glucosyl)-*p*-coumaroyl) glucoside] 5-*O*-(6⁗ -*O*-malonyl) glucoside, whose structure is shown in Fig. 4a. This molecule had four sugar residues; in addition to 3-*O*-glucose and 5-*O*-glucose, a xylose residue attached to the C2-position of the 3-*O*-glucose and a glucose residue attached to the *p*-position of the coumaroyl moiety. Considering the results of up-regulated genes and flavonoid profiles in the roots

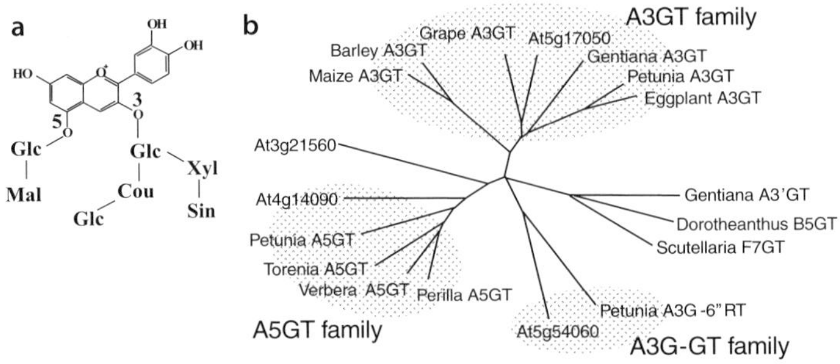

Fig. 4. **a** A major anthocyanin in *PAP1*-overexpressing *Arabidopsis* leaves. **b** Molecular phylogenetic tree of the amino acid sequences of the flavonoid glycosyltransferases. The amino acid sequences were aligned using the multiple sequence alignment CLUSTALW (http://clustalw.genome.ad.jp/). The GenBank accession numbers for the sequences are as follows; eggplant A3GT (X77369); petunia A3GT (AB027454); gentiana A3GT (D85186); grape A3GT (AF000371); barley A3GT (X15694); maize A3GT (X13501); petunia A5GT (AB027455); torenia A5GT (AB076698); verbena A5GT (BAA36423); perilla A5GT (AB013596); petunia A3G 6″RT (Z25802); *Dorotheanthus* B5GT (CAB56231); *Scutellaria* F7GT (BAA83484)

of *PAP1*-overexpressing lines, four candidate glucosyltransferase genes could be assigned to specific functions. Reverse genetics approach and in vitro enzymatic assay using recombinant gene products have proved the predicted functions of At5g17050 and At4g14090 as flavonoid 3-*O*-glucosyltransferase and anthocyanin 5-*O*-glucosyltransferase, respectively (Tohge et al. 2005).

6 Conclusions

In the present studies on sulfur metabolism and anthocyanin production, we could integrate metabolomics and transcriptomics and predict comprehensively gene function especially in secondary metabolism. Concerning the production of secondary metabolites, the regulation at the transcriptional level may be dominant over other regulation at translational and enzymatic activity levels, and hence the transcript profile may determine directly the metabolite profile. We believe that almost all genes involved in the secondary metabolism of interest can be identified by the approach presented in this article. This type of functional genomics can be applied to novel biosynthetic pathway in non-model plants, crops and medicinal plants by using transcriptome analysis such as cDNA-AFLP and cDNA subtraction as substitutions for DNA array.

Acknowledgements. We are grateful to our co-workers whose names are in our recent publications on metabolomics and transcriptomics. Part of the authors' study was supported by Core Research for Evolutional Science and Technology of Japan Science and Technology Agency, and by Grants-in-Aids from the Ministry of Education, Science, Culture, Sports and Technology, Japan.

References

Abe T, Kanaya S, Kinouchi M, Ichiba Y, Kozuki T, Ikemura T (2003) Informatics for unveiling hidden genome signature. Genome Res 13:693–702

Bloor SJ, Abrahams S (2002) The structure of the major anthocyanin in *Arabidopsis thaliana*. Phytochemistry 59:343–346

Borevitz JO, Xia Y, Blount J, Dixon RA, Lamb C (2000) Activation tagging identifies a conserved MYB regulator of phenylpropanoid biosynthesis. Plant Cell 12:2383–2393

Graham TL (1998) Flavonoid and flavonol glycoside metabolism in *Arabidopsis*. Plant Physiol Biochem 36:135–144

Hirai MY, Fujiwara T, Awazuhara M, Kimura T, Noji M, Saito K (2003) Global expression profiling of sulfur-starved *Arabidopsis* by DNA macroarray reveals the role of *O*-acetyl-L-serine as a general regulator of gene expression in response to sulfur nutrition. Plant J 33:651–663

Hirai MY, Yano M, Goodenowe DB, Kanaya S, Kimura T, Awazuhara M, Arita M, Fujiwara T, Saito K (2004) Integration of transcriptomics and metabolomics for understanding of global responses to nutritional stresses in *Arabidopsis*. Proc Natl Acad Sci USA 101:10205–10210

Hirai MY, Klein M, Fujikawa Y, Yano M, Goodenowe DB, Yamazaki Y, Kanaya S, Nakamura Y, Kitayama M, Suzuki H, Sakurai N, Shibata D, Tokuhisa J, Reichelt M, Gershenzon J, Papenbrock J, Saito K (2005) Elucidation of gene-to-gene and metabolite-to-gene networks in *Arabidopsis* by integration of metabolomics and transcriptomics. J Biol Chem 280:25590–25595

Kanaya S, Kinouchi M, Abe T, Kudo Y, Yamada Y, Nishi T, Mori H, Ikemura T (2001) Analysis of codon usage diversity of bacterial genes with a self-organizing map (SOM): characterization of horizontally transferred genes with emphasis on the E. coli O157 genome. Gene 276:89–99

Kohonen T (1990) The self-organizing map. Proc IEEE 78:1464–1480

Kohonen T, Oja E, Simula O, Visa A, Kangas J (1996) Engineering applications of the self-organizing map. Proc IEEE 84:1358–1384

Saito K (2004) Sulfur assimilatory metabolism. The long and smelling road. Plant Physiol 136:2443–2450

Tohge T, Nishiyama Y, Hirai MY, Yano M, Nakajima J, Awazuhara M, Inoue E, Takahashi H, Goodenowe DB, Kitayama M, Noji M, Yamazaki M, Saito K (2005) Functional genomics by integrated analysis of metabolome and transcriptome of *Arabidopsis* plants overexpressing an MYB transcription factor. Plant J 42:218–325

Veit M, Pauli GF (1999) Major flavonoids from *Arabidopsis thaliana* leaves. J Nad Prod 62:1301–1303

III.3 Targeted Profiling of Fatty Acids and Related Metabolites

T.R. Larson and I.A. Graham[1]

1 Introduction

1.1 Plant Lipids and Society

Plant oils make up as much as 25% of the calorific intake in the human diet (Broun et al. 1999), and the balance of saturated and unsaturated fatty acids has profound effects on human health, particularly in the amelioration of the metabolic syndrome (Hulbert et al. 2005). In addition, novel oils containing unusual fatty acids or fatty acids of a specified composition in high yield are valuable as industrial feedstocks (Carole et al. 2004). To address these interests, the factors that control triacylglycerol (TAG) yield (oil quantity per seed or hectare) and quality (fatty acids present, oil-soluble impurities) in oilseed crops have been studied for many years by plant breeders, and more recently by biotechnologists and molecular geneticists (Thelen and Ohlrogge 2002). Metabolic engineering approaches presume an understanding of how fatty acid metabolism is regulated, particularly with respect to fatty acid metabolism during seed development. Unfortunately this understanding is far from complete, although in excess of 600 genes involved in plant lipid metabolism have been identified by expressed sequence tag (EST) analysis (Beisson et al. 2003). These interests have provided a significant driving force behind technologies developed to profile lipids and their metabolites from plant tissues.

1.2 Biochemistry of Plant Fatty Acid Metabolism

There are in excess of 200 different fatty acid species present in plants (van de Loo et al. 1993). Fatty acids have multiple roles in plant metabolism; they are the building blocks for structural membrane lipids (phospho-, galacto-, sphingo-, sulfolipids), are stored for energy as TAG, and are the precursors for signalling molecules in wounding and pathogenic response pathways (Li et al. 2005; Schneider et al. 2005). This diversity is the result of acyl chain elongation, desaturation and other modifications that occur during lipid synthesis, which is compartmentalized between the cytosol and the plastid in both leaf and seed tissues (Ohlrogge et al. 1991; for an excellent recent review, see Wallis

[1] CNAP, Department of Biology (Area 7), University of York, PO Box 373, York YO10 5YW, UK, e-mail: trl1@york.ac.uk

Biotechnology in Agriculture and Forestry, Vol. 57
Plant Metabolomics (ed. by K. Saito, R.A. Dixon, and L. Willmitzer)

and Browse 2002). The mechanisms that control the partitioning of fatty acids between structural and storage lipid synthesis and catabolism remain unclear, and much biochemical and molecular genetic investigation over the last two decades has not given any appreciable insight into this question. This is in part due to the redundancy of biochemical routes involved in lipid synthesis and the differential expression of these routes at different developmental stages and in different tissues. These complex interactions have increased the importance of comprehensive analytical techniques for profiling plant lipids and their metabolites in studies that seek to understand the regulatory mechanisms of lipid synthesis in oilseeds (Abbadi et al. 2004).

Within the plastid (Fig. 1, step 1), fatty acids are synthesized de novo from acetyl-CoA in a series of condensation and elongation reactions to form a pool of mostly 16:0- and 18:0-acyl carrier proteins (ACPs), of which a portion are desaturated to monounsaturated acyl-ACPs. These acyl-ACPs are cleaved by specific thioesterases to free fatty acids and then converted to acyl-CoAs by acyl-CoA synthetases at or near the outer plastid envelope so as to be available for passage to the cytosol (Pollard and Ohlrogge 1999). Export to the cytosol or entry to the "eukaryotic" pathway diverts acyl chains away from membrane lipid synthesis in the plastid (the "prokaryotic" pathway; Fig. 1, step 8), and makes them available for incorporation into TAGs (Fig. 1, steps 4, 5, 6). Acyl-chain flux back into the plastid (probably as free fatty acids) and the prokaryotic pathway also occurs from the cytosolic acyl-CoA pool (Fig. 1, steps 6, 7). This flux is probably in the form of free fatty acids. Within the cytosol further modification of the acyl chains occurs through the complex interaction of desaturation, elongation, and phospholipid/acyl-CoA exchange mechanisms (Fig. 1, steps 4, 5, 6). A portion of the acyl-CoA pool may be catabolised through β-oxidation, which is the normal route for storage-oil derived fatty

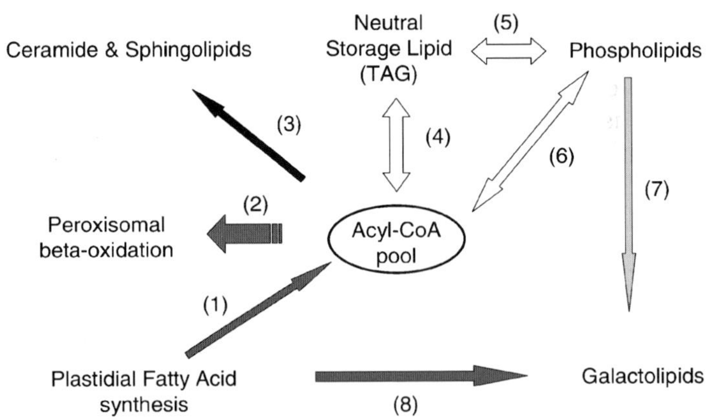

Fig. 1. The central role of the acyl-CoA pool in plant lipid metabolism. *Arrows* represent directional fluxes of cytosolic acyl-CoAs in a general model representing all plant tissues. *Numbers* refer to biochemical routes and genes referenced in the text

acid breakdown during seed germination (Fig. 1, step 2). The extra carbon required for acyl chain elongation commonly observed in the brassicaceae originates from a distinct cytosolic acetyl-CoA pool, probably derived from mitochondrial citrate metabolism (Fatland et al. 2005). The interplay of these mechanisms ultimately controls the final fatty acid composition, and together with consideration of the rate of de novo fatty acid synthesis, yield of the TAGs formed from the condensation of diacylglycerol and acyl-CoA pools in the final step of the Kennedy pathway. In fact, the acyl-CoA pool is central to lipid metabolism and is potentially a sensitive indicator of perturbations in the acyl-chain flux through any of the pathways indicated in Fig. 1.

The complexity of fatty acid synthesis and catabolism in plants, and the wide range of lipid-related intermediates involved in these processes, emphasise the need for a wide range of metabolite profiling techniques. The development and use of these techniques is discussed in the following sections.

2 Metabolite Profiling Techniques Used to Study Plant Lipid Metabolism

2.1 Fatty Acid Methyl Esters

Seed oil profiling, specifically the measurement of fatty acids derived from TAGs as their methyl esters (FAMEs), is one of the oldest metabolite profiling techniques used in plant functional genomics studies. This technique has also been used to examine the acyl chain composition of many other common plant glycerolipids in addition to TAGs, such as the galactolipids that make up chloroplast membranes. Lipids, being extremely hydrophobic, can rapidly and easily be partitioned away from other cell components during sample work-up. The extracted lipids can then be quantitatively converted to FAMEs using a range of transesterification reagents. FAMEs can even be routinely prepared by the direct transmethylation of plant material with no preparative lipid extraction (Browse et al. 1986). This simple preparation assists in large-scale screening projects for alterations in fatty acid quality or content in plant lipids. The equipment used for separating and detecting plant-derived FAMEs is the gas chromatograph (GC) with flame-ionization detection (FID). A large range of commercial GC columns are available for FAME analysis, and commercial reference standards are also available for their identification and quantification. FAME profiling has the potential to become extremely rapid with the recent introduction of fast GC technology, which allows analysis times of a few minutes per sample (Mondello et al. 2003). FID detectors have a response directly proportional to the carbon mass eluted from the column, and they benefit from having a linear quantification range greater than most other types of detector, including mass spectrometers. This enables the absolute quantification of novel or unknown FAMEs for which no standards are available. Some of the first arabidopsis mutants in fatty acid synthesis were discovered using genetic forward screening techniques on mutagenised populations, where leaf and seed material was analysed by FAME profiling. Using this technology, leaf lipids from an arabidopsis population of only 2000 M2 mutagenised lines uncovered seven mutants and five loci encoding desaturases (Browse et al. 1985). FAME profiling was subsequently used for the first example of map-based cloning in plants, used to isolate the arabidopsis plastidic desaturase, *FAD3* (Arondel et al. 1992), and has subsequently been used to isolate several desaturase mutants (for review, see Wallis and Browse 2002).

The examples mentioned above used qualitative differences in FAME profiles to identify lesions in fatty acid biosynthesis. However, GC profiling can also be used to obtain quantitative oil yield data, by summing the total amounts of fatty acids present per unit weight or per seed. Quantitative data is also required to calculate changes in the overall flux of carbon into storage lipids, which may occur as a result of upregulated biosynthesis, and/or reduced fatty acid catabolism during seed filling. For model plants such as arabidopsis, eicosenoic acid (20:1n9) is a useful marker for TAG when yield information is required, as this fatty acid is only present in storage oil and is absent from leaf lipids. However, alternative technologies such as NMR (Rutar 1989) are perhaps more suited to rapid screening of total seed yields, when compositional information is not as important. Despite the analytical options available for measuring lipid concentrations, the study of oil yields is fraught with difficulties, as estimation of such a useful agronomic trait is also dependent on the number of seeds produced per plant and other growth parameters, which makes it difficult to pinpoint discrete genetic factors responsible for increasing the amount of oil produced per hectare of a planted crop.

For the identification of new or novel fatty acids in plants, secondary derivatization techniques are available to assist in the calculation of carbon number and double bond position in FAMEs using GC coupled to mass spectrometry (GCMS) (for an excellent summary of these techniques, see Christie's comprehensive website at http://www.lipidlibrary.co.uk). These techniques are especially useful for the identification of novel fatty acids produced by the heterologous expression of candidate biosynthetic genes in model systems. For example, a putative $\Delta 11$ desaturase isolated from a marine alga and expressed in yeast resulted in the synthesis of a new fatty acid, for which no standards were available. This fatty acid was subsequently identified as $16{:}1^{\Delta 11}$ by examination of the dimethyldisulphide adduct by GCMS (Tonon et al. 2004).

2.2 Intact Glycerolipids

Triacylglycerols, free fatty acids, phospholipids and galactolipids are all converted to FAMEs using acidic transmethylation techniques. However, the major disadvantage of these techniques is that the derivatized FAMEs originate from several different lipids. Alkaline transmethylation techniques are slightly more specific in that free fatty acids are not transmethylated (Christie 1982), but the problem remains that FAME measurements can not distinguish between storage and membrane lipids. In addition, some important regulatory lipids, notably sphingolipids, cannot be detected by transmethylation/GC methods, and require more specialised extraction and analysis techniques (Sperling et al. 1998). The fatty acid composition of storage and membrane lipid species in the model plant arabidopsis is similar, with the exception that 16:3n3 is found only in galactolipids and 20:1n9 in TAGs. In this case, these two FAMEs can be used as markers for changes in membrane and TAG lipid pools, respectively, but their levels can not be used to infer the compartmentalised regulation of other fatty acids present in both pools, such as 18:3n3. This problem can be somewhat circumvented by careful choice of the tissues analysed; i. e. leaves will contain mostly galactolipids and seeds TAGs. These limitations can be overcome by preparative techniques that separate lipid classes before derivatization to FAMEs; for example by the use of thin-layer or column chromatography (TLC).

Preparative scale techniques have been refined for high-throughput screening of changes in glycerolipid species. For example, Benning has developed a robotic system to spot leaf lipid extracts onto TLC plates to screen for mutants in galactolipid biosynthesis (Benning 2004). Polar lipid extracts from the green alga *Chlamydomonas reinhardtii* were used to develop the system before the technique was used for arabidopsis lipid analysis. Cells or leaf material in 96-well plates are robotically harvested and extracted with chloroform/methanol, and an aliquot spotted by the robot onto TLC plates. After further robotic addition of developing solvents, lipid bands develop as small concentric rings around the spotted sample, which can then be visualised with various headgroup-specific reagents. This approach was successfully used to screen 25,000 M2 arabidopsis lines in a forward screen, which subsequently isolated 25 *dgd1* suppressor mutants that accumulated di-, tri-, or tetra-galactosyldiacylglycerols (Benning 2004).

However, the most comprehensive techniques for lipid class profiling involve infusing plant extracts into a mass spectrometer. Infusion has the advantage of being rapid without relying on a lengthy chromatographic separation step. Lipid classes are identified by tandem MS fragmentation patterns specific for different headgroups, with quantification achieved by the infusion of relevant standards. This is referred to as "two-dimensional" fingerprinting, and has found use in animal (Han et al. 2004) and plant systems (Welti and Wang 2004). In arabidopsis, over 120 polar lipids (galacto- and phospholipids) can be routinely profiled using simple sample preparation, electrospray ionisation, and detection by precursor ion/neutral loss scanning (Welti and Wang 2004). This approach was used to determine that phosphatidylcholine, rather than other phospholipid species, is the major substrate for the most abundant phospholipase D activated during cold treatment in arabidopsis (Welti et al. 2002). A further refinement of rapid fingerprinting techniques is the use of matrix assisted laser desorption ionization (MALDI) coupled with high-resolution mass spectrometry. MALDI has been used to fingerprint phospholipids compositions in whole cells, such as in bacterial or yeast suspensions (Jones et al. 2004). A variation of this technique, laser desorption/ionization mass spectrometry (i. e. without added matrix) has been used to obtain qualitative TAG fingerprints from commercial olive oil samples (Calvano et al. 2005).

Lipids and their derivatives are popular candidate molecules in studies of plant disease resistance responses. In arabidopsis, some genes associated with the induction of systemic acquired resistance (SAR) and salicylic acid signalling are involved in lipid metabolism. A suppressor of the positive regulator gene *NPR1* in the salicylic acid signalling pathway, designated *ssi2* (*s*uppressor of *s*alicylate *i*nsensitivity), was found to induce pathogenesis-related (PR) genes though a salicylic acid dependent but NPR1-independent pathway (Shah et al. 2001). *SSI2* was subsequently cloned and, with expression analysis in *E. coli*, shown to encode a stearoyl-ACP desaturase (Kachroo et al. 2001). In *ssi2* plants, GCMS analysis of leaf FAMEs indicated substantially increased levels of 18:0 compared to 18:1, with minor reductions in 16:3, 18:2 and 18:3. How this alteration in leaf fatty acid composition might specifically regulate or interact with plant defence response pathways is unclear, although the authors did propose that an increased 18:0/18:1 ratio might activate lipid signalling events that would induce the pathogenesis response pathway, or that the ratio of saturated and unsaturated fatty acids might alter protein phosphatase activities. More recent work showed that increasing plastidic 18:1 levels (probably 18:1-ACP) by decreasing acylation into lipids via decreased glycerol-3-phosphate acyltransferase activity modified the defence signalling pathway (Kachroo et al. 2003, 2004). A second gene related to plant defence and involved in glycerolipid metabolism, denoted *SFD1* (suppressor of fatty acid desaturase deficiency1) was described by Nandi et al. (2004). *SFD1* was shown to encode a putative dihydroxyacetone phosphate (DHAP) reductase, essential for the production of plastidic glycerolipids. LCMS direct-infusion lipid profiling was used to determine that *sfd1* plants had lower levels of 16:3 in monogalactosyldiacylglycerol (MGDG), but increased levels of 18:3 in both MGDG and digalactosyldiacylglycerol (DGDG). In contrast, the non-plastidic phospholipid pools did not change. Although *sfd1* plants are compromised in the activation of SAR, the signalling link between an altered plastidic glycerolipid profile and the SAR response is unknown.

Infusion-MS technology allows polar lipid classes to be identified and the acyl chains to be measured in terms of the total number of carbons and double bonds. However, the structure (number of carbons, number and position of double bonds) of the individual fatty acyl chains in these lipids cannot be readily determined using this technique; neither can the regiospecificity of acyl chain attachment be determined. Additional structural information is potentially very useful because it reflects the pathway of lipid assembly via different possible desaturase, elongase, and acyltransferase mechanisms in plants. This may be important when selecting or considering particular plant lines or species for the production of beneficial fatty acids in TAGs. For example, heterologous expression of a *Pythium irregulare* Δ6 desaturase in *Brassica juncea* led to a preferential accumulation of gamma linolenic acid (18:3n6) in the *sn*-2 position of TAGs compared to endogenous fatty acids, as determined by pancreatic lipase treatment (Hong et al. 2002). TAG positional information is especially relevant for studies on human nutrition, where studies suggest that fatty acids on the *sn*-2 position may be more bioavailable (Jensen et al. 1994; Christensen et al. 1995).

TAGs cannot be easily profiled by infusion-MS techniques, because there is no unique headgroup present to use as an identity marker in neutral loss scans, and regioisomers cannot be distinguished by parent ion mass. In fact, the infusion and MALDI techniques have been developed with phospho- and other lipids with polar headgroups in mind, which are relatively easy to characterize using tandem MS techniques. In addition, the loss of signal from trace-level components in background noise and ion suppression through matrix effects makes quantitative lipid profiling in the absence of a full set of standards problematic. Therefore, while the infusion technique is useful for profiling glycerolipids involved in leaf metabolism, different techniques are needed to determine identity and positional location of individual fatty acids on these lipids and on TAGs. Some of the problems associated with infusion can be overcome by using chromatographic techniques, such as liquid chromatography (LC) coupled to MS (for review, see Buchgraber et al. 2004). GCMS can be used for TAG separation and detection, but the high boiling point of TAGs and the relatively poor selectivity of high-temperature GC columns for lipid isomers has excluded the widespread adoption of this technique. The LC separation step can use normal or reversed-phase columns, or even a combination of the two to achieve maximum molecular species separation (Houjou et al. 2005). This technology could be used to characterize the multitude of TAG species present in model species such as arabidopsis, and thus provide a tool for examining subtle changes in lipid remodelling and acyltransferase specificities that are not possible with simple FAME or infusion-MS of intact lipid profiling.

For example, LCMS analysis of seed lipid extracts from a range of putative arabidopsis lipase mutants revealed an increase in TAGs containing 18:1 in one line, suggesting that an alteration in lipid composition took place during seed maturation (Fig. 2). There is circumstantial evidence that TAG remodelling occurs during seed development. It has been reported that arabidopsis TAG levels decrease by as much as 28% in the final stages of seed maturation (Baud et al. 2002), and by approximately 10% in *Brassica napus* (Chia et al. 2005). This decrease in TAGs is probably not due to a specific activation of lipid catabolism, as β-oxidation genes and enzymes are known to be expressed throughout seed development (Eastmond and Graham 2001). This suggests that acyltransferase and/or lipase activities may change in the final stages of seed development to either prepare the mature seed for dormancy or prime it for germination and reserve mobilization. The development of better tools to examine closely changes in TAG composition during seed development will undoubtedly help our understanding of how these processes are regulated.

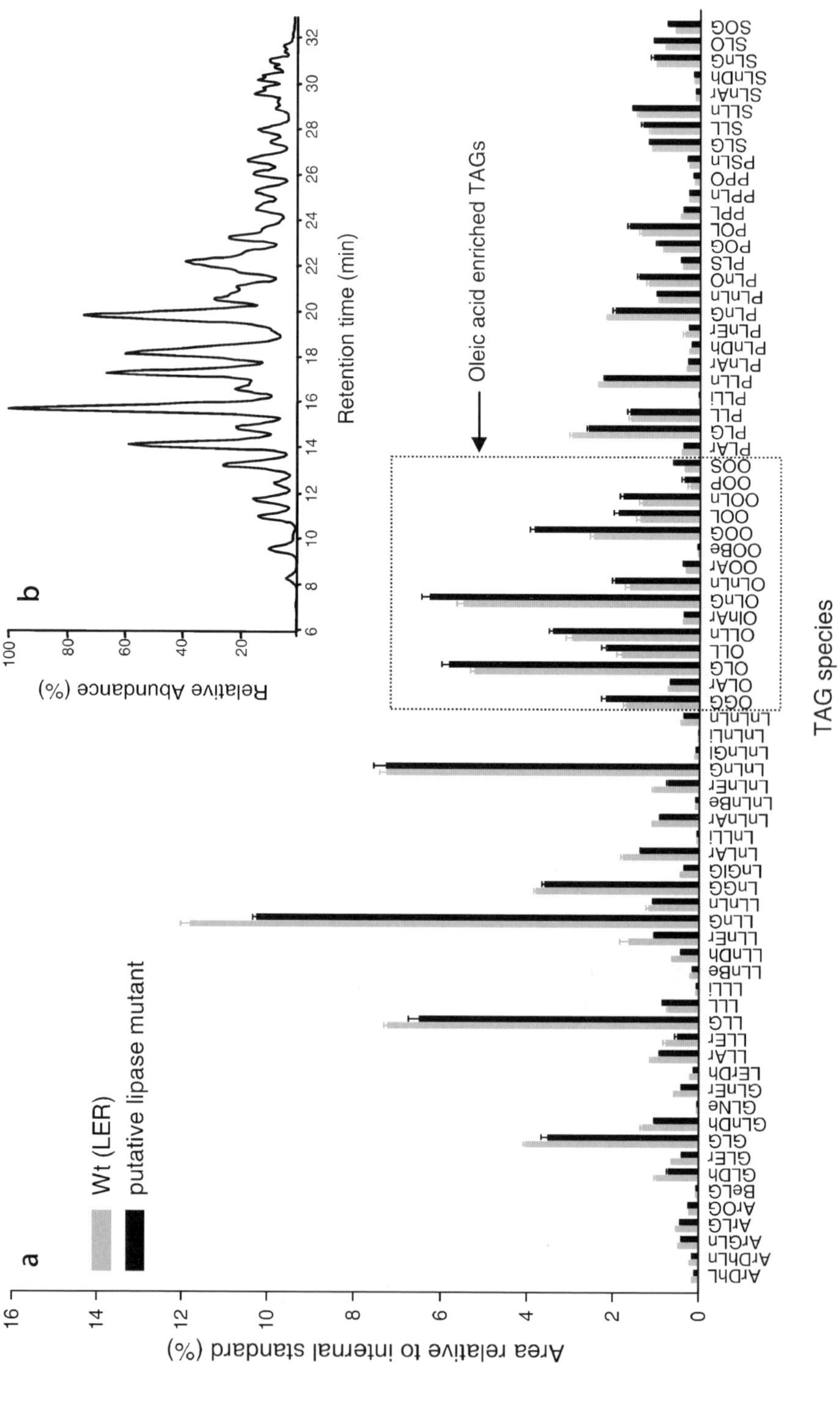
a
Wt (LER)
putative lipase mutant
Area relative to internal standard (%)
TAG species
ArDhL ArDhLn ArGLn ArLG ArOG BeLG GLDh GLEr GLG GLnDh GLNe GLnEr LErDh LLAr LLEr LLG LLL LLLi LLnBe LLnDh LLnEr LLnG LLnLn LnGG LnGlG LnLAr LnLLi LnLnAr LnLnBe LnLnEr LnLnG LnLnGl LnLnLi LnLnLn OGG OLAr OLG OLL OLLn OlnAr OLnG OLnLn OOAr OOBe OOG OOL OOLn OOP OOS PLAr PLG PLL PLLi PLLn PLnAr PLnDh PLnEr PLnG PLnLn PLnO PLS POG POL PPL PPLn PPO PSLn SLG SLL SLLn SLnAr SLnDh SLnG SLO SOG
Oleic acid enriched TAGs
b
Relative Abundance (%)
Retention time (min)

◀ **Fig. 2.** TAG profiling in dry arabidopsis seeds: **a** TAGs were extracted in hexane:isopropanol from 10 mg of both Ler seeds and a putative TAG lipase knockout mutant in the same background; **b** TAGs were resolved by HPLC and quantified by MS. More than 70 TAG species can be routinely resolved and identified using MS/MS with this method; results from the putative lipase mutant (**b**) suggest that oleic acid is preferentially incorporated into TAG during seed development in this line

Despite the rich positional information that can be obtained using lipid class profiling techniques, a major hurdle that needs addressing is obtaining quantitative information. This is especially important when concentrations must be calculated to determine regulatory points in biochemical pathways. FAME analysis using GC-FID is ideal for this purpose, because detector response is directly proportional to mass and is also linear over several orders of magnitude. Unfortunately, the same cannot be said of intact lipid profiling by infusion into a mass spectrometer, or by HPLC followed by MS or evaporative-light scattering detection (ELSD). Although both MS and ELSD detectors are now routinely used for detecting lipid species, response factors can vary tenfold, being sensitive to changes in head groups, chain length, and degree of unsaturation (Christie 1985; Holcapek et al. 2003; Schaefer et al. 2003). Thus, it is vital to have access to representative standards for quantitative purposes, and this may not be possible for plant extracts that have complex lipid profiles.

A second analytical constraint that needs addressing is that complete positional analysis of acyl chains on glycerolipids is currently tedious and low-throughput. Established techniques require lipid isolation by TLC, partial hydrolysis with Grignard reagents (and/or lipases), derivatization to chiral urethanes and separation by HPLC, further derivatization to FAMEs and analysis by GC-FID, and finally back-calculation to determine which fatty acid species were present at each position on the glycerol backbone (Christie et al. 1991). These procedures, although tedious, have been necessarily used in studies of lipid synthesis, especially in inferring the mechanisms responsible for determining the positional specificity of fatty acids in TAGs. However, such techniques are not appropriate for functional genomics studies or forward-screen designs, where analytical automation and high-throughput are prerequisites. Thus new techniques are required for the analysis of lipid regioisomers. One technique that shows promise is the partial or complete separation of triacylglycerol regioisomers by HPLC, with subsequent determination of acyl chain attachment specificity by tandem MS (Kusaka et al. 1996; Fauconnot et al. 2004). Tandem MS has also been demonstrated as a useful technique to determine both acyl chain double bond position and regioisomer structure in wheat flour galactolipids (Kim et al. 2001). For TAGs, these methods make use of characteristic diacylglycerol fragment intensity patterns to determine *sn*-2 and *sn*-1,3 attachment points. It is already possible to separate seed oil TAGs with different acyl chain components using this technique (Fig. 2), and also to identify some regioisomers. Chromatographic resolution may be further improved by the use of recently introduced stable silver ion HPLC columns, where Ag complexation with unsaturated double bonds increases selectivity in TAG isomer separation (Adlof and List 2004). However, tandem MS cannot yet differentiate between *sn*-1 and *sn*-3 acyl chains; this requires the classical Grignard/lipase positional analysis technique.

2.3 Intermediates and Trace-level Components in Lipid Metabolism

Most functional genomics studies on plant lipids have used measurements of storage or structural lipids to indirectly infer upstream modifications in metabolism. Ideally, the immediate substrates or products of the metabolic step under investigation should be measured, as this gives more direct evidence of the specific gene product involved in pathway modification. Control points in lipid synthesis can be elucidated by metabolic profiling, when combined with other biochemical or molecular data. For example, Perry et al. (1999) demonstrated that diacylglycerol accumulated relative to other intermediates in the Kennedy pathway during TAG biosynthesis in rape seed, and diacylglycerol acyltransferase (DGAT) activity remained low, suggesting that this acyltransferase may control flux into storage oils. Further work with arabidopsis identified the *TAG1* locus, encoding for a DGAT. Two allelic knockout mutants in this locus, designated *as11* and *abx45*, displayed the phenotype of reduced amounts of fatty acids associated with TAGs during seed maturation (Routaboul et al. 1999).

The list of metabolic intermediates involved in lipid metabolism is extensive, and includes: acyl-ACPs in the plastid; free fatty acids; acyl-CoAs in the cytosol, mitochondrion, and peroxisome; partially acylated glycerolipid intermediates in the cytosol (diacylglycerol, phospholipids); and fatty acids attached to intra- and intercellular binding or transfer proteins. These compounds are more difficult to extract or measure compared to structural or storage lipids because they are, by their nature as metabolic intermediates, present at relatively low concentrations. This excludes techniques that rely on FAME measurements unless rigorous extraction and preparative procedures can separate them from potentially contaminating structural and storage lipids. Specific metabolic derivatives of lipids, such as hydroxylated free fatty acids produced by lipoxygenase activity, have unique chemical properties and can therefore be successfully separated and measured by reverse or chiral-phase HPLC (Feussner et al. 1995).

One approach to locate the activity of an enzyme or gene product involved in lipid metabolism is to feed radio- or isotopically-labelled substrates in vivo or in vitro. Such studies are very useful for estimating carbon fluxes in lipid metabolism (Schwender et al. 2004), but are not an ideal way to screen for mutants.

Several techniques have been developed to determine directly the concentration of metabolic intermediates involved in lipid metabolism. Acyl-ACPs have been quantified using western blotting techniques to determine the regulatory role of acetyl-CoA carboxylase in spinach and pea chloroplasts (Post-Beittenmiller et al. 1992). Similarly, there have been attempts to monitor key proteins, such as plant acyl-CoA binding protein (ACBP) (Brown et al. 1998; Engeseth et al. 1998) and lipid transfer protein (LTP) (Sohal et al. 1999). These techniques belong to the field of proteomics rather than metabolomics, and will not be dealt with in detail here. Nevertheless, it would be useful to have some way to measure the pool sizes for binding proteins, as the acyl-protein complexes will have a direct impact on the metabolically active pool of acyl chains. There is evidence to suggest that most of the intracellular long-chain acyl-CoA pool in plants is not free, but bound to ACBP (Brown et al. 1998; Engeseth et al. 1998).

Binding proteins themselves are useful tools for determining concentrations of free (i. e. not bound to endogenous proteins) acyl-CoAs and non-esterified (free) fatty acids. One method has been developed using a recombinant bovine ACBP linked to a fluorescent reporter (Wadum et al. 2002). This method is useful in determining the amount of free vs ACBP-bound long-chain acyl CoAs in biological samples. Such a method would be particularly useful if it could be adapted to determine the intracellular location of different acyl-CoA pools, as there is circumstantial evidence that alterations in the cytosolic vs peroxisomal pool sizes have different effects on peroxisome size and biogenesis. In the peroxisomal ABC transporter mutant, *cts,* acyl-CoAs accumulate during lipid catabolism in germinating seeds, presumably in the cytosol (Footit et al. 2002), whereas in several β-oxidation mutants, such as the thiolase mutant, *kat2*, acyl-CoAs also accumulate, but presumably in the peroxisome (Germain et al. 2001). In the case of *cts*, peroxisomes have a normal morphology, whereas in *kat2* they are enlarged. The location of the accumulated pools could be identified by careful subcellular fractionation, but with the many caveats associated with these isolation techniques. A far better solution would be to label and visualise the acyl-CoA pools in vivo. An immunological method has been developed where long-chain acyl CoAs associated with membranes can be detected using ELISA techniques (Maneta-Peyret et al. 1998), and more recently, a related immunogold localization method has also been published (Diakou et al. 2002).

In isolated spinach and pea chloroplasts incubated in the light, Post-Beittenmiller et al. (1992) demonstrated that the short-chain acyl-CoA pool consisted almost entirely of acetyl-CoA with much lower amounts of free CoA and malonyl-CoA, suggesting that acetyl-CoA carboxylase is an important regulatory step in fatty acid synthesis. However, measurement of plant acyl-CoA pool components is a decidedly difficult task in plants, made difficult by their low micromolar concentrations and the presence of interfering compounds that interfere with UV detection of the CoA moiety at 260 nm. One method has been developed to measure the components of the acyl-CoA pool, specifically in plant extracts (Larson and Graham 2001; Larson et al. 2002). This technique uses fluorescent derivatization and HPLC separation to identify and quantify individual acyl-CoA species. The technique measures individual components of the entire acyl-CoA pool, releasing any protein-bound acyl-CoA during the extraction procedure. Using this method, acyl-CoA profiling has been extensively used to characterize lesions in lipid catabolism, where TAG mobilisation may be stopped or slowed during seed germination, and acyl-CoAs retained. In this case, acyl-CoA profiles do more than mirror fatty acid profiles; they also provide information on where exactly the lesion in lipid catabolism is occurring. For example, analysis of acyl-CoA profiles in germinating seeds of several arabidopsis acyl-CoA oxidase (ACOX) deletion mutants showed that lines lacking short- and medium-chain ACOX activities had distinctive acyl-CoA profiles compared to wild-type lines (Rylott et al. 2003). The short- and medium-chain ACOX deletion lines accumulated short- and medium-chain acyl-CoAs, which they could not efficiently metabolize in peroxisomal β-oxidation.

During lipid synthesis, acyl-CoA accumulation can also be used to evaluate the efficiency of the acyltransferase enzymes, such as DGAT, that use acyl-CoAs as substrates for TAG synthesis. An accumulation of a specific acyl-CoA may indicate poor incorporation into TAG. In *B. napus* transformed with a California Bay medium chain thioesterase, developing seed acyl-CoA profiles were used as evidence to conclude that poor accumulation of medium chain fatty acids in the transgenic seed oils indicated a poor acyltransferase specificity rather than insufficient medium chain fatty acid synthesis (Larson et al. 2002). Similarly, acyl-CoA analysis of plants metabolically engineered to produce polyunsaturated fatty acids in their TAGs has been used to determine if the required acyl-CoA intermediates were available for PUFA synthesis (Abbadi et al. 2004).

In summary, gene over-expression and reverse genetics approaches in arabidopsis have revealed that the acyl-CoA pool composition and size is altered by manipulating gene expression at steps 1 and 2 in Fig. 1. Acyl-CoAs are subject to catabolism by β-oxidation, primarily in the peroxisome in addition to feeding the synthesis of neutral storage lipid as well as glycerolipids. Thus, acyl-CoAs act as a pivotal point in both anabolic and catabolic lipid metabolism. There is good evidence for metabolic "bottlenecks" and branch points impacting on the biosynthesis of TAG (Abbadi et al. 2004; Napier et al. 2004). For example, the substrate dichotomy of microsomal glycerolipid desaturation and cytosolic acyl-CoA elongation represents a bottleneck in heterologous long-chain polyunsaturated fatty acid (LC-PUFA) biosynthesis, requiring enhanced acyl-exchange (Fig. 1, steps 4, 5, 6) to increase the flux between the two compartments (microsome, cytosol). In particular, the composition and quantity of the acyl-CoA pool may play an important role in the homeostatic regulation of lipid metabolism. This is based on the knowledge that acyl-CoAs not only are incorporated into phosphatidylcholine and TAG via the Kennedy pathway, but are also the primary inputs into ceramide and sphingolipid biosynthesis (Fig. 1, step 3). It is well known that acyl-CoAs have profound regulatory roles in mammalian lipid metabolism (Faergeman and Knudsen 1997), and these roles are only just starting to be investigated in plants.

Finally, it should be stressed that understanding how plant lipid synthesis is regulated will require a wider metabolite profiling effort that is not limited to measuring the biosynthetic intermediates between acetyl-CoA and acylglycerols. The source and flux of carbon to acetyl-CoA requires further investigation, and efforts using stable isotope analysis of sugars, amino acids, and lipid-derived fatty acids with GCMS analysis of isotopomers in oilseed feeding studies have begun to address this issue. For example, Schwender and Ohlrogge (2002) demonstrated that over 90% of the carbon for fatty acid synthesis originates from plastidic glycolysis, but carbon derived from amino acids is used to provide the cytosolic acetyl-CoA pool necessary from acyl-CoA elongation and the production of fatty acids with greater than 18 carbons. Understanding how these processes are regulated will assist in metabolic engineering attempts to increase the yield of very long-chain fatty acids in oilseeds. In addition, regulatory genes have been identified that alter seed oil yields, such as *WRINKLED1* (Focks and Benning 1998) and *APETALA2* (Jofuku et al. 2005; Ohto et al. 2005), both of which appear to control carbohydrate partitioning in developing seeds. How this control is exerted is uncertain; however, it is clear that broad or global-scale metabolite profiling will be necessary to understand more fully which biochemical pathways are being affected.

3 Future Developments

Data collection from multiple lipid profiling techniques will eventually be stored as lipidomic datasets, where information on thousands of lipids and their intermediates will be deposited. The large number of data-points generated will require annotation into lipid classes, identification of the acyl chains, and accurate quantification. Automation of data analysis using specifically designed computer algorithms is under development for lipidomic datasets (Hermansson et al. 2005), and will be necessary for application in functional genomic studies.

One emerging need for the utilisation of metabolomic datasets is their integration with other 'omic data, particularly mRNA transcript data (transcriptomics). For model plants such as arabidopsis, rice, and tomato, there is already a wealth of transcriptomic data available that can be analysed and modelled to identify genes that may be involved in particular areas of metabolism, by evidence of correlative expression with genes of known function. There have been efforts to catalogue genes involved in lipid metabolism (Beisson et al. 2003), and several attempts to model general transcriptomic data. It is a logical step to make the same correlations within metabolomic data sets, and also between metabolomic and transcriptomic datasets. This would be especially valuable, given that 40% of predicted proteins in transcriptomic analysis have an unknown function (Benning 2004), and of the 600 identified lipid related genes, two-thirds have not been functionally characterized (Beisson et al. 2003).

One potential use of lipidomic datasets will be in the discovery of new regulatory pathways involved in plant glycerolipid metabolism. This could be achieved by the integration of rapidly expanding transcriptomic datasets with new metabolomic data. In particular, new insights could be gained using 'omic data obtained from different developmental time-points or tissues that reflect specific biosynthetic or catabolic stages in lipid metabolism. For example, major changes in transcriptional activity accompany the progression of seed development from embryogenesis to maturation and the onset of storage reserve accumulation. A preliminary study of the Arabidopsis developmental atlas using data derived from the Affymetrix GeneChip ATH1 Genome Array has recently revealed that two main opposing expression trends operate during the progression into seed maturation (Schmid et al. 2005). Approximately 800 transcripts are induced and 1500 transcripts are repressed during this period. These cover a broad range of cellular processes. In a more focussed transcript profiling study Ruuska et al. (2002) generated data on 3500 genes previously identified from a developing seed EST collection. Among a number of interesting observations they found that genes related to biosynthesis of storage compounds showed several distinct temporal expression patterns and some of these were similar to the patterns of putative regulatory genes. This work also included transcript profiling of seed tissue from the *wrinkled 1* (*wri1*) mutant which is disrupted in a putative AP2/EREBP transcription factor which results in severe reduction of seed oil accumulation (Cernac and Benning 2004).

Marker-assisted plant breeding programs will also be assisted by new analytical technologies that enable rapid metabolite profiling of plant lipids. This approach, together with forward genetic screening methodologies, requires the rapid analysis of many thousands of individual samples in order to map to genes or loci of interest. In arabidopsis, variations in seed lipid composition have been identified by FAME profiling of hundreds of different ecotypes (O'Neill et al. 2003), and production of recombinant inbred lines have been used to identify quantitative trait loci (QTL) associated with increased oil yields (Hobbs et al. 2004). If intact glycerolipids and their intermediates could additionally be rapidly profiled during seed development, many more loci involved in TAG synthesis could be identified using QTL techniques and plant breeding could also be more finely tailored to achieve optimum seed compositions in oilseed crops.

Acknowledgements. We thank Dr Peter Eastmond for supplying seed material for triacylglycerol analysis.

References

Domergue F, Bauer J, Napier JA, Welti R, Zähringer U, Cirpus P, Heinz E (2004) Biosynthesis of very-long-chain polyunsaturated fatty acids in transgenic oilseeds: constraints on their accumulation. Plant Cell 16:2734–2748

Adlof R, List G (2004) Analysis of triglyceride isomers by silver-ion high-performance liquid chromatography. Effect of column temperature on retention times. J Chromatogr A 1046:109–113

Arondel V, Lemieux B, Hwang S, Gibson S, Goodman HR, Somerville CR (1992) Map-based cloning of a gene controlling omega-3 fatty acid desaturation in Arabidopsis. Science 258:1353–1355

Baud S, Boutin J, Miquel M, Lepiniec L, Rochat C (2002) An integrated overview of seed development in *Arabidopsis thaliana* ecotype WS. Plant Physiol Biochem 40:151–160

Beisson F, Koo AJK Russka S, Schwender J, Pollard M, Thelen JJ, Paddock T, Salas JJ, Savage L, Milcamps A, Mhaske VB, Cho Y, Ohlrogge JB (2003) Arabidopsis genes involved in acyl lipid metabolism A 2003 consensus of candidates a study of the distribution of expressed sequence tags in organs and a web-based database. Plant Physiol 132:681–697

Benning C (2004) Genetic mutant screening by direct metabolite analysis. Anal Biochem 332:1–9

Broun P, Gettner S, Somerville C (1999) Genetic engineering of plant lipids. Annu Rev Nutr 19:197–216

Brown AP, Johnson P, Rawsthorne S, Hills MJ (1998) Expression and properties of acyl-CoA binding protein from *Brassica napus*. Plant Physiol Biochem 9:629–635

Browse J, McCourt C, Somerville C (1985) A mutant of Arabidopsis lacking a chloroplast-specific lipid. Science 227:763–765

Browse J, McCourt PJ, Somerville C (1986) Fatty acid composition of leaf lipids determined after combined digestion and fatty acid methyl ester formation from fresh tissues. Anal Biochem 152:141–145

Buchgraber M, Ulberth M, Emons H, Anklam E (2004) Triacylglycerol profiling by using chromatographic techniques. Eur J Lipid Sci Technol 106:621–648

Calvano CD, Palmisano F, Zambonin CG (2005) Laser desorption/ionization time-of-flight mass spectrometry of triacylglycerols in oils. Rapid Comm Mass Spectrom 19:1315–1320

Carole TM, Pellegrino J, Paster MD (2004) Opportunities in the industrial biobased products industry. Appl Biochem Biotechnol 115:871–886

Cernac A, Benning C (2004) WRINKLED1 encodes an AP2/EREB domain protein involved in the control of storage compound biosynthesis in Arabidopsis. Plant J 40:575–585

Chia TYP, Pike MJ, Rawsthorne S (2005) Storage oil breakdown during embryo development of *Brassica napus* (L.). J Exp Bot 56:1285–1296

Christensen MS, Mullertz A, Hoy CE (1995) Absorption of triglycerides with defined or random structure by rats with biliary and pancreatic diversion. Lipids 30:521–526

Christie WW (1982) A simple procedure for rapid transmethylation of glycerolipids and cholesteryl esters. J Lipid Res 23:1072–1075

Christie WW (1985) Rapid separation and quantification of lipid classes by high performance liquid chromatography and mass (light-scattering) detection. J Lipid Res 26:507–512

Christie WW, Nikolova-Damyanova B, Laaska P Herslof B (1991) Stereospecific analysis of triacyl-sn-glycerols via resolution of diastereomeric diacylglycerol derivatives by HPLC on silica. J Am Oil Chem Soc 68:695–701

Diakou P, Fedou S, Carde JP, Cassagne C, Moreau P, Maneta-Peyret L (2002) Immunolocalization of long-chain acyl-CoAs in plant cells. Biochim Biophys Acta 1583:85–90

Eastmond PJ, Graham IA (2001) Re-examining the role of the glyoxylate cycle in oilseeds. Trends Plant Sci 6:72–77

Engeseth NJ, Pakovsky RS, Newman T, Ohlrogge JB (1998) Characterization of an acyl-CoA-binding protein from *Arabidopsis thaliana*. Arch Biochem Biophys 331:55–62

Faergeman NJ, Knudsen J (1997) Role of long-chain fatty acyl-CoA esters in the regulation of metabolism and in cell signalling. Biochem J 323:1–12

Fatland BL, Nikolau BJ, Wurtele ES (2005) Reverse genetic characterization of cytosolic acetyl-CoA generation by ATP-citrate lyase in Arabidopsis. Plant Cell 17:182–203

Fauconnot L, Hau J, Aeschlimann J, Fay L, Dionisi F (2004) Quantitative analysis of triacylglycerol regioisomers in fats and oils using reverse-phase high-performance liquid chromatography and atmospheric pressure chemical ionization mass spectrometry. Rapid Comm Mass Spectrom 18:218–224

Feussner I, Wasternack C, Kindl H, Kuhn H (1995) Lipoxygenase-catalyzed oxygenation of storage lipids is implicated in lipid mobilization during germination. Proc Natl Acad Sci USA 92:11849–11853

Focks N, Benning C (1998) *Wrinkled1*: a novel low-seed-oil mutant of Arabidopsis with a deficiency I the seed-specific regulation of carbohydrate metabolism. Plant Physiol 118:91–101

Footitt S, Slocombe SP, Larner V, Kurup S, Wu Y, Larson T, Graham I, Baker A, Holdsworth M (2002) Control of germination and lipid mobilization by COMATOSE the Arabidopsis homologue of human ALDP. EMBO J 21:2912–2922

Germain V, Rylott EL, Larson TR, Sherson SM, Bechtold N, Carde JP, Bryce JH, Graham IA, Smith SM (2001) Requirement for 3-ketoacyl-CoA thiolase-2 in peroxisome development fatty acid beta-oxidation and breakdown of triacylglycerol in lipid bodies of Arabidopsis seedlings. Plant J 28:1–12

Han X, Yang J, Cheng H, Hongping Y, Gross RW (2004) Toward fingerprinting cellular lipidomes directly from biological samples by two-dimensional electrospray ionization mass spectrometry. Anal Biochem 330:317–331

Hermansson M, Uphoff A, Kakela R, Somerharju P (2005) Automated quantitative analysis of complex lipidomes by liquid chromatography/mass spectrometry. Anal Chem 77:2166–2175

Hobbs DH, Flintham JE, Hills MJ (2004) Genetic control of storage oil synthesis in seeds of Arabidopsis. Plant Physiol 136:3341–3349

Holcapek M, Jandera P, Zderadika P, Hrubá L (2003) Characterization of triacylglycerol and diacylglycerol composition of plant oils using high-performance liquid chromatography – atmospheric pressure chemical ionization mass spectrometry. J Chromatogr A 1010:195–215

Hong H, Datla N, Reed DW, Covello PS, MacKenzie SL, Qiu X (2002) High-level production of γ-linolenic acid in *Brassica juncea* using a Δ-6 desaturase from *Pythium irregulare*. Plant Physiol 129:354–362

Houjou T, Yamatani K, Imagawa M, Shimizu T, Taguchi R (2005) A shotgun tandem mass spectrometric analysis of phospholipids with normal-phase and/or reverse-phase liquid chromatography/electrospray ionization mass spectrometry. Rapid Comm Mass Spectrom 16:654–666

Hulbert AJ, Turner N, Storlien LH, Else PL (2005) Dietary fats and membrane function: implications for metabolism and disease. Biol Rev Camb Philos Soc 80:155–169

Jensen MM, Christensen MS, Hoy CE (1994) Intestinal absorption of octanoic decanoic and linoleic acids: effect of triglyceride structure. Ann Nutr Metab 38:104–116

Jofuku KD, Omidyar PK, Gee Z, Okamuro JK (2005) Control of seed mass and seed yield by the floral homeotic gene *APETALA2*. Proc Natl Acad Sci USA 102:3117–3122

Jones JJ, Stump MJ, Fleming RC, Lay JO, Wilkins CL (2004) Strategies and data analysis techniques for lipid and phospholipid chemistry elucidation by intact cell MALDI-FTMS. J Am Soc Mass Spectrom 15:1665–1674

Kachroo A, Lapchyk L, Fukushige H, Hildebrand D, Klessig D, Kachroo P (2003) Plastidial fatty acid signaling modulates salicylic acid- and jasmonic acid-mediated defense pathways in the Arabidopsis *ssi2* mutant. Plant Cell 15:2952–2965

Kachroo A, Venugopal SC, Lapchyk L, Falcone D, Hildebrand D, Kachroo P (2004) Oleic acid levels regulated by glycerolipid metabolism modulate defense gene expression in Arabidopsis. Proc Natl Acad Sci USA 101:5152–5157

Kachroo P, Shanklin J, Shah J, Whittle EJ, Klessig DF (2001) A fatty acid desaturase modulates the activation of defense signaling pathways in plants. Proc Natl Acad Sci USA 98:9448–9453

Kim YH, Gil JH, Hong J, Yoo JS (2001) Tandem mass spectrometric analysis of fatty acyl groups of galactolipid molecular species from wheat flour. Microchem J 68:143–155

Kusaka T, Ishihara S, Sakaida M, Mifune A, Nakano Y, Tsuda K, Ikeda M, Nakano H (1996) Composition analysis of normal plant triacylglycerols and hydroperoxidized *rac*-1-stearoyl-2-oleoyl-3-linoleoyl-*sn*-glycerols by liquid chromatography-atmospheric pressure chemical ionization mass spectrometry. J Chromatogr A 730:1–7

Larson TR, Graham IA (2001) Technical advance: a novel technique for the sensitive quantification of acyl-CoA esters from plant tissues. Plant J 25:115–125

Larson TR, Edgell T, Byrne J, Dehesh K, Graham IA (2002) Acyl-CoA profiles of transgenic plants that accumulate medium-chain fatty acids indicate inefficient storage lipid synthesis in developing oilseeds. Plant J 32:519–527

Li C, Schilmiller AL, Liu G, Lee GI, Jayanty S, Sageman C, Vrebalov J, Giovannoni JJ, Yagi K, Kobayashi Y, Howe GA (2005) Role of β-oxidation in jasmonate biosynthesis and systemic wound signaling in tomato. Plant Cell 17:971–986
Maneta-Peyret L, Sturbois-Balcerzak B, Cassagne C, Moreau P (1998) Antibodies to long-chain acyl-CoAs. A new tool for lipid biochemistry. Biochim Biophys Acta 1389:50–56
Mondello L, Tranchida PQ, Costa R, Casilli A, Dudo P, Cotroneo A, Dugo G (2003) Fast GC for the analysis of fats and oils. J Sep Sci 26:1467–1473
Nandi A, Welti R, Shah J (2004) The *Arabidopsis thaliana* dihydroxyacetone phosphate reductase gene *suppressor of fatty acid desaturase deficiency1* is required for glycerolipid metabolism and for the activation of systemic acquired resistance. Plant Cell 16:465–477
Napier JA, Sayanova O, Baixiu Q, Lazarus CM (2004) Progress toward the production of long-Chain polyunsaturated fatty acids in transgenic plants. Lipids 39:1067–1075
Ohlrogge JB, Browse J, Somerville CR (1991) The genetics of plant lipids. Biochim Biophys Acta 1082:1–26
Ohto M, Fischer RL, Goldberg RB, Nakamura K, Harada JJ (2005) Control of seed mass by *APETALA2*. Proc Natl Acad Sci USA 102:3123–3128
O'Neill CM, Gill S, Hobbs D, Morgan C, Bancroft I (2003) Natural variation for seed oil composition in *Arabidopsis thaliana*. Phytochemistry 64:1077–1090
Perry HJ, Bligny R, Gout E, Harwood JH (1999) Changes in Kennedy pathway intermediates associated with increased triacylglycerol synthesis in oil-seed rape. Phytochemistry 52:799–804
Pollard M, Ohlrogge J (1999) Testing models of fatty acid transfer and lipid synthesis in spinach leaf using in vivo oxygen-18 labeling. Plant Physiol 121:1217–1226
Post-Beittenmiller D, Roughan G, Ohlrogge JB (1992) Regulation of plant fatty acid biosynthesis. Analysis of acyl-coenzyme A and acyl-acyl carrier protein substrate pools in spinach and pea chloroplasts. Plant Physiol 100:923–930
Routaboul J, Benning C, Bechtold N, Caboche M, Lepiniec L (1999) The *TAG1* locus of Arabidopsis encodes for a diacylglycerol acyltransferase. Plant Physiol Biochem 37:831–840
Rutar V (1989) Magic angle sample spinning NMR spectroscopy of liquids as a nondestructive method for studies of plant seeds. J Agric Food Chem 37:67–70
Ruuska SA, Girke T, Benning C, Ohlrogge JB (2002) Contrapuntal networks of gene expression during Arabidopsis seed filling. Plant Cell 14:1191–1206
Rylott EL, Rogers CA, Gilday AD, Edgell T, Larson TR, Graham IA (2003) Arabidopsis mutants in short- and medium-chain acyl-CoA oxidase activities accumulate acyl-CoAs and reveal that fatty acid β-oxidation is essential for embryo development. J Biol Chem 278:21370–21377
Schaefer A, Kuchler T, Simat TJ, Steinhart H (2003) Migration of lubricants from food packagings Screening for lipid classes and quantitative estimation using normal-phase liquid chromatographic separation with evaporative light scattering detection. J Chromatogr A 10178:107–116
Schmid M, Davison TS, Henz SR, Pape UJ, Demar M, Vingron M, Scholkopf B, Weigel D, Lohmann JU (2005) A gene expression map of *Arabidopsis thaliana* development. Nat Genet 37:501–506
Schneider K, Kienow L, Schmelzer E, Colby T, Bartsch M, Miersch O, Wasternack C, Kombrink E, Stuible HP (2005) A new type of peroxisomal acyl-coenzyme A synthetase from *Arabidopsis thaliana* has the catalytic capacity to activate biosynthetic precursors of jasmonic acid. J Biol Chem 280:13962–13972
Schwender J, Ohlrogge JB (2002) Probing in vivo metabolism by stable isotope labeling of storage lipids and proteins in developing *Brassica napus* embryos. Plant Physiol 130:347–361
Schwender J, Ohlrogge J, Shachar-Hill Y (2004) Understanding flux in plant metabolic networks. Curr Opin Plant Biol 7:309–317
Shah J, Kachroo P, Nandi A, Klessig DF (2001) A recessive mutation in the Arabidopsis *SSI2* gene confers SA- and *NPR1*-iindependent expression of *PR* genes and resistance against bacterial and oomycete pathogens. Plant J 25:563–574
Sohal A, Love A, Cecchini E, Covey S, Jenkins G, Milner J (1999) Cauliflower mosaic virus infection stimulates lipid transfer protein gene expression in Arabidopsis. J Exp Bot 50:1727–1733

Sperling P, Zahringer U, Heinz E (1998) A sphingolipid desaturase from higher plants Identification of a new cytochrome b5 fusion protein. J Biol Chem 273:28590–28596
Thelen JJ, Ohlrogge JB (2002) Metabolic engineering of fatty acid biosynthesis in plants. Metabolic Engineering 4:12–21
Tonon T, Harvey D, Qing R, Li Y, Larson TR, Graham IA (2004) Identification of a fatty acid Delta11-desaturase from the microalga *Thalassiosira pseudonana*. FEBS Lett 563:28–34
Van de Loo FJ, Fox BG, Somerville C (1993) Unusual fatty acids. In: Moore T (ed) Plant lipids. CRC Press, Boca Raton, Florida, pp 91–126
Wadum MC, Villadsen JK, Feddersen S, Moller RS, Neergaard TB, Kragelund BB, Hojrup P, Faergeman NJ, Knudsen J (2002) Fluorescently labelled bovine acyl-CoA-binding protein acting as an acyl-CoA sensor: interaction with CoA and acyl-CoA esters and its use in measuring free acyl-CoA esters and non-esterified fatty acids. Biochem J 365:165–172
Wallis JG, Browse J (2002) Mutants of Arabidopsis reveal many roles for leaf lipids. Prog Lipid Res 41:254–278
Welti R, Wang X (2004) Lipid species profiling: a high-throughput approach to identify lipid compositional changes and determine the function of genes involved in lipid metabolism and signaling. Curr Opin Plant Biol 7:337–344
Welti R, Li W, Li M, Sang Y, Biesiada H, Zhou H, Rajashekar CB, Williams TD, Wang X (2002) Profiling membrane lipids in plant stress responses. Role of phospholipase Dα in freezing-induced lipid changes in Arabidopsis. J Biol Chem 277:31994–32002

III.4 Metabolic Profiling and Quantification of Carotenoids and Related Isoprenoids in Crop Plants

P.D. Fraser and P.M. Bramley[1]

1 Introduction

Metabolomics has been defined as "a comprehensive analysis in which all the metabolites of an organism are defined and quantified" (Fiehn 2002). It is estimated that within the plant kingdom 200,000 different metabolites exist (Pickersky and Gang 2000), while in *Arabidopsis* leaf tissue alone, 5000 different metabolites are estimated of which about 10% have been annotated using current technologies. Numerous analytical techniques have now been applied to metabolomic analysis including NMR, MS and hyphenated-MS, such as LC-MS/MS, GC-MS and CE-MS (reviewed by Halket et al. 2005). Although chemical derivatisation can be exploited to alter ionization and chromatographic properties of compounds it is typically the chemical nature of the metabolites that determines extraction, separation, and detection methodologies. Carotenoids are important metabolites involved in many biological processes, are essential components of human diet (Fraser and Bramley 2004), and are present in most fruits and vegetables (O'Neill et al. 2001). Over 600 have been identified (Britton et al. 2004). In this chapter we will describe targeted metabolite profiling of carotenoids and related isoprenoids, as well as application of the technology and its integration within metabolomic studies.

1.1 Abundance and Biological Functions of Carotenoids

Carotenoids represent one of the largest groups of natural pigments found in nature and their presence is widespread throughout the plant kingdom (Harborne 1991). All photosynthetic tissues must contain carotenoids in order to function and most of the yellow, orange and red colours of fruits and flowers are due to the presence of carotenoids. Functionally, carotenoids perform a variety of roles and are involved in numerous biological processes. For example, they are essential for photosynthesis as they act as ancillary light harvesting pigments. They are also potent antioxidants acting as protectants against environmental and metabolically generated free radicals (Miller et al. 1996). Carotenoids also provide membrane stability and act as precursors

[1] School of Biological Sciences, Royal Holloway, University of London, Egham, Surrey, TW20 0EX, UK, e-mail: p.bramley@rhul.ac.uk

Biotechnology in Agriculture and Forestry, Vol. 57
Plant Metabolomics (ed. by K. Saito, R.A. Dixon, and L. Willmitzer)

for phytohormones such as abscisic acid, as well as recently elucidated apocarotenoid signalling molecules (Booker et al. 2004).

1.2 Health Aspects of Carotenoids

Carotenoids are essential dietary components for humans. β-Carotene is the most potent precursor of vitamin A, while other carotenoids alleviate age-related diseases such as macular degeneration (zeaxanthin and lutein; Seddon et al. 1994) and prostate cancer (lycopene; Giovannucci 1999). Although carotenoids have been shown to have positive effects on human health, relatively large amounts need to be consumed in order to gain noticeable benefits. Typically, nutritional advice is to eat large quantities of fruits and vegetables ('five-a-day') that contain health-promoting phytochemicals such as carotenoids. The perceived health benefits of carotenoids have been reviewed by Fraser and Bramley (2004).

1.3 Biotechnological Importance of Carotenoids

One strategy that has been employed to increase levels of health promoting carotenoids in fruits and vegetables for human and animal consumption is genetic modification. Perhaps the best-publicised example is that of Golden rice, which contains β-carotene (provitamin A) as a result of transformation with carotenoid biosynthetic genes (Ye et al. 2000). Other workers have produced tomato fruit with elevated levels of lycopene and β-carotene (Römer et al. 2000; Rosati et al. 2000; Fraser et al. 2002). However, these varieties have yet to be commercialised, largely due to the resistance of consumers and public bodies to the introduction of GM crops into Europe. One concern of consumers and certain public bodies is that the metabolic changes to such crops are not only unpredictable and unknown, but cannot be detected by the use of the substantial equivalence approach, used by regulatory bodies, prior to commercialisation. The desire of the consumer for health-based traits in crop plants, but not via a GM approach, has led to a resurgence of conventional breeding programs to screen for high carotenoid phenotypes. Consequently, germplasm resources such as the *Lycopersicon pennellii* introgression lines (see Sect. 3.2) and ecotype collections have been generated (Gur et al. 2004).

Besides their health benefits, carotenoids are commercially important natural products used in the food, feed, pharmaceutical and cosmetic industries. The world market for carotenoids in 1999 was about US$ 800 million and projections estimate this will increase to US$ 1 billion in 2005. Although chemical synthesis is the most often used method to produce carotenoids industrially, production from plants can offer a more cost effective option (Ausich 1997).

1.4 Chemical Properties of Carotenoids

Carotenoids are predominately C_{40} hydrocarbon molecules possessing a polyene chain. A series of conjugated double bonds of varying length is a fundamental feature of the carotenoid molecule (Fig. 1). Cyclisation of acyclic carotenes results in the introduction of β-ionone and/or ε-ionone rings at the ends of the carotenoid molecule. Introduction of hydroxyl and/or oxygen moieties into ring structures results in xanthophylls. It is these structural features that have a direct bearing on the metabolite profiling methodologies employed. For example, the C_{40} hydrocarbon polyene chain results in an extremely hydrophobic molecule that is not soluble in aqueous or polar solvents. The series of conjugated double bonds give rise to a chromophore and the spectral properties of the molecule. The fine spectra can be modified by the presence of ring structures and oxygen moieties. Collectively, these structural properties result in characteristic UV and visible spectra for each carotenoid molecule. In addition, the number of conjugated double bonds, end groups and oxygen moieties affect polarity and thus chromatographic properties. Finally carotenoids are sensitive to light, oxygen, acid and in some cases alkali; thus, careful consideration must be made when designing and implementing metabolomic approaches to carotenoids. A comprehensive list of carotenoids and their structures can be found in Britton et al. (2004).

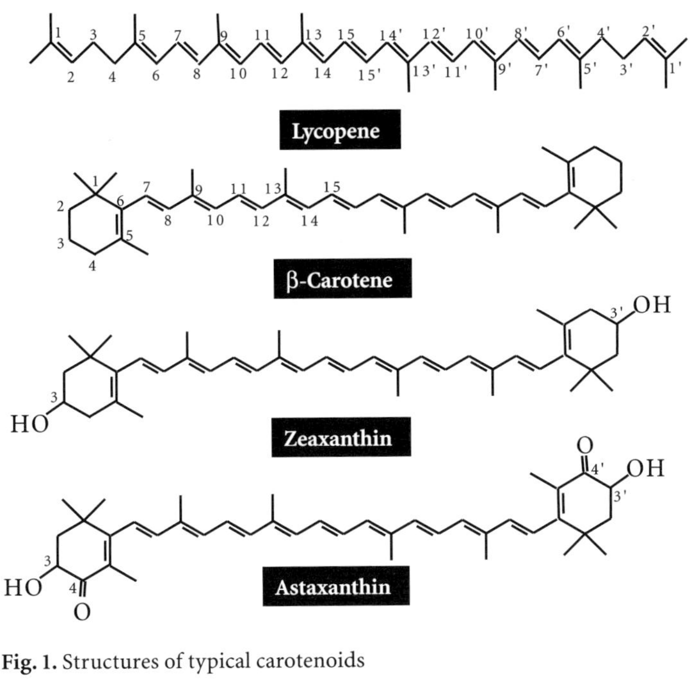

Fig. 1. Structures of typical carotenoids

1.5 Biosynthesis of Carotenoids and Related Isoprenoids

Carotenoids are isoprenoids and therefore biosynthetically related to other classes of isoprenoid compounds via the common C_5 precursor isopentenyl diphosphate (IPP). A schematic overview of the biosynthetic pathway (Fig. 2) reveals the large number of common intermediates, the varied nature of the end products and also the highly branched nature of the pathway itself. In addition, the branches of the pathway are located in different subcellular compartments. All the carotenogenic reactions occur within the plastid, in common with those forming plastoquinone, tocopherols and gibberellins, but the formation of sterols occurs in the cytoplasm. One should note, however, that these pathways are not completely separate from each other, as transport of at least IPP occurs across the plastid envelope (Bick and Lange 2003). Taken together, these complexities of metabolism and subcellular location suggest a sophisticated series of regulatory interactions with the coordinated flux of isoprenoid units into each branch of the pathway. Thus, the likelihood of metabolic cross talk is high, and perturbations of the flux by genetic modification, mutations or seasonal effects may result in unintended effects on individual isoprenoids. Further details

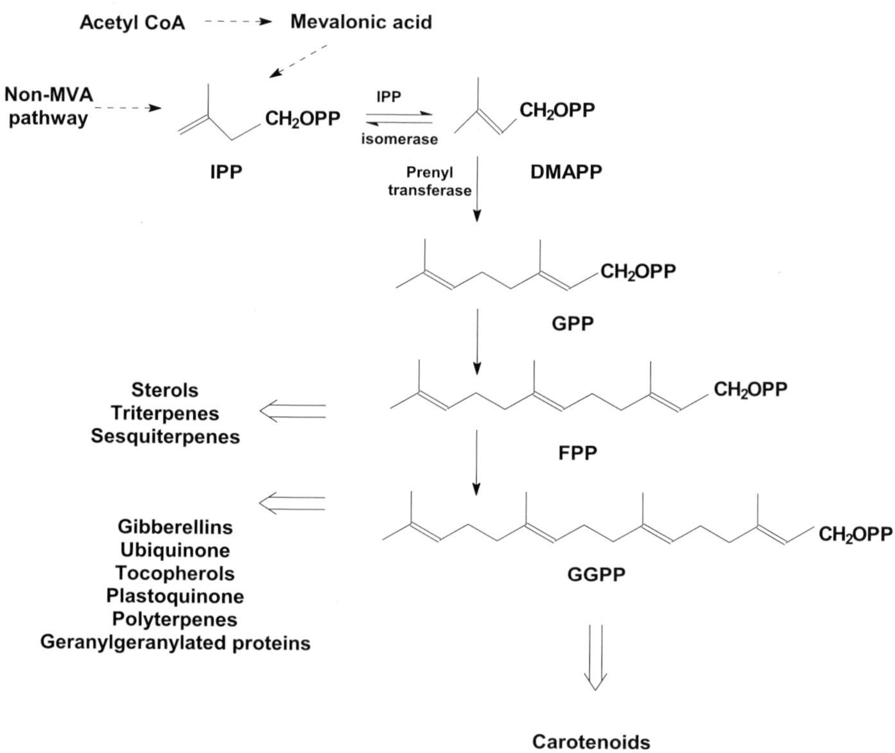

Fig. 2. Overview of isoprenoid biosynthesis

of the biosynthetic pathways can be found in several reviews (Hirschberg 2001; Rodriguez-Concepcion and Boronat 2002; Fraser and Bramley 2004), whilst aspects of the regulation of isoprenoid biosynthesis are described elsewhere (Bramley 2002).

2 Analytical Methodologies Employed in the Analysis of Carotenoids

2.1 Harvesting and Sample Preparation

It is essential to standardise the analytical system adopted, so that phenotypic variation between samples can be determined accurately. This necessitates minimising variations in environmental factors by standardising growth regimes. In the case of crop plants, metabolite changes due to seasonal variation must be considered. Growth plots should be randomised and the adequate number of the correct controls interspersed in order to minimise intra and inter-plant variability. If genetically modified lines are grown, the appropriate controls (e. g. an azygous or empty vector line) must be included.

Another factor that must be considered is the developmental stage at which the tissue is harvested. For example, in the case of tomato fruit, seven days post breaker is commonly used for carotenoid analysis and can be reproducibly defined by "tagging" at the breaker stage, i. e. the time at which green fruit begin to change colour. The paradigm growth stages are also an ideal standardization when using *Arabidopsis* tissues (Boyes et al. 2001).

Once the plant tissue has been harvested (ideally at the same daily time point) the optimal storage and preparation conditions need to be determined. Typically with plant tissue destined for carotenoid analysis, freezing in liquid nitrogen or at −70 °C is adequate and prolonged frozen storage (several years) does not affect the carotenoids present. Repeated freeze thawing, however, does reduce the levels of carotenoids and should be avoided. Lyophilisation is an effective method of preparation that facilitates ease of storage and extraction. It is essential that lyophilisation is complete, as incomplete freeze-drying results in loss of carotenoids.

As plant material typically comprises several tissue types of differing textures it is essential to employ homogenisation procedures that yield homogeneous material and eliminate intra-sample variation. In the case of tomato, where the skin tissue is not amenable to many homogenisation procedures, we recommend the use of a freezer mill, which provides vigorous homogenisation at low temperatures and can be standardised in terms of power and time.

2.2 Extraction Procedures

As carotenoids and most isoprenoids are non-polar molecules they require organic solvents for extraction. The literature describes numerous carotenoid extraction procedures with a variety of organic solvents. However, no one solvent is optimal for all carotenoids and differential extraction can occur resulting in misleading determinations. It is therefore important to know which carotenoids are present and optimise the extraction accordingly. Figure 3 illustrates the variability that can arise from using different extraction solvents. It can be seen that, with tomato fruit, methanol is very poor at extracting the very hydrophobic carotenoids such as lycopene, whilst hexane and chloroform show differential extraction of β-carotene and lycopene. Typically a greater recovery of carotenoids is achieved with freeze-dried material rather than fresh tissue, and freeze-dried material is also easier to handle practically. Recoveries must be assessed and optimised, and re-extraction carried out to ensure complete recovery of the metabolites, especially if the compounds are at high concentrations that may saturate the solvent during the first extraction procedure. Unfortunately, a wide range authentic carotenoids are not available commercially for use as internal standards for monitoring extraction efficiency. Therefore, carotenoids that are exclusive to some microorganisms such as canthaxanthin, astaxanthin and echinenone have been used to monitor extraction efficiency of higher plant tissues.

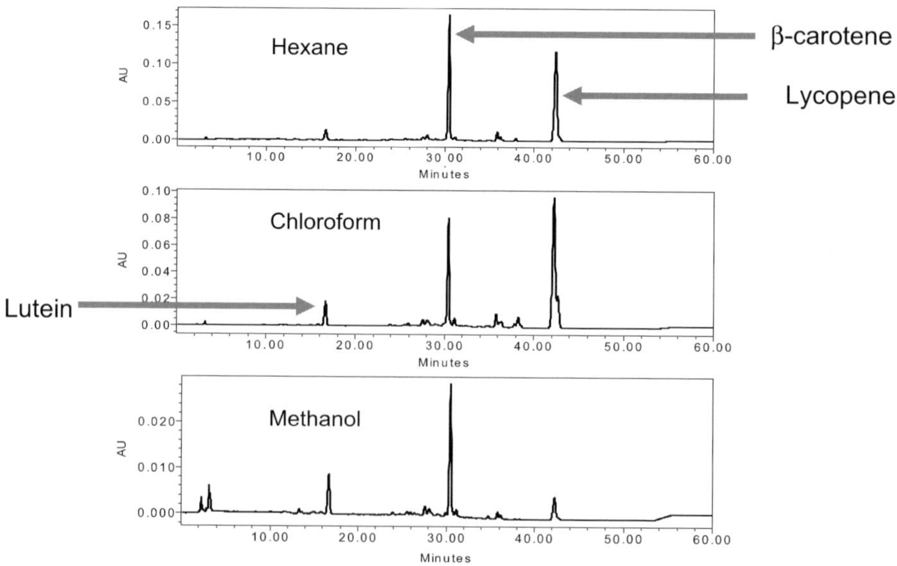

Fig. 3. HPLC chromatograms of freeze dried tomato fruit extracted with hexane, methanol or chloroform showing differential extraction of individual carotenoids

2.3 Separation Procedures

Modern liquid chromatographic methods are highly sophisticated, allowing precise maintenance of low flow rates, automation and data handling. More traditional approaches, however, are often useful in preliminary separations of carotenoids. For example, thin layer chromatography (TLC) is useful for the initial profiling of carotenoids in crude extracts. The hydrophobic nature of carotenoids means they are suited to most silica or alumina-based stationary phases, and mobile phases comprised of hexane or light petroleum (Davies 1976; Britton 1991). Numerous samples can be loaded onto the thin layers and run simultaneously, thus providing a high throughput screening protocol (e. g., Ralley et al. 2004). Since most carotenoids are coloured, visual detection is straightforward. For colourless carotenoids such as phytoene, staining the chromatogram with iodine can be used (Davies 1976).

Carotenoids are not compatible with separation by gas chromatography unless they have been hydrogenated (Taylor and Ikawa 1971). Thus, HPLC has become the method of choice for separation, identification and quantification. Reverse-phase C_{18} columns are the most popular matrix for separating carotenoids. Typically, a methanol or acetonitrile based mobile phase will be used with modifiers such as water or ethyl acetate (reviewed Bramley 1992; Craft 1992). Normal phase silica columns are another alternative. Mobile phases compatible with normal phase separation are usually hexane based, with ethyl acetate used as the modifier (reviewed Bramley 1992). Traditionally, reverse or normal phase separation systems have been used to analyse a specific carotenoid or class of carotenoids. For example, normal phase columns have been used principally applied for the separation of xanthophylls, while reverse phase columns have been used extensively in the analysis of acyclic and cyclic carotenes. More recently, C_{30} reverse-phase columns have been utilised to profile a range of carotenoids with diverse polarities (Fraser et al. 2000) as well as numerous other isoprenoids such as tocopherols, ubiquinones and plastoquinone. The C_{30} reverse-phase matrix is also ideal for the separation of geometric isomers (Breitenbach et al. 2001).

2.4 Identification and Quantification

The number of conjugated double bonds, the nature of the cyclic end groups and oxygen moieties present in the carotenoid molecule (Fig. 1) give rise to characteristic UV/VIS spectra. The ability of in-line photodiode array detectors (PDA) to record spectra simultaneously across the whole spectrum makes them ideal for carotenoid identification. Co-chromatography and comparison of spectral characteristics with authentic standards enable conclusive identification (Fig. 4). Carotenoid standards can be either purchased commercially or purified from known biological sources, and compared to their properties documented in the literature (e. g. Britton 1995). The weak and

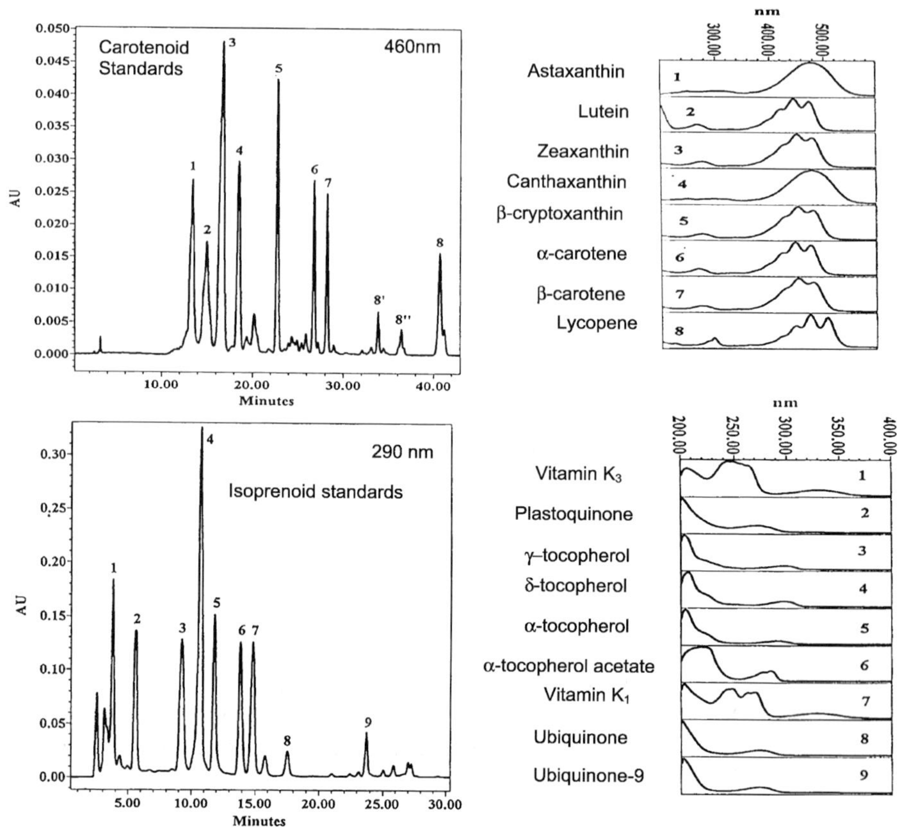

Fig. 4. Typical HPLC separations of isoprenoids, with corresponding spectra captured in line with a photodiode array detector

differentialtion of carotenoids makes detection by mass spectrometry (MS) poor. However, where the spectra of two or more carotenoids are very similar (e. g. keto group-containing carotenoids) MS can be useful to distinguish functional groups. In these cases MS detection by APCI-MS (van Breemen 1997) or MALDI-Tof/MS (Fraser and Bramley, unpublished) has been used effectively.

Quantitation of carotenoids separated by HPLC can be achieved by the construction of dose-response curves prepared from authentic standards. For accurate determination, it is advantageous to prepare a curve for each carotenoid and record the chromatographic area at the λ_{max} for each carotenoid. If an authentic standard is unavailable, a carotenoid with similar chromatographic properties and λ_{max} can be used (e. g. β-carotene for α- or δ-carotene).

3 Examples of Carotenoid/isoprenoid Profiling

In this section we describe the methodologies used to separate, identify and evaluate the transgenic lines synthesizing novel carotenoids as an example of metabolite profiling for novel metabolites, as well as the analysis of introgression lines of tomato.

3.1 Determination of High-value Carotenoids in Transgenic Plants

Astaxanthin, canthaxanthin and zeaxanthin are high-value carotenoids used industrially as colourants and feed supplements. Higher plants (with the exception of *Adonis* flowers) do not produce astaxanthin or canthaxanthin but possess the precursors zeaxanthin and β-carotene. In order to generate a readily available source of ketocarotenoids the biosynthetic genes responsible for ketocarotenoid formation in the bacterium *Paracoccus* (Misawa et al. 1995) have been expressed simultaneously in *Nicotiana tabacum* and *Lycopersicon esculentum* (Ralley et al. 2004). This gives raise to the potential production of at least seven unique carotenoids not endogenously present in the plant.

No visible differences were apparent between control and transgenic tobacco plants containing the ketocarotenoid biosynthetic genes. In order to rapidly screen the transgenic population for novel carotenoids a TLC system was devised. Crude chloroform extracts were prepared from freeze-dried material. The sensitivity of ketocarotenoids to alkali prevents removal of chlorophylls by saponification. In order to make accurate comparison between lines, the amount of tissue, extraction volumes and loading onto the TLC plates were standardised. The most effective TLC system comprised a silica stationary phase and mobile phase of ethyl acetate/hexane (40:60 v/v). Control extracts were run concurrently with transgenic samples. Visual comparisons between control and transgenic lines identified several unique bands with R_f values similar to those of known products of the astaxanthin biosynthetic pathway. This TLC approach enabled the rapid detection of novel carotenoids in crude extracts containing abundant endogenous pigments. In order to identify these products conclusively, co-chromatography was performed with authentic samples by HPLC-PDA (Ralley et al. 2004).

The nectary flower tissue of these transgenic lines exhibited a more intense and reddish colour in comparison to controls. Crude extracts were prepared and analysed on a C_{30} reverse-phase column using an unbiased system with a broad carotenoid polarity range. The chromatogram from the transgenic line extract contained 30 peaks with carotenoid spectra (Fig. 5). Six of the peaks showed quantitative increases compared to the control while four peaks represented novel components. The spectra of these novel peaks identified them as astaxanthin, 4-ketozeaxanthin, 3′-OH-echinenone and canthaxanthin and allowed quantification (Fig. 5).

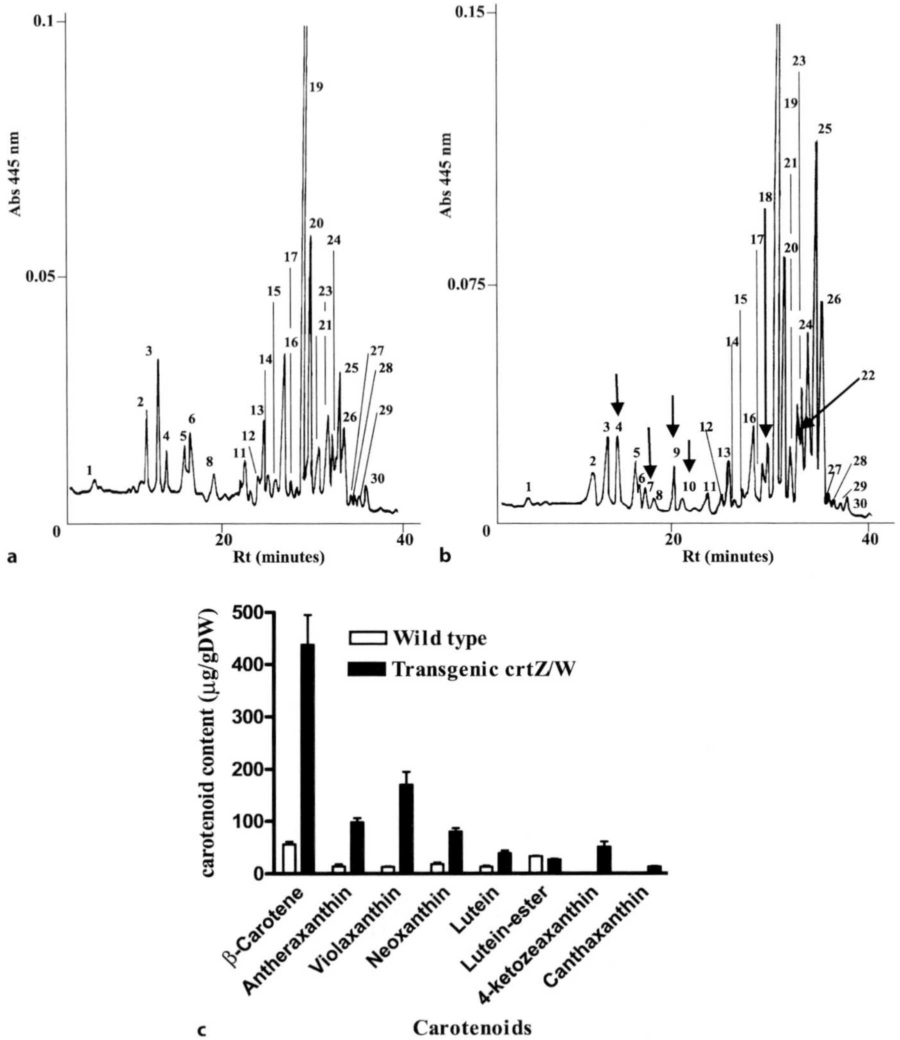

Fig. 5. Analysis of carotenoids in transgenic tobacco expressing the *Paracoccus* genes *crtZ* and *crtW*: **a** control nectary tissue; **b** transgenic tissue showing additional carotenoids (shown with an *arrow*); **c** levels of carotenoids in nectary tissue. Taken in part from Ralley et al. (2004)

▶ **Fig. 6.** Separation and identification of carotenoids from two introgression lines of tomato (4–3–2 and 6–2). Each chromatogram was monitored in line at three wavelengths: 286, 350 and 450 nm. Peaks: *1* – violaxanthin, *2* – lutein, *3* – zeaxanthin, *4* – α-carotene, *5* – β-carotene, *6* – *cis*-β-carotene, *7* – *cis* lycopene, *8* – δ-carotene, *9* – *cis*-lycopene, *10* – *cis*-lycopene, *11* – *cis*-lycopene, *12* – all *trans*-lycopene, *13* – *cis*-lycopene, *14* – *cis*-phytofluene, *15* – *cis*-phytofluene, *16* – *cis*-ζ-carotene, *17* – *cis*-ζ-carotene, *18* – γ-tocopherol, *19* – α-tocpherol, *20* – *cis*-phytoene, *21* – all *trans*-phytoene, *22* – *cis*-phytoene and *23*– ubquinone. The spectra of lycopene (peak 12), β-carotene (peak 5) and lutein (peak 2) are shown

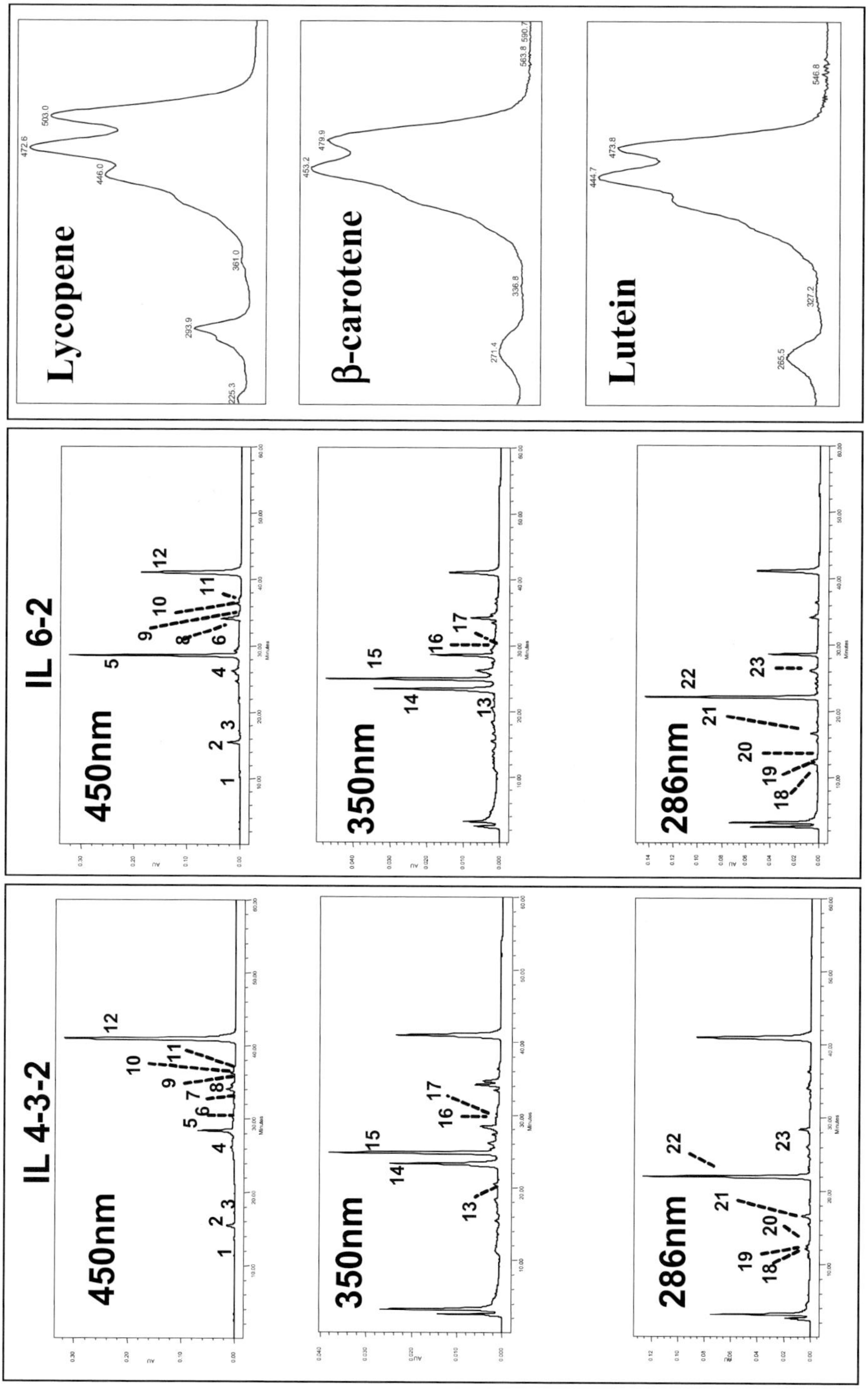
IL 4-3-2
450nm
350nm
286nm
IL 6-2
450nm
350nm
286nm
Lycopene
225.3
293.9
361.0
446.0
472.6
503.0
β-carotene
271.4
336.8
453.2
479.9
563.8
590.7
Lutein
265.5
327.2
444.7
473.8
546.8

3.2 Identification of Carotenoids in Fruit of Tomato Introgression Lines

It has been shown that the carotenoid content of tomato can be altered and increased by the application of a GM approach (e. g., Römer et al. 2000). However, many consumers are not willing to accept GM foods. The production of recombinant inbred lines (introgression lines) with genetically defined regions that represent inherent genetic variation offer an approach for breeding tomato varieties with improved quality traits (Gur et al. 2004). In order to define the regions of the genome that confer colour or health traits due to altered carotenoid content, metabolite profiling is required. In Fig. 6 the application of metabolite profiling to the elucidation of ILs with altered carotenoid content is illustrated. The respective profiles show that IL 3–2–4 contains a greater proportion of β-carotene (provitamin A) and many of the pathway intermediates have been altered.

4 Conclusions

The value of carotenoids to health and biotechnology make them an important class of metabolites that require qualitative and quantitative profiling. Their chemical properties are not amenable to GC-MS profiling or NMR fingerprinting. Therefore, targeted pathway profiling using HPLC and spectral data is the best alternative approach. As the need for the profiling of different plant species and tissues increases, standardisation of the nomenclature used to document the components will be required. The approach described in Bino et al. (2004) could be utilised as a foundation and incorporation into databases.

References

Ausich RL (1997) Commercial opportunities for carotenoid production by biotechnology. Pure Appl Chem 69:2169–2173

Bick JA, Lange BM (2003) Metabolic cross talk between cytosolic and plastidial pathways of isoprenoid biosynthesis: unidirectional transport of intermediates across the chloroplast envelope membrane. Arch Biochem Biophys 415:146–154

Bino RJ, Hall RD, Fiehn O, Kopka J, Saito K, Draper J, Nikolau BJ, Mendes P, Roessner-Tunali U, Beale MH, Trethewey RN, Lange BM, Syrkin Wurtele E, and Sumner LW (2004) Potential of metabolomics as a functional genomics tool. Trends Plant Sci 9:418–425

Booker J, Aulderidge M, Wills S, McCarty D, Klee HJ, Leyser O (2004) MAX 3 is a carotenoid cleavage dioxygenase required for the synthesis of a novel plant signalling molecule. Curr Biol 14:1232–1238

Boyes DC, Zayed A, Ascenzi R, MacCaskill AJ, Hoffman NE, Davis KR, Gorlach J (2001) Growth stage-based phenotypic analysis of *Arabidopsis*: a model for high throughput functional genomics in plants. Plant Cell 13:1499–1510

Bramley PM (1992) Analysis of carotenoids by high-performance liquid-chromatography and diode-array detection. Phytochem Anal 3:97–104

Bramley PM (2002) Regulation of carotenoid formation during tomato fruit ripening and development. J Exp Bot 53:2107–2113

Breitenbach J, Braun G, Steiger S, Sandmann S (2001) Chromatographic performance on a C-30 bonded stationary phase of monohydroxycarotenoids with variable chain length or degree of desaturation and of lycopene isomers synthesized by various carotene desaturases. J Chromatogr A 936:59–69

Britton G (1991) Carotenoids. In: Dey PM, Harborne JB (eds) Methods in plant biochemistry, vol 7. Academic Press, London, pp 473–518

Britton G (1995) UV/Vis spectroscopy. In: Britton G, Liaaen-Jensen S, Pfander H (eds) Carotenoids, vol 1B. Spectroscopy. Birkhäuser, Basel, pp 13–63

Britton G, Liaaen-Jensen S, Pfander H (2004) Carotenoids handbook. Birkhauser, Basel

Craft NE (1992) Carotenoid reversed-phase high performance liquid-chromatography methods-reference compendium. Methods Enzymol 213:185–205

Davies BH (1976) Carotenoids. In: Goodwin TW (ed) Chemistry and biochemistry of plant pigments, vol 2. Academic Press, London, pp 38–165

Fiehn O (2002) Metabolomics-the link between genotypes and phenotypes. Plant Mol Biol 48:155–171

Fraser PD, Bramley PM (2004) The biosynthesis and nutritional uses of carotenoids. Prog Lipid Res 43:228–265

Fraser PD, Pinto MES, Holloway DE, Bramley PM (2000) Application of high-performance liquid chromatography with photodiode array detection to the metabolic profiling of plant isoprenoids. Plant J 24:551–558

Fraser PD, Römer S, Shipton CA, Mills PB, Kiano JW, Misawa N, Drake RG, Schuch W, Bramley PM (2002) Evaluation of transgenic tomato plants expressing an additional phytoene synthase in a fruit specific-manner. Proc Natl Acad Sci USA 99:1092–1097

Giovannucci E (1999) Tomatoes, tomato-based products, lycopene and prostate cancer: review of the epidemiologic literature-response. J Natl Cancer Inst 91:317–331

Gur A, Semel Y, Cahaner A, Zamir D (2004) Real time QTL of complex phenotypes in tomato interspecific introgression lines. Trends Plant Sci 9:107–109

Halket JM, Watermann D, Przyborowska AM, Patel RKP, Fraser PD, Bramley PM (2005) Chemical derivatisation and mass spectral libraries in metabolic profiling by GC-MS and LC-MS/MS. J Exp Bot 56(410):219–243

Harborne JB (1991) Recent advances in the ecological chemistry of plant terpenoids. In: Harborne JB, Tomas-Barberan RA (eds) Ecological chemistry and biochemistry of plant terpenoids. Clarendon Press, Oxford, pp 399–426

Hirschberg J (2001) Carotenoid biosynthesis in flowering plants. Curr Opin Plant Biol 4:210–218

Miller NJ, Sampson J, Candeias LP, Bramley PM, Rice-Evans C (1996) Antioxidant activities of carotenes and xanthophylls. FEBS Lett 384(3):240–242

Misawa N, Satomi Y, Kondo K, Yokoyama A, Kaijwara S, Saito T, Ohtani T, Miki W (1995) Structure and functional analysis of a marine bacterial carotenoid biosynthesis gene cluster and astaxanthin biosynthetic pathway proposed at the gene level. J Bacteriol 177:6575–6584

O'Neill ME, Carroll Y, Corridan B, Granado F, Blanco I, Van den Berg H, Hinnger I, Rousell AM, Chopra M, Southon S, Thurnham DI (2001) European carotenoid database to assess carotenoid intakes and its use in a five-country comparative study. Br J Nutr 85(4):499–507

Pickersky E, Gang D (2000) genetics and biochemistry of secondary metabolites: an evolutionary perspective. Trends Plant Sci 5:439–445

Ralley L, Enfissi EMA, Misawa N, Schuch W, Bramley PM, Fraser PD (2004) Metabolic engineering of ketocarotenoid formation in higher plants. Plant J 39:477–486

Rodriguez-Concepcion M, Boronat A (2002) Elucidation of the methylerythritol phosphate pathway for isoprenoid biosynthesis in bacteria and plastids. A metabolic milestone achieved through genomics. Plant Physiol 130:1079–1089

Römer S, Fraser PD, Kiano JW, Shipton CA, Misawa N, Schuch W, Bramley PM (2000) Elevation of the provitamin A content of transgenic tomato plants. Nature Biotechnol 18:666–669

Rosati C, Aquilani R, Dharmapuri S, Pallara P, Marusic C, Tavazza R, Bouvier F, Camara B, Giuliano G (2000) Metabolic engineering of beta-carotene and lycopene content in tomato fruit. Plant J 24:413–419

Seddon JM, Ajani UA, Sperduto RD, Hiller R, Blair N, Burton TC, Farber MD, Gragoudas ES, Haller J, Miller DT, Yannuzzi LA, Willett W (1994) Dietary carotenoids, vitamin A, vitamin C and vitamin E and advanced age-related macular degeneration. J Am Med Assoc 272(18):1413–1420

Taylor RF, Ikawa M (1971) Gas chromatography of carotenoids. Anal Biochem 44:623–627

Van Breemen RB (1997) Liquid chromatography mass spectrometry of carotenoids. Pure Appl Chem 69:2061–2066

Ye X, Al-Babili S, Kloti A, Zhang J, Lucca P, Beyer P, Potrykus I (2000) Engineering the provitamin A (β-carotene) biosynthetic pathway into (carotenoid-free) rice endosperm. Science 287:303–305

III.5 Metabolomics and Gene Identification in Plant Natural Product Pathways

R.A. Dixon, L. Achnine, B.E. Deavours, and M. Naoumkina[1]

1 Introduction

Collectively, plants produce more than 100,000 natural products (also known as secondary metabolites). The underlying genetic basis of their chemical elaboration appears at first to be dauntingly complex. However, the rich diversity of many chemical structures found in the plant kingdom arises from a number of chemical scaffolds (of many types in terpene biosynthesis, of a much more limited number in flavonoid biosynthesis) modified by a limited number of chemical substitution types (hydroxylation, glycosylation, acylation, prenylation, *O*-methylation, etc.) (Fig. 1). Much of the chemical diversity is brought about by the substrate- and/or regio-specificities of the substitution enzymes. Functional genomics of plant natural product pathways therefore centers in large part on identifying genes encoding the substitution enzymes that determine the chemical complexity of a given plant species.

This chapter highlights the problems of how to assign metabolic function to gene sequences that appear to encode enzymes of secondary metabolism. We argue that metabolomic analysis is an essential complement to "genomic" approaches for functional annotation of genes involved in plant natural product biosynthesis (see Fig. 2 for a summary of the concept and potential strategies). However, developments in this field have lagged far behind those for gene discovery per se.

2 Gene Discovery – Past and Present Strategies

The classical biochemical approach for characterization of metabolic pathway genes relied on assay-directed purification of the protein followed by protein sequence determination, synthesis of gene-specific oligonucleotides based on the protein sequence, and screening of cDNA or genomic libraries (or use of polymerase chain-reaction (PCR) approaches) to clone the corresponding gene. This approach has been very successful in the past (see Kutchan 2002 for an excellent summary in relation to alkaloid biosynthesis), but is somewhat laborious and "low throughput". Manipulation of proteins is often more

[1] Plant Biology Division, Samuel Roberts Noble Foundation, 2510 Sam Noble Parkway, Ardmore, OK 73401, USA, e-mail: radixon@noble.org

Biotechnology in Agriculture and Forestry, Vol. 57
Plant Metabolomics (ed. by K. Saito, R.A. Dixon, and L. Willmitzer)

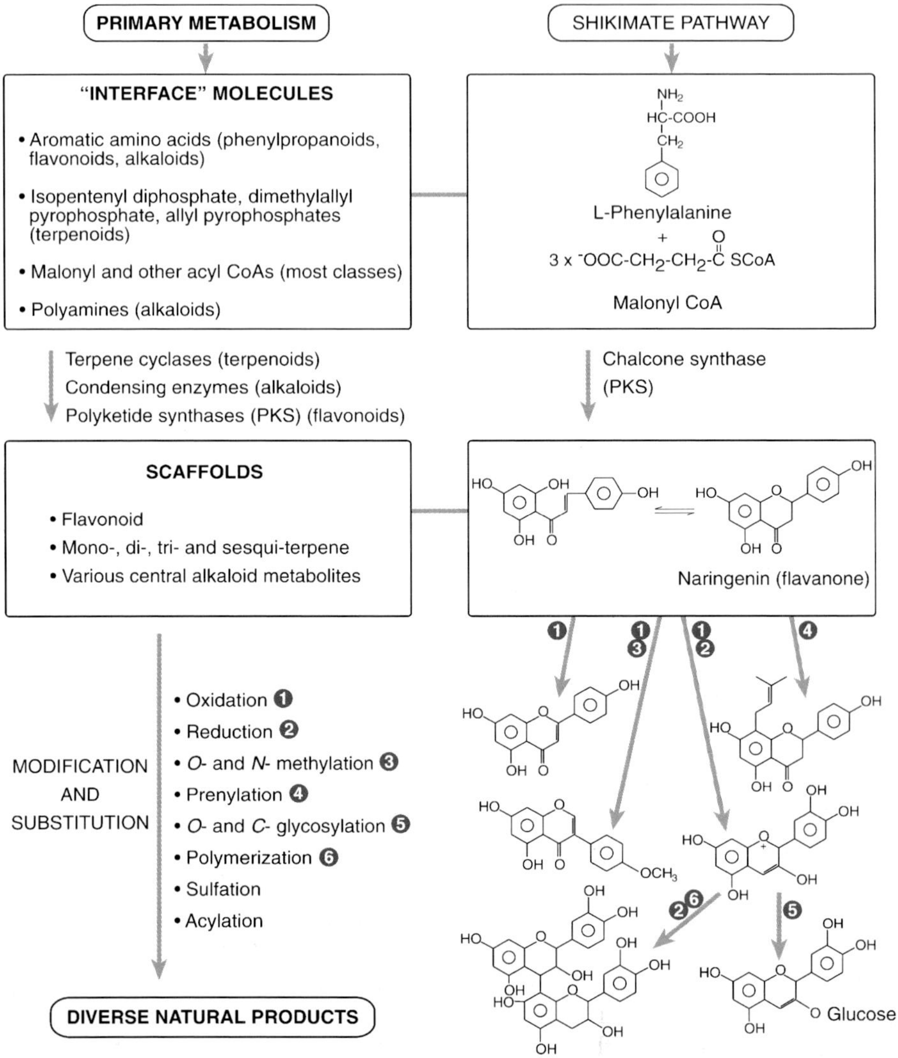

Fig. 1. Simplified outline of plant natural product biosynthesis, showing the interface between primary and secondary metabolism and the intermediacy of common scaffolds. In flavonoid biosynthesis, the great diversity of chemical structures depends largely on the modification and substitution of the scaffold, often catalyzed by the products of members of large gene families (an example is given on the *right*). Note that in terpenoid biosynthesis, much of the diversity arises from the formation of different scaffolds arising from differential cyclization products formed by different but closely related terpene cyclase enzymes, and further changes most often involve oxidation and reduction reactions

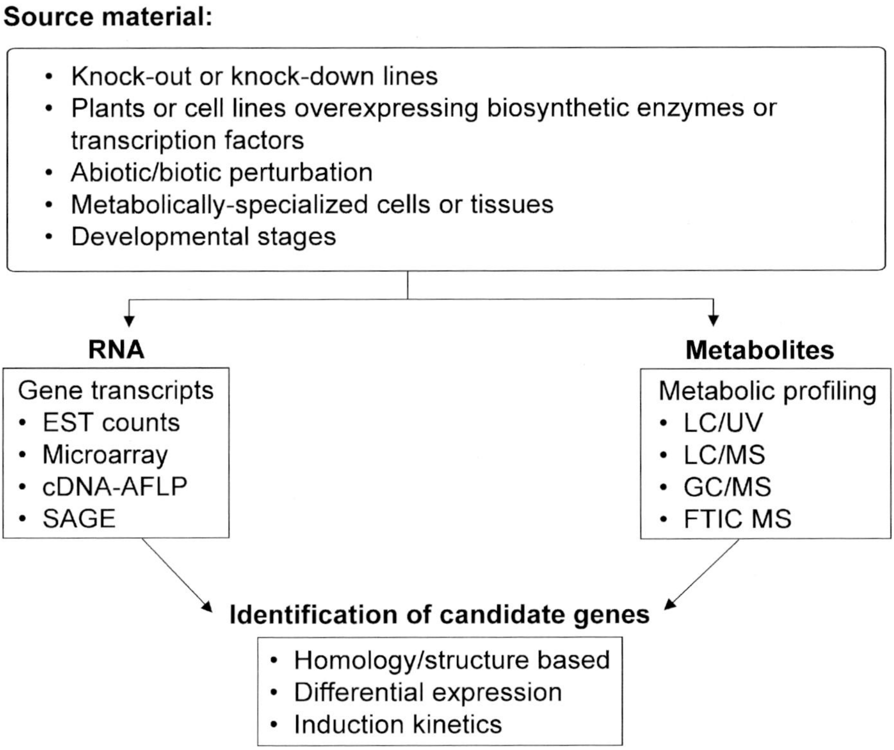

Fig. 2. Strategies for the application of metabolomics to assign gene function in plant natural product biosynthesis

difficult, and less intuitive, than that of DNA, and low abundance membrane proteins with poor stability, such as some cytochrome P450s (Kochs and Grisebach 1986), pose particular problems. The advantage of the approach is that it leads directly and unequivocally to a gene encoding a protein of known catalytic activity. However, that catalytic activity has already been defined by the experimenter, and may not always reflect the in vivo activity of the enzyme. Furthermore, it is still necessary to confirm function of the gene by expression in *E. coli* or an alternative heterologous system such as yeast or insect cells.

The availability of large collections of cDNAs (e. g., expressed sequence tag [EST] libraries), which can be prepared in "Gateway" vectors for direct transfer, via site-specific recombination, into a variety of "destination vectors" for heterologous expression (Karimi et al. 2002), efficiently by-passes the need for initial protein purification if methods are available for selecting candidates for a particular catalytic activity. DNA sequence-based comparison and annotation of EST clones can give an overall list of genes potentially encoding enzymes of a specific class, although, as the annotation in the databases is based purely on sequence comparisons, it may sometimes be incorrect. For example, some genes annotated as serine proteases are now known to be acyltransferases involved in plant secondary metabolism (Li and Steffens 2000; Shirley et al. 2001), and genes annotated as encoding chalcone synthase (CHS, the polyketide synthase (Fig. 1) at the entry point of the flavonoid pathway) may encode related polyketide synthases (Tropf et al. 1994; Schröder 1997; Schröder et al. 1998).

Generally speaking, sequence-based functional predictions have to be tested by heterologous expression followed by enzyme assay. This becomes problematic when the gene is a member of a large gene family, such as a cytochrome P450 or glycosyltransferase (GT), and the experimenter still has to make the decision as to which substrates will be tested, a decision influenced by the often limited availability of potential substrates, or their instability in vitro. Furthermore, the substrate specificity of the recombinant enzyme may differ from that of the enzyme purified from the plant as a result of in planta post-translational modifications (Vogt 2004).

The most rigorous, unbiased approach to gene function is the analysis of knock-out lines in the gene of interest, possible for Arabidopsis where comprehensive collections of such lines exist covering almost the entire genome (Alonso et al. 2003), but this is not possible for most other species. In some cases, knock-outs or knock-downs in secondary metabolic pathways have clear biochemical and visible developmental phenotypes (Chapple et al. 1992; Franke et al. 2002). In other cases there may be a strong developmental phenotype but the biochemical basis for this may be unclear (Woo et al. 1999). Even in Arabidopsis with its small genome, genetic redundancy is a problem for functional identification. For example, knock-outs in several Arabidopsis GTs with ascribed in vitro activities fail to yield a clear phenotype (D.J. Bowles, personal communication). Lack of a discovered phenotype in knock-out lines may also reflect subtle biochemical changes in the mutant that escape targeted metabolite profiling.

3 Enzyme Promiscuity in Natural Product Pathways

In vitro biochemical analyses provide information on substrate preference and catalytic properties determined for the substrates selected by the experimenter. An enzyme with a relatively high K_m or low K_{cat}/K_m value may have more favorable kinetic constants for other substrates unknown at the time. Furthermore, the exact in vitro conditions chosen may, in some cases, have profound effects on relative substrate specificity (Lukacin et al. 2004). In cases where enzymes are promiscuous, a range of alternative in vivo substrates may exist. For example, developing strawberry fruits contain lignified achenes and vascular bundles, and an *O*-methyltransferase was cloned from the fruit with substrate specificity for ortho-diphenols including caffeic acid and caffeoyl CoA (Wein et al. 2002). It was thought likely that this enzyme is involved in lignification in the fruit. However, the enzyme was also active with the vanillin precursor protocatechuic aldehyde, and could also methylate 2,5-dimethyl-4-hydroxy-3(2*H*)-furanone (DMHF). The latter two compounds are involved in flavor production in strawberry. In spite of the relatively low K_{cat}/K_m value for DMHF, it was concluded, from temporal and spatial examination of expression patterns, that the promiscuous OMT may play an important role in flavor production (Wein et al. 2002). Clearly, for any cell type within the fruit, knowing the presence, absence or relative levels of the potential OMT substrates would be instructive for assigning a biochemical function.

A recent example from mammalian cells nicely illustrates the problems of enzyme promiscuity, and how these may be addressed by a metabolomics approach. Untargeted LC/MS was used to analyze lipophilic compounds extracted from the brains and spinal cords of wild-type and transgenic mice in which the enzyme fatty acid amide hydrolase (FAAH) had been knocked-out (Saghatelian et al. 2004). Peaks seen in the knock-out but not in the wild type samples were in fact FAAH substrates. Surprisingly, the relative hydrolytic activity of FAAH shown for lipid metabolites in vitro was not necessarily indicative of the specific contribution of this enzyme in vivo (Saghatelian et al. 2004).

4 Examples of the Use of Metabolomics in the Elucidation of Gene Function

It will be clear from the above sections that ascribing gene function in secondary metabolism is not straightforward because activity in vitro may not always reflect activity in vivo. Particularly for promiscuous enzymes, knowledge of the cellular levels of all potential substrates may be critical for assigning in vivo function. Unfortunately, the depth of transcriptomic analyses (which can be close to genome wide for model species such as Arabidopsis, rice, *Medicago* and poplar) far exceeds that of metabolomic analyses at the present time. This is particularly true in natural product biosynthesis, where compounds tend to be identified on an "at need basis", rather than globally.

4.1 The Isoflavonoid Pathway in *Medicago*

Isoflavonoids comprise a class of plant natural products with important biological activities including health promotion in humans and antimicrobial activity against plant pathogens (Dixon 1999, 2004; Dixon and Ferreira 2002). They are found primarily in leguminous plants, where they function as pre-formed or inducible antimicrobial or anti-insect compounds, as inducers of the nodulation genes of symbiotic *Rhizobium* bacteria, or as allelopathic agents (Dixon 1999). Isoflavonoids originate from a flavanone intermediate (either liquiritigenin or naringenin, Fig. 3a) that is ubiquitously present in plants. For entry into the isoflavonoid pathway, flavanone undergoes migration of the B-ring to the 3-position followed by hydroxylation at the 2-position, catalyzed by the cytochrome P450 enzyme CYP93C1 (2-hydroxyisoflavanone synthase, commonly termed isoflavone synthase [IFS]). The resulting 2-hydroxyisoflavanone is dehydrated to the corresponding isoflavone (Kochs and Grisebach 1986) (Fig. 3a), which is then modified by substitution, reduction, ring cyclization and glycosylation to yield the range of isoflavone, isoflavanone and pterocarpan compounds illustrated in Fig. 3a.

Most, but not all, enzymes in the pathway to medicarpin are known (Dixon 1999). Several of these characterized enzymes are encoded by large multigene families; these include P450s such as IFS, OMTs and isoflavone reductases (Dixon et al. 2002). These genes were first discovered using classical biochemical approaches, such that only one member of the family was initially isolated and functionally characterized. Questions exist as to whether the other family members provide redundancy, tissue-specificity, or even encode enzymes with different catalytic properties. One way to address these questions is to link gene-specific transcript analysis with metabolic profiling that targets products and intermediates of the pathway in different tissues and/or in tissues in which the pathway is induced in response to biotic or abiotic stimuli. This approach has recently proved instructive for addressing gene function in sulfur metabolism (including glucosinolate biosynthesis) in Arabidopsis (Hirai et al. 2004) and in pyridine alkaloid biosynthesis in tobacco (Goossens et al. 2003).

▶ **Fig. 3.** Integration of transcriptomics and metabolomics for gene identification in the isoflavonoid pathway: **a** scheme for isoflavonoid biosynthesis in *Medicago sativa* (alfalfa) and *M. truncatula*. Enzymes are: PAL, L-phenylalanine ammonia-lyase; C4H, cinnamate 4-hydroxylase; 4CL, 4-coumarate: CoA ligase; CHS, chalcone synthase; CHI, chalcone isomerase; CHR, "chalcone reductase"; IFS, "isoflavone synthase"; HI4'OMT, 2-hydroxyisoflavanone 4′-*O*-methyltransferase; DH, dehydratase; I2'H, isoflavone 2′-hydroxylase; IFR, isoflavone reductase; VR, vestitone reductase; GT, glycosyltransferase; MT, malonyltransferase; P450, cytochrome P450; OMT, *O*-methyltransferase. *Boxed structures* show compounds for which enzymes remain uncharacterized in Medicago. MG, medicarpin glucoside; FGM, formononetin glucoside malonate. Compounds marked with a * have been observed in metabolomic analysis; **b** *color-coded panels* show DNA microarray analysis of transcripts encoding functionally assigned genes of the isoflavonoid pathway (only one gene family member is shown for each enzyme apart from CHS) and candidates for genes encoding predicted *O*-methyltransferase, cytochrome P450 and GT genes involved in isoflavone modification, based on co-induction with identified metabolites. *Color coding* reflects relative expression level at the times shown (hours) after exposure to elicitor or water (control). Normalization was preformed using GeneTraffic software. Signal intensities between two fluorescent images (Cy3 reference, Cy5 experimental) were normalized using LOWESS subgrid normalization. The *color scale* indicates normalized signal intensities (log 2 ratio of fold change between experimental and reference samples); **c,d** HPLC-UV profiling of (iso)flavonoids in alfalfa cell suspension cultures 48 h after exposure to water (control, C) or yeast elicitor (D); **e** a hydrolyzed extract from cells that had been fed with labeled liquiritigenin (L), with label incorporation in formononetin (F), 2′hydroxyformononetin (2'HF) and medicarpin (M) indicating de novo synthesis in response to elicitation. IL, isoliquiritigenin

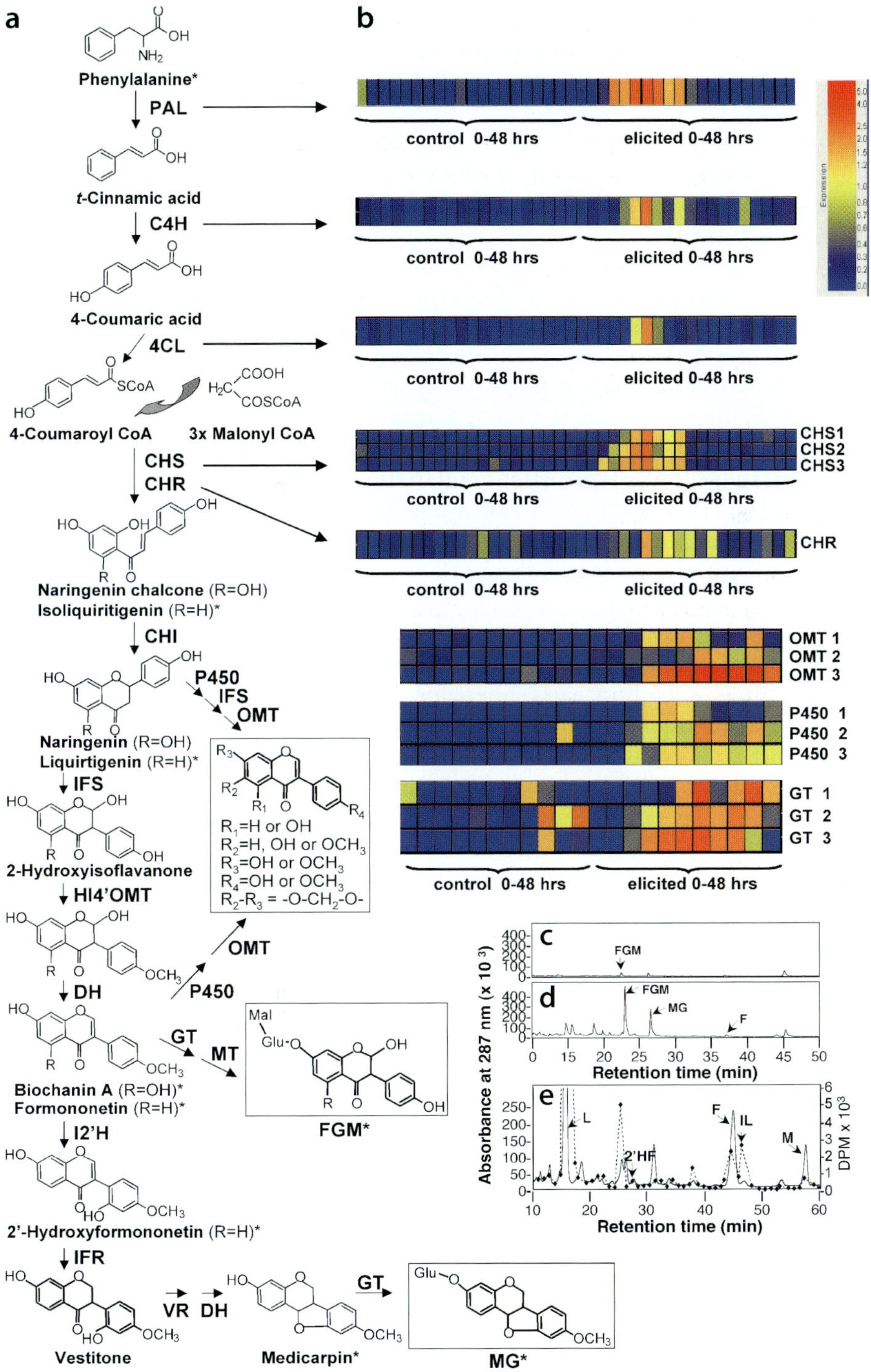
a
Phenylalanine*
PAL
t-Cinnamic acid
C4H
4-Coumaric acid
4CL
4-Coumaroyl CoA
3x Malonyl CoA
CHS
CHR
Naringenin chalcone (R=OH)
Isoliquiritigenin (R=H)*
CHI
P450
IFS
OMT
Naringenin (R=OH)
Liquirtigenin (R=H)*
IFS
2-Hydroxyisoflavanone
HI4'OMT
DH
Biochanin A (R=OH)*
Formononetin (R=H)*
I2'H
2'-Hydroxyformononetin (R=H)*
IFR
VR DH
GT
MT
FGM*
Vestitone
Medicarpin*
MG*
R1=H or OH
R2=H, OH or OCH3
R3=OH or OCH3
R4=OH or OCH3
R2-R3 = -O-CH2-O-
b
control 0-48 hrs
elicited 0-48 hrs
Expression
CHS1
CHS2
CHS3
CHR
OMT 1
OMT 2
OMT 3
P450 1
P450 2
P450 3
GT 1
GT 2
GT 3
c
d
e
FGM
MG
F
L
2'HF
IL
M
Absorbance at 287 nm (x 10 3)
Retention time (min)
DPM x 10³

Targeted metabolite profiling in alfalfa (*Medicago sativa*) (Fig. 3c–e) and *Medicago truncatula* cell suspension cultures reveals induction of the same major isoflavonoid metabolites following exposure of the cultures to an elicitor from yeast cell walls (Liu and Dixon 2001; Suzuki et al. 2005; M. Farag and L.W. Sumner, unpublished results). The profiling methods developed in the past for flavonoids/isoflavonoids, and shown in Fig. 3b–d, rely primarily on HPLC with UV/visible detection (Graham 1991). This is a simple and reliable method for cases where metabolites have already been characterized and their properties (HPLC retention times and UV spectra) are known and authentic standards are available, e. g., for fingerprinting specific metabolites. However, mass spectrometry offers many advantages as a detection system for a more unbiased metabolomic approach, including greatly improved sensitivity and better structural resolution, particularly when employing tandem MS (Fiehn et al. 2000a,b; Sumner et al. 2003). Thus, LC/MS/MS analysis of elicited *M. truncatula* cell suspension cultures resulted in identification not only of the group of isoflavones, pterocarpans and their glycosides revealed by UV/visible analysis (Fig. 3a,c–e), but also of a number of additional isoflavones with unexpected A-ring methylation and methelenedioxy substitution, as well as isoflavene and aurone metabolites (M. Farag and L.W. Sumner, unpublished results).

DNA microarray analysis of *M. truncatula* cell cultures harvested at a range of times post-elicitation revealed induction of several members of the multigene families encoding early phenylpropanoid pathway, flavonoid branch and isoflavonoid-specific branch pathway enzymes (Fig. 3b). Integration of metabolite and transcript data from such experiments in an interrogable database allows in silico comparison of transcript induction kinetics with appearance of specific metabolites. For example, yeast elicitor induced accumulation of a range of methylated isoflavones, suggesting the need for multiple isoflavone OMTS, and potential candidate genes can be identified from the microarray dataset (Fig. 3a). Likewise, it is possible to identify a number of candidate *cytochrome P450* and *GT* genes potentially involved in the formation of the various glycosylated, oxidized isoflavone derivatives accumulating in response to elicitation (Fig. 3b).

4.2 Deciphering the Triterpene Pathway in *Medicago*

Triterpene saponins are a class of plant natural products with a wide range of bioactivities, from allelopathic and anti-microbial to anticancer and anticholesterolemic (Waller et al. 1993; Behboudi et al. 1999; Haridas et al. 2001; Osbourn 2003), and are important components of a number of herbal medicines (Xu 2001; Chan et al. 2002). Most of the genes involved in the biosynthesis of these complex molecules remain uncharacterized.

The saponins of *M. truncatula* and alfalfa exist as glycosides of at least five different triterpene aglycones; medicagenic acid, hederagenin, soyasapogenol B, soyasapogenol E and bayogenin (Huhman and Sumner 2002). These compounds are derived from β-amyrin, the cyclization product of 2,3-oxidosqualene (Kushiro et al. 1998; Suzuki et al. 2002). The downstream reactions in the biosynthesis of *M. truncatula* saponins include a number of cytochrome P450 dependent hydroxylations/oxidations and several glycosyl transfer reactions catalyzed by uridine diphosphate-dependent GTs. Based on current EST and partial genome sequence information, *M. truncatula* contains P450 and GT supergene families each with approximately 300 members. It is more than likely that most of the genes involved in triterpene biosynthesis in *Medicago* are already physically present in EST and genomic library collections, and approaching the identification from these resources, rather than taking a protein purification approach, is attractive in view of the relative instability of P450s and GTs, and the insoluble nature of the former. However, the numbers of potential candidate genes is problematic. An integrated approach involving comparison of transcript and metabolite behavior in response to a metabolic perturbation represents one way to address this problem.

Exposure of *M. truncatula* cell suspension cultures to methyl jasmonate (MeJA) induces triterpene saponin accumulation preceded by induction of the triterpene cyclase β-amyrin synthase (β-AS) (Suzuki et al. 2002, 2005). By coupling DNA array approaches to profile transcripts corresponding to all 100-plus GTs in MeJA-induced and control cell cultures with metabolite (saponin) profiling and in silico expressed sequence tag (EST) data mining, two GTs, designated UGT71G1 and UGT73K1, were selected and subsequently found to be active with *Medicago* triterpene aglycones (Achnine et al. 2005). The basis of the selection was to compare tissue-specificity (in nearly 40 cDNA libraries used for EST sequencing) and induction kinetics (assuming co-induction) of candidate GTs compared to the triterpene cyclase β-AS. A genomics approach has also recently been used to identify diterpene GTs from *Stevia rebaudiana* (Richman et al. 2005).

Of more than 40 potential triterpene and phenolic acceptor molecules tested in in vitro assays, UGT73K1 only showed activity against three triterpenes (soyasapogenol E, soyasapogenol B, and hederagenin), consistent with the appearance of glucosides of these compounds in the MeJA-treated cell cultures (Achnine et al. 2005). However, on the basis of in vitro kinetic measurements, UGT71G1 appeared to have a clear preference for flavonoid compounds (quercetin and 5-hydroxyisoflavones) as compared to triterpenes (Fig. 4a). The K_{cat}/K_m ratio of UGT71G1 was more than 30-fold higher for quercetin than for hederagenin. This raised the question of whether the identification of UGT71G1 as a triterpene GT was erroneous, and simply a reflection of in vitro enzyme promiscuity. However, comparisons of the levels of UGT71G1 transcripts, triterpenes, and isoflavones in *M. truncatula* cell cultures responding to MeJA or yeast elicitor (a strong inducer of isoflavones) suggested a lack of association of UGT71G1 with isoflavone glycoside formation, and quercetin glycosides were not detected in the cultures (Achnine et al. 2005) (Fig. 4b). Thus, metabolomic analysis resolved the issue of in vivo substrate specificity

a

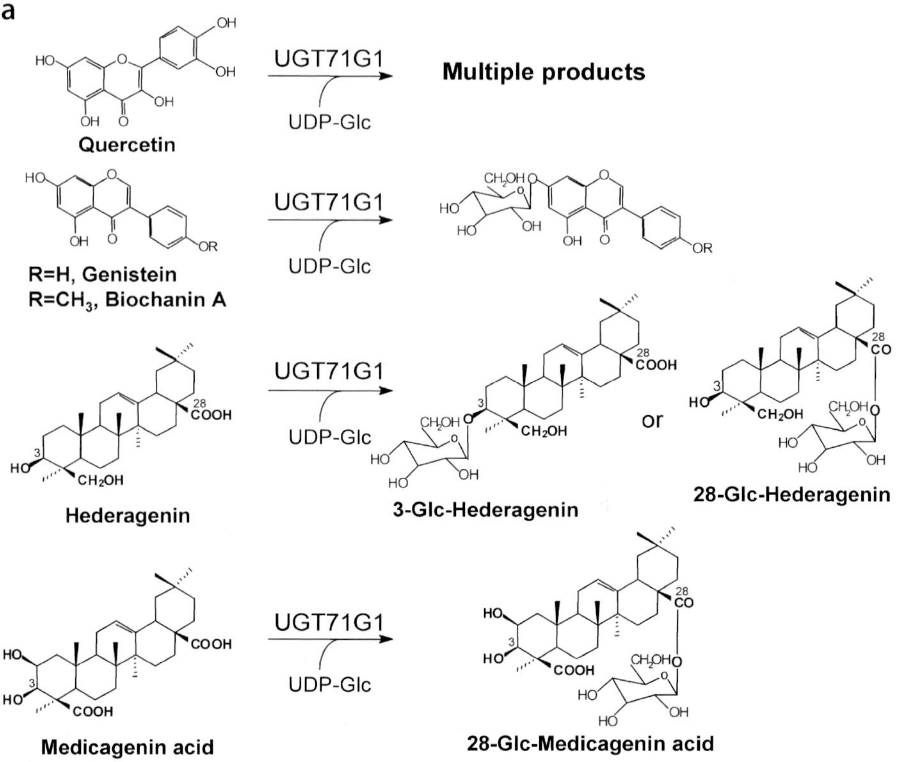

b

	Yeast elicitor	Methyl Jasmonate
UGT7G1	–	++
Quercetin glucosides	–	–
Genistein-7-*O*-glucoside	++	–
Biochanin A-7-*O*-glucoside	–	–
3-Glc-Hederagenin	–	++
28-Glc-Hederagenin	–	++
3-Glc-28-Glc-Medicagenic acid	–	++

Fig. 4. Metabolomics for gene identification in the triterpene pathway: **a** substrates and products of UGT71G1 in vitro; **b** summary of transcript and metabolite analyses in vivo. Yeast elicitor induces the formation of genistein glucoside, but not UGT71G1 transcripts, whereas methyl jasmonate induces hederagenin glucosides and UGT71G1. Other GTs are likely involved in the formation of isoflavone glucosides in yeast-elicited cells

in this cell culture. Even so, a full understanding of the specificity of UGT71G1 in vivo must await metabolomic analysis of plants in which its gene is selectively down-regulated.

5 Single Cell or Isolated Tissue Metabolomics

The above examples make a strong case for the argument that knowledge of which compounds are, and which are not, present in a particular plant tissue may be essential for ascribing function to a cloned gene. This statement also implies that it is necessary to be able to correlate presence of an enzyme or its transcripts with those metabolites made in the same cells or cell types.

Many of the plant EST projects currently accessible through publicly available web sites, such as the TIGR Plant Gene Indices (http:www.tigr.org/tdb/tgi/plant.shtml), provide expression data as EST counts in a wide range of cDNA libraries prepared from different tissue, developmental stages, and responses to various biotic and abiotic stresses and stimuli. A list of currently available *M. truncatula* EST libraries is given in Table 1. It is informative to contrast this list with the paucity of information, and almost non-existent tissue resolution, on the *M. truncatula* metabolome. A survey of the literature, including a comprehensive phytochemical dictionary (ILDIS 1994), can provide a list of secondary metabolites found in different legume species, but the conditions of the tissues from which the metabolites were identified is generally poorly defined, and, without exception, the studies reported were targeted and non-comprehensive. A similar situation holds for those other species, such as Arabidopsis, rice, corn, soybean and poplar, for which extensive genomic and EST sequence information is available.

Furthermore, the degree of tissue and treatment resolution currently found in EST databases such as indicated in Table 1 is of itself insufficient to provide the kind of integration necessary to allow for meaningful correlations between transcriptome and metabolome. For example, proanthocyanidins are found in alfalfa in specific regions of the seed coat (Debeaujon et al. 2003) and in the heads of glandular trichomes (Aziz et al. 2005), but have not been shown to be present in other parts of the plant. As described above, several studies have utilized cell suspension cultures to obtain a more homogeneous cellular background for integrated transcript and metabolome profiling (Goossens et al. 2003; Achnine et al. 2005; Suzuki et al. 2005). However, although the kinetics of changes in many metabolites can be shown to cluster with changes in transcript levels, most of the metabolites remain unknown. Spatially resolved metabolomic and transcriptomic analysis is technically challenging but is an essential approach for understanding the relationship between gene expression and metabolism in whole plants. Laser capture microdissection (LCM) techniques can now provide tissue samples for such analyses (Kerek et al. 2003), and techniques are available for the construction of cDNA libraries from minute tissue samples (Belyavsky et al. 1989).

Trichomes provide an excellent system with which to develop technologies for integrated transcriptomics/metabolomics on a tissue or even single cell level. Trichomes are epidermal appendages found on the aerial organs of many plants. Glandular trichomes have a high capacity to synthesize, store and secrete secondary metabolites that help protect the plant against insect

Table 1. *M. truncatula* EST libraries as of May 1st, 2003 (TIGR Release 7.0), with a total of 189,714 EST sequences (http://www.tigr.org/tigr-scripts/tgi/T_index.cgi?species=medicago)

Tissue	EST library	EST #
1. Leaf		
Non-challenged	Developing leaf[a]	9415
Biotic challenged	Insect herbivore attacked leaf	10,309
	Colletotrichum trifolii infected leaf	6003
	Phoma medicaginis infected leaf	3281
Abiotic challenged	Phosphate starved leaf	10,188
2. Root		
Non-challenged	Developing root, no symbiosis[a]	3054
	KV0, non-nodulated root	2752
	MtRHE, root hair-enriched	899
Biotic challenged		
	KV1, root – four day post-nodulation	2840
	KV2, root – two days post-nodulation	3330
	KV3, root – three days post-nodulation	4315
	MtBB, root – four days post-nodulation	7807
	GVN, one-month-old root nodules	6468
	GVSN, senescent nodules	2788
	R108, young root nodules	438
	Nodulated root, mixed	3299
	MHAM, *Glomus versiforme* infected root	7368
	MtBC, mycorrhizal root	8601
	DSIR, fungus-elicited root	2463
	BNIR, nematode-infected root	3154
Abiotic challenged	MtBA, nitrogen starved root	7939
	Rootphos(−), phosphate starved root	1967
	MHRP-, phosphate starved root	2658
	MGHG, β-glucan-elicited root	2687
	HOGA, oligogalacturonide-elicited root	2861
3. Stem		
Non-challenged	Developing stem[a]	10,783
4. Seed		
Non-challenged	GESD, developing seed[a]	2672
	GLSD, developing seed	2944
	Germinating seed	6619
5. Flower		
Non-challenged	Developing flower[a]	6724
6. Cell culture		
Biotic challenged	Yeast elicited cell culture[a]	9859
Abiotic challenged	Methyl jasmonate-induced cell culture[a]	6900
7. Pods		
Non-challenged	Developing pod[a]	1915

Table 1. (continued)

Tissue	EST library	EST #
8. Mixed tissues		
Non-challenged	Cotyledon and leaf	2143
Abiotic challenged	Drought stressed seedlings	9520
	UV Irradiated seedling	6748

[a] These tissue types are the only ones to date for which metabolomic analysis has been initiated. Preliminary results have led to identification of approximately 200 primary metabolites, and a significantly smaller number of secondary metabolites from each tissue source (ILDIS 1994; Huhman and Sumner 2002; Achnine et al. 2005; Broeckling et al. 2005; Suzuki et al. 2005; L.W. Sumner, C. Broeckling, D.V. Huhman and M. Farag, unpublished results)
For EST libraries descriptions, refer to CEDA (Comparative EST Data Analysis in *M. truncatula*) at http://bionfo.noble.org/CEDA.htm

predation and other biotic challenges (Wagner 1991; Ranger and Hower 2001; Wagner et al. 2004). For example, the peltate glandular trichomes of mint produce a suite of defensive monoterpenes that are the major components of, and give the characteristic smell and flavor to, mint oil (McCaskill et al. 1992; Voirin and Bayet 1996); trichomes from tomato species collectively produce a number of insecticidal sesquiterpenes, acyl sugars and methyl ketones (Li et al. 1999; Antonious 2001; Maluf et al. 2001); and tobacco trichomes produce diterpenes and acyl sugars (Kandra et al. 1990; Guo and Wagner 1995).

cDNA libraries have been constructed from trichomes of mint (Lange et al. 2000), sweet basil (Gang et al. 2001), and wild and cultivated tomatoes (http:www.tigr.org/tigr-scripts/tgi/T{_}index.cgi?species=tomato). The mint and tomato trichomes show a strong preponderance of transcripts (represented by ESTs) encoding enzymes of terpene metabolism. Because of the highly specialized biosynthetic functions of the trichomes from these two species, considerable biosynthetic information was obtained by sequencing only a relatively small number of ESTs (1000–2500). This is likely to be the case for other species that produce biochemically-specialized trichomes or other secretory cell types, such as hops (Hirosawa et al. 1995) and vanilla orchid (Joel et al. 2003). Combining in-depth EST sequencing and metabolite profiling will provide a powerful approach for gene discovery that, because of the specialized nature of the trichome, will directly address bioactive secondary metabolites.

6 Concluding Remarks

We argue for the wider use of metabolomics in the context of natural product pathway gene discovery. In depth profiling of plant tissues for natural products, coupled to parallel analysis of gene transcript levels using EST count, microarray, cDNA-AFLP or SAGE analysis, is a powerful tool when applied to a biological system in which "differential display" of metabolites and gene transcripts can be visualized and compared. Such systems include knock-out mutants, plants or cell lines over-expressing biosynthetic enzymes or transcription factors, and plants, tissues or cell lines exposed to biotic or abiotic perturbations. Comparisons of tissues at different developmental stages should also be addressed. In this context, it would be very valuable to obtain in depth metabolite profiles for all tissue and treatment types that have been sampled in EST sequencing projects, thereby providing a resource for initial in silico analysis pairing enzymes with potential substrates and products as a tool to assist functional annotation. The major problem is that many metabolites are unknown and, where they can be predicted (e.g., as intermediates in a complex pathway, Fig. 4a), standard compounds are generally unavailable.

Acknowledgements. We thank David Huhman and Drs Lloyd Sumner and Mohamed Farag for assistance with mass spectrometry, sharing results prior to publication, and useful discussions. Work in the authors' laboratory was supported by grants to RAD from the National Science Foundation (#DBI 0109732), Forage Genetics International, the Oklahoma Center for the Advancement of Science and Technology Health Sciences Program, and the Samuel Roberts Noble Foundation.

References

Achnine L, Huhman DV, Sumner LW, Blount JW, Dixon RA (2005) Genomics-based selection and functional characterization of triterpene glycosyltransferases from the model legume Medicago truncatula. Plant J 41:875–887

Alonso JM, Stepanova AN, Leisse TJ, Kim CJ, Chen H, Shinn P, Stevenson DK, Zimmerman J, Barajas P, Cheuk R, Gadrinab C, Heller C, Jeske A, Koesema E, Meyers CC, Parker H, Prednis L, Ansari Y, Choy N, Deen H, Geralt M, Hazari N, Hom E, Karnes M, Mulholland C, Ndubaku R, Schmidt I, Guzman P, Aguilar HL, Schmid M, Weigel D, Carter DE, Marchand T, Risseeuw E, Brogden D, Zeko A, Crosby WL, Berry CC, Ecker JR (2003) Genome-wide insertional mutagenesis of *Arabidopsis thaliana.* Science 301:653–657

Antonious GF (2001) Production and quantification of methyl ketones in wild tomato accessions. J Environ Sci Health B36:835–848

Aziz N, Paiva NL, May GD, Dixon RA (2005) Profiling the transcriptome of alfalfa glandular trichomes. Planta 221:28–38

Behboudi S, Morein B, VillacresEriksson MC (1999) Quillaja saponin formulations that stimulate proinflammatory cytokines elicit a potent acquired cell-mediated immunity. Scand J Immunol 50:371–377

Belyavsky A, Vinogradova T, Rajewsky K (1989) PCR-based cDNA library construction: general cDNA libraries at the level of a few cells. Nucleic Acids Res 17:2919–2933

Broeckling CD, Huhman DV, Farag M, Smith JT, May GD, Mendes P, Dixon RA, Sumner LW (2005) Metabolic profiling of *Medicago truncatula* cell cultures reveals effects of biotic and abiotic elicitors on primary metabolism. J Exp Bot 56:323–336

Chan RYK, Chen WF, Dong A, Guo D, Wong MS (2002) Estrogen-like activity of ginsenoside Rg1 derived from *Panax notoginseng*. J Clin Endocrinol Metab 87:3691–3695

Chapple CCS, Vogt T, Ellis BE, Somerville CR (1992) An Arabidopsis mutant defective in the general phenylpropanoid pathway. Plant Cell 4:1413–1424

Debeaujon I, Nesi N, Perez P, Devic M, Grandjean O, Caboche M, Lepiniec L (2003) Proanthocyanidin-accumulating cells in Arabidopsis testa: regulation of differentiation and role in seed development. Plant Cell 15:2514–2531

Dixon RA (1999) Isoflavonoids: biochemistry, molecular biology and biological functions. In: Sankawa U (ed) Comprehensive natural products chemistry, vol 1. Elsevier, Amsterdam, pp 773–823

Dixon RA (2004) Phytoestrogens. Annu Rev Plant Biol 55:225–261

Dixon RA, Ferreira D (2002) Molecules of Interest: Genistein. Phytochemistry 60:205–211

Dixon RA, Achnine L, Kota P, Liu C-J, Reddy MS, Wang L (2002) The phenylpropanoid pathway and plant defense – a genomics perspective. Mol Plant Pathol 3:371–390

Fiehn O, Kopka J, Dormann P, Altmann T, Trethewey RN, Willmitzer L (2000a) Metabolite profiling for plant functional genomics. Nature Biotechnol 18:1157–1161

Fiehn O, Kopka J, Trethewey RN, Willimitzer L (2000b) Identification of uncommon plant metabolites based on calculation of elemental compositions using gas chromatography and quadrupole mass spectrometry. Anal Chem 72:3573–3580

Franke R, Humphreys JM, Hemm MR, Denault JW, Ruegger MO, Chapple C (2002) The Arabidopsis *REF8* gene encodes the 3-hydroxylase of phenylpropanoid metabolism. Plant J 30:33–45

Gang DR, Wang J, Dudareva N, Nam KH, Simon JE, Lewinsohn E, Pichersky E (2001) An investigation of the storage and biosynthesis of phenylpropenes in sweet basil. Plant Physiol 125:539–555

Goossens A, Häkkinen S, Laakso I, Seppänen-Laakso T, Biondi S, de Sutter V, Lammertyn F, Nuutila AM, Söderlund H, Zabeau M, Inzé D, Oksman-Caldentey KM (2003) A functional genomics approach toward the understanding of secondary metabolism in plant cells. Proc Natl Acad Sci USA 100:8595–8600

Graham TL (1991) A rapid, high resolution high performance liquid chromatography profiling procedure for plant and microbial aromatic secondary metabolites. Plant Physiol 95:584–593

Guo Z, Wagner GJ (1995) Biosynthesis of cembratrienols in cell-free extracts from trichomes of *Nicotiana tabacum*. Plant Sci 110:1–10

Haridas V, Higuchi M, Jayatilake GS, Bailey D, Mujoo K, Blake ME, Arntzen CJ, Gutterman JU (2001) Avicins: triterpenoid saponins from *Acacia victoriae* (Bentham) induce apoptosis by mitochondrial perturbation. Proc Natl Acad Sci USA 98:5821–5826

Hirai MY, Yano M, Goodenowe DB, Kanaya S, Kimura T, Awazuhara M, Arita M, Fujiwara T, Saito K (2004) Integration of transcriptomics and metabolomics for understanding of global responses to nutritional stresses in *Arabidopsis thaliana*. Proc Natl Acad Sci USA 101:10205–10210

Hirosawa T, Saito T, Tanaka T, Matsushima H (1995) SEM observation and HPLC analysis of the accumulation of alpha- and beta-acids in the fresh developing hop (*Humulus lupulus* L.) peltate glandular trichomes. J Electron Microsc 44:145–147

Huhman DV, Sumner LW (2002) Metabolic profiling of saponin glycosides in *Medicago sativa* and *Medicago truncatula* using HPLC coupled to an electrospary ion-trap mass spectrometer. Phytochemistry 59:347–360

ILDIS (1994) Phytochemical dictionary of the leguminosae. Plants and their constituents, vol 1. Chapman and Hall, London

Joel DM, French JC, Graft N, Kourteva G, Dixon RA, Havkin-Frenkel D (2003) A hairy tissue produces vanillin. Israel J Plant Sci 51:157–159

Kandra G, Severson R, Wagner GJ (1990) Modified branched-chain amino acid pathways give rise to acyl acids of sucrose esters exuded from tobacco leaf trichomes. Eur J Biochem 188:385–391

Karimi M, Inze D, Depicker A (2002) GATEWAY™ vectors for *Agrobacterium*-mediated plant transformation. Trends Plant Sci 7:193–195

Kerek NM, Ceserani T, Tausta SL, Sussex IM, Nelson TM (2003) Laser capture microdissection of cells from plant tissues. Plant Physiol 132:27–35

Kochs G, Grisebach H (1986) Enzymic synthesis of isoflavones. Eur J Biochem 155:311–318

Kushiro T, Shibuya M, Ebizuka Y (1998) β-Amyrin synthase. Cloning of oxidosqualene cylase that catalyzes the formation of the most popular triterpene among higher plants. Eur J Biochem 256:238–244

Kutchan TM (2002) Sequence-based approaches to alkaloid biosynthesis gene identification. In: Romeo JT, Dixon RA (eds) Phytochemistry in the genomics and post-genomics eras. Elsevier Science, Oxford, pp 163–178

Lange BM, Wildung MR, Stauber EJ, Sanchez C, Pouchnik D, Croteau R (2000) Probing essential oil biosynthesis and secretion by functional evaluation of expressed sequence tags from mint glandular trichomes. Proc Natl Acad Sci USA 97:2934–2939

Li AX, Steffens JC (2000) An acyltransferase catalyzing the formation of diacylglucose is a serine carboxypeptidase-like protein. Proc Natl Acad Sci USA 97:6902–6907

Li AX, Eannetta N, Ghangas GS, Steffens JC (1999) Glucose polyester biosynthesis. Purification and characterization of a glucose acyltransferase. Plant Physiol 121:453–460

Liu C-J, Dixon RA (2001) Elicitor-induced association of isoflavone O-methyltransferase with endomembranes prevents formation and 7-O-methylation of daidzein during isoflavonoid phytoalexin biosynthesis. Plant Cell 13:2643–2658

Lukacin R, Matern U, Specker S, Vogt T (2004) Cations modulate the substrate specificity of bifunctional class 1 *O*-methyltransferase from *Ammi majus*. FEBS Lett 577:367–370

Maluf WR, Campos GA, das Gracas CM (2001) Relationships between trichome types and spider mite (*Tetranychus evansi*) repellence in tomatoes with respect to foliar zingiberene contents. Euphytica 121:73–80

McCaskill D, Gershenzon J, Croteau R (1992) Morphology and monoterpene biosynthetic capabilities of secretory cell clusters isolated from glandular trichomes of peppermint (*Mentha piperita* L.). Planta 187:445–454

Osbourn AE (2003) Molecules of interest. Saponins in cereals. Phytochemistry 62:1–4

Ranger CM, Hower AA (2001) Glandular morphology from a perennial alfalfa clone resistant to the potato leafhopper. Crop Sci 41:1427–1434

Richman A, Swanson A, Humphrey T, Chapman R, McGarvey BD, Pocs R, Brandle J (2005) Functional genomics uncovers three glucosyltransferases involved in the synthesis of the major sweet glucosides of *Stevia rebaudiana*. Plant J 41:56–67

Saghatelian A, Trauger S, Want E, Hawkins E, Siuzdak G, Cravatt B (2004) Assignment of endogenous substrates to enzymes by global metabolite profiling. Biochemistry 43:14332–14339

Schröder J (1997) A family of plant-specific polyketide synthases: facts and predictions. Trends Plant Sci 2:373–378

Schröder J, Raiber S, Berger T, Schmidt A, Schmidt J, Soares-Sello AM, Bardshiri E, Strack D, Simpson TJ, Veit M, Schröder G (1998) Plant polyketide synthases: a chalcone synthase-type enzyme which performs a condensation reaction with methylmalonyl-CoA in the biosynthesis of C-methylated chalcones. Biochemistry 37:8417–8425

Shirley AM, McMichael CM, Chapple C (2001) The sng2 mutant of Arabidopsis is defective in the gene encoding the serine carboxypeptidase-like protein sinapoylglucose: choline sinapoyltransferase. Plant J 28:83–94

Sumner LW, Mendes P, Dixon RA (2003) Plant metabolomics: large-scale phytochemistry in the functional genomics era. Phytochemistry 62:817–836

Suzuki H, Achnine L, Xu R, Matsuda SPT, Dixon RA (2002) A genomics approach to the early stages of triterpene saponin biosynthesis in *Medicago truncatula*. Plant J 32:1033–1048

Suzuki H, Reddy MSS, Naoumkina M, Aziz N, May GD, Huhman DV, Sumner LW, Blount JW, Mendes P, Dixon RA (2005) Methyl jasmonate and yeast elicitor induce differential genetic and metabolic re-programming in cell suspension cultures of the model legume *Medicago truncatula*. Planta 220:698–707

Tropf S, Lanz T, Rensing SA, Schröder J, Schröder G (1994) Evidence that stilbene synthases have developed from chalcone synthases several times in the course of evolution. J Mol Evol 38:610–618

Vogt T (2004) Regiospecificity and kinetic properties of a plant natural product *O*-methyltransferase are determined by its N-terminal domain. FEBS Lett 561:159–162

Voirin B, Bayet C (1996) Developmental changes in the monoterpene composition of *Mentha X piperita* leaves from individual peltate trichomes. Phytochemistry 43:573–580

Wagner GJ (1991) Secreting glandular trichomes: more than just hairs. Plant Physiol 96:675–679

Wagner GJ, Wang E, Shepherd RW (2004) New approaches for studying and exploiting an old protuberance, the plant trichome. Ann Bot 93:3–11

Waller GR, Jurzysta M, Thorne RLZ (1993) Alleopathic activity of root saponins from alfalfa (*Medicago sativa* L.) on weeds and wheat. Bot Bull Acad Sin 34:1–11

Wein M, Lavid N, Lunkenbein S, Lewinsohn E, Schwab W, Kaldenhoff R (2002) Isolation, cloning and expression of a multifunctional O-methyltransferase capable of forming 2,5-dimethyl-4-methoxy-3(2H)-furanone, one of the key aroma compounds in strawberry fruits. Plant J 31:755–765

Woo HH, Orbach MJ, Hirsch AM, Hawes MC (1999) Meristem-localized inducible expression of a UDP-glycosyltransferase gene is essential for growth and development in pea and alfalfa. Plant Cell 11:2303–2315

Xu Y (2001) Perspectives on the 21st century development of functional foods: bridging Chinese medicated diet and functional foods. Int J Food Sci Technol 36:229–242

III.6 Metabolomic Analysis of *Catharanthus roseus* Using NMR and Principal Component Analysis

H.K. Kim, Y.H. Choi, and R. Verpoorte[1]

1 Introduction

The ultimate goal of plant metabolomics is to map all metabolites in a plant both qualitatively and quantitatively. Detection of all plant metabolites seems impossible due to the large number, the chemical complexity of the metabolites and their different characteristics such as solubility and polarity. A proper analytical method should be selected in order to be able to detect as many compounds as possible in a plant. A number of analytical methods have been proposed and applied to profile the plant metabolome (Sumner et al. 2003). Basically, two types of methods can be distinguished – methods based on a chromatographic separation, e.g. HPLC, GC and TLC, and methods based on a physical characteristic of the metabolites, e.g. MS (molecular weight) and NMR (resonance of magnetic nuclei, e.g. ^{1}H or ^{13}C in a strong magnetic field). To obtain maximum selectivity, both methods can also be combined. Chromatographic methods are based on the relative behavior of the individual metabolites in a system with a mobile phase (gas or liquid) and a stationary phase. This allows a selective separation; however, reproducibility is very much dependant on the quality of both phases. For the mobile phase this is reasonable; however, for the stationary phase this is different. Many different stationary phases exist for GC and HPLC, and "improved" stationary phase are regularly introduced. Long term reproducibility is thus difficult. With respect to the requirements of speed and reproducibility, nuclear magnetic resonance (NMR) and mass spectrometry (MS) based approaches score very well if compared to chromatography. The MS based metabolomic analysis shows high separation efficiency and sensitivity, and easy coupling with chromatographic methods. These characteristics of MS analysis in the plant metabolomics allow the detection of a larger number of metabolites if compared to NMR (e.g. 3000 metabolites) (Aharoni et al. 2002; Fiehn 2002). However, there are some inevitable limitations in the MS based methods. These are mainly in terms of quantitation. Each compound will show different sensitivity which may also be different by the matrix in which it is analyzed. For absolute quantitation calibration curves are needed for each single compound. For relative occurrence of a certain compound in different materials this is of no importance. The range of metabolites covered by gas chromatography (GC)-MS is restricted more or

[1] Division of Pharmacognosy, Section Metabolomics, Institute of Biology, Leiden University, P.O. Box 9502, 2300RA, Leiden, The Netherlands, e-mail: verpoort@chem.leidenuniv.nl

Biotechnology in Agriculture and Forestry, Vol. 57
Plant Metabolomics (ed. by K. Saito, R.A. Dixon, and L. Willmitzer)

less to small and volatile metabolites having a molecular weight less than 400 (e. g. mono or disaccharides, amino acids, or organic acids). Consequently, this excludes the detection of unstable and non-volatile plant secondary metabolites such as glycosides. MS can be coupled with high performance liquid chromatography (HPLC) using soft ionization methods such as electrospray ionization (ESI) or matrix assisted laser desorption ionization (MALDI) (Huhman and Sumner 2002; Tolstilkov and Fiehn 2002). It is suited for all kinds of metabolites and can also detect polar or high molecular weight metabolites; however, lack of fragmentation makes it difficult to identify the metabolites – an MS/MS system is required to obtain structural information.

Despite its low sensitivity, the recent advances in NMR methods offer some advantages compared to chromatography and MS methods. The range of compounds is not limited by their volatility, presence of chromophores or polarity. The broad range of metabolites detected by NMR makes it the optimum choice for macroscopic metabolomics, a total representative view of all metabolites present. Moreover, NMR has the great advantage that the spectra are highly reproducible as it concerns a physical characteristic of a compound. In other words, the NMR data obtained at different places or time can be compared with each other. It is also possible to elucidate the structure of unknown metabolites, particularly secondary metabolites. Another advantage is that the signals of NMR spectra are based on molar concentration and can directly be compared while the intensity of metabolites in MS is highly affected by the ionization level.

Despite the advantages of NMR in plant metabolomics, the spectral complexity, lower sensitivity and costly combination with chromatography have made researchers hesitant to apply it as a tool of metabolomics. Most applications of NMR to plant metabolomics are in food quality control, e. g. identification of origin of wine (Brescia et al. 2002), coffee (Charlton et al. 2002), juice (Vogels et al. 1996) and beer (Duarte et al. 2002).

Here we will discuss several factors which should be considered for plant metabolomics using NMR. For plant materials in biological experiments, aspects such as harvesting, extraction and choice of NMR solvent are important factors to be considered in metabolomic studies. We will also show some applications of various NMR methods using several plants, and in particular *Catharanthus roseus*, as a model.

2 Experimental Consideration for Metabolomics Using NMR

2.1 Harvesting Plant Material

When plants are harvested, a plant might recognize itself as being damaged and immediate wound reactions will occur. In the short term (starting immediately after wounding) this self-defense mechanism of plants results in oxidation or hydrolysis of metabolites. These reactions will continue and in the longer term (e. g. 12–72 h) even phytoalexin biosynthesis will produce novel compounds if the material is stored at room temperature for a certain time. To avoid these reactions, all biochemical reactions in the plant material should be stopped immediately at harvesting, for example, by freezing in liquid nitrogen followed by storage at −80 °C. Stopping biochemical reactions can also be done by heating, or by adding organic solvents or strong acid. Heating will stop enzyme activities involved in the defense reaction, but might cause decomposition of metabolites. Microwave treatment might be helpful as all material is heated to 100 °C in very short time, where conventional heating will result in a temperature gradient, in which defense reactions still may happen. To extract secologanin from *Symphoricarpos albus*, a few minutes of microwave treatment in water solution inhibited enzymatic degradation by β-glucosidase and resulted in a higher yield of secologanin (Kim et al. 2004).

The next factor to consider is the drying of the plant material. Fresh plants contain approximately 70–80% of water. As this variable water content will mix with extraction solvent, it causes inaccuracy in the ratio of the extraction solvents which results in a lower reproducibility of the metabolomic profile. Moreover, in an aqueous environment the various enzymes involved in defense will be active, whereas in a dry material these metabolites are not likely to occur. For example, the extraction efficiency of sinigrin, a well known glucosinolate of *Brassica nigra* leaves, is drastically reduced in fresh material compared to that of dried ones (unpublished results). It might be due to the fact that sinigrin is degraded by myrosinase in fresh material. For this reason, freeze dried plant material would be preferable due to higher extraction reproducibility and less degradation of metabolites.

2.2 Extraction

A number of solvents can be considered for extraction from non-polar to highly polar. In fact, there are various metabolites in plants with diverse polarity such as alkaloids, amino acids, carbohydrates, fatty acids, lipids, steroids and terpenoids. It is impossible to extract all these metabolites using one single solvent. Thus, the choice of an optimum solvent is one of the most important factors in plant metabolomics. Figure 1 shows ^{1}H NMR spectra of *Catharanthus roseus* leaves extracted with different polar solvents: methanol, 0.1% trifluoroacetic acid (TFA) and methanol-KH_2PO_4 buffer (pH 6.0). The phenolic region in the ^{1}H NMR spectra shows different profiles of metabolites. In particular, the signals from alkaloids (catharanthine and vindoline) are changeable depending on the extraction solvents. In another experiment, various solvents were tested for *Arabidopsis thaliana* and *Brassica rapa* (unpublished data). More than 90% of metabolites extracted by chloroform and *n*-hexane were fatty acids or lipids. Adenosine, cytosine, phenylpropanoids, flavonoids and terpenoids were abundant metabolites extracted by methanol, acetone or

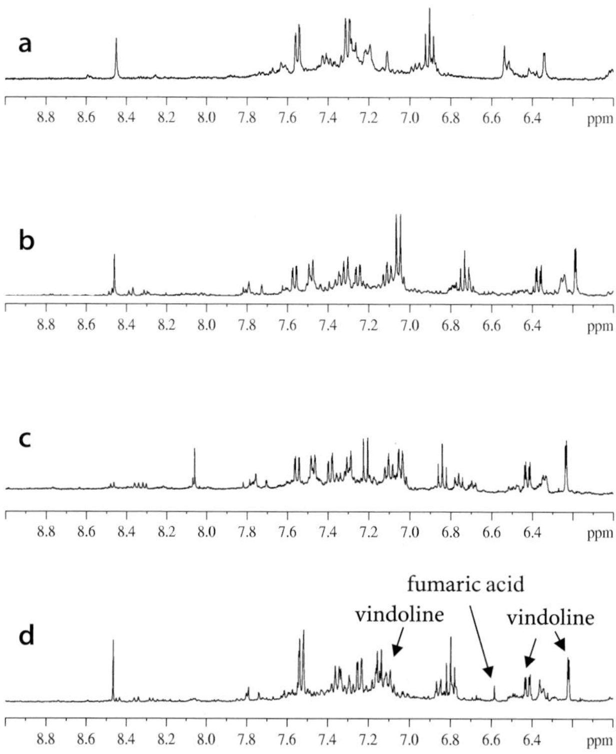

Fig. 1. ^{1}H NMR spectra (δ 9.0–6.0) of *Catharanthus roseus* leave extracted by different solvents: **a** 0.1% TFA; **b** MeOH; **c** MeOH+TFA; **d** MeOH:KH_2PO_4 buffer (1:1) pH 6.0. Leaves were extracted directly with corresponding NMR solvents (unpublished data)

acetonitrile. Amino acids and carbohydrates are well extracted by water. The extraction with the mixture of methanol-water (1:1) showed a broader range of extracted metabolites if compared to other solvents. A mixture of chloroform-methanol-water (2:1:1) was employed for several plants including *Nicotiana tabacum* (Choi et al. 2004a), *Cannabis sativa* (Choi et al. 2004b), *Catharanthus roseus* (Choi et al. 2004c), and *Ilex* species (Choi et al. 2005). In this mixture, the chloroform fraction contained a high level of fatty components, steroids, terpenoids and alkaloids (caffeine, theobromine, theophylline) while carbohydrates, phenylpropanoid glycosides and saponins were major metabolites in the water fraction. This two-phase extraction method provided a wide range of metabolites compared to a single solvent extraction. However, even with this two-phase extraction method, the extraction efficiency of medium polar metabolites such as indole alkaloids (catharanthine and ajmalicine in *Catharanthus roseus* leaves and roots) and aglycones of flavonoids was relatively low. For extraction of alkaloids it is necessary to adjust the pH of extraction solvent.

To handle hundreds of samples at one time, direct extraction with NMR solvents will be very helpful to reduce elaborate extraction procedures. It also minimizes the degradation or loss of metabolites, which occurs during elaborate extraction procedures.

For all these reasons, direct extraction using MeOD:KH_2PO_4 buffer (pH 6.0) is now routinely used for our work.

2.3 Solvent for NMR

Although NMR provides reproducible signals based on physical properties of molecules compared to other analytical methods, the signals in NMR (chemical shifts) are quite dependent on NMR solvents. Several factors should be considered to choose the solvents of NMR in plant metabolomics. The pH of the solution and concentration may influence the reproducibility of NMR spectra. Since the pH is known to affect shifts in ionizable compounds such as alkaloids (Schripsema et al. 1987), a controlled pH is thus required for metabolomic analysis. The pH control can be done using a buffer or simply adding acid. Commonly used buffers are acetate (pH range 3.7–5.6) and phosphate (pH range 5.0–7.4), in the concentration of 10–50 mmol/L. As an example, Fig. 2 shows the effect of pH on the chemical shifts of malic acid in a plant extract (*Senecio aquaticus*). The chemical shift of malic acid is highly affected not only by the pH of the NMR solvent but also by the sort of buffer. It indicates clearly the importance of pH in order to obtain reproducible chemical shifts.

For alkaloids the addition of acid can control the pH of extracts. Trifluoroacetic acid (TFA) has been used as a pH modifier to adjust pH of the NMR solvent. Several indole alkaloids including icajine, brucine, strychnine and vomicine from *Strychnos* species were analyzed by ^{1}H NMR using 1% TFA in methanol-d_4 (Frédérich et al. 2004).

Even under controlled pH, ^{1}H NMR signals of some metabolites are largely affected by their concentration in the solution. Recently, we encountered this problem with ungeremine (Rhee et al. 2004). Lower concentration of ungeremine in the solution resulted in the downfield shift of each proton by 0.1–0.01 ppm. Other examples are organic acids such as citric acid and malic acid. These acids can be found in the plant as major products of the TCA cycle. They show characteristic signals in the range of δ 2.5–δ 3.0. As shown in Fig. 3, the chemical shifts of the organic acids are also largely changed by their concentration.

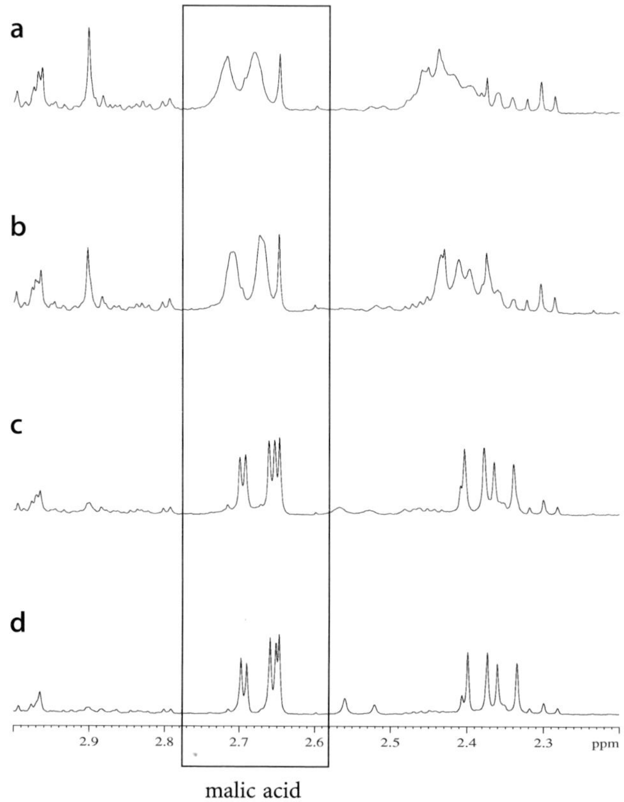

Fig. 2. ^{1}H NMR spectra of aqueous extract of *Senecio* measured in different solvents: **a** pH 6.0 in acetate buffer; **b** pH 6.0 in phosphate buffer; **c** pH 7.0 in phosphate buffer; **d** pH 8.0 in phosphate buffer. Note that the resolution of malic acid is affected by different solvent and pH (unpublished data)

3 Application of NMR for Plant Metabolome

3.1 ^{1}H NMR

Because of its relatively high sensitivity and the universal occurrence of protons in organic metabolites, ^{1}H NMR is a good starting tool for a metabolomic study. ^{1}H NMR spectroscopy has been shown to provide a wealth of information about the main metabolites in plants. From 10 to 50 mg of dried plant material, a ^{1}H NMR spectrum can be generated within 10 min. The spectrum covers approximately 50–100 metabolites, of which 10–20 compounds are easily identified. Basically the identification of the metabolites is possible by means of chemical shifts and coupling constants.

By visual inspection of the ^{1}H NMR spectrum, one has a first view of the whole metabolome of the plant material. Figures 4 and 5 show the ^{1}H NMR

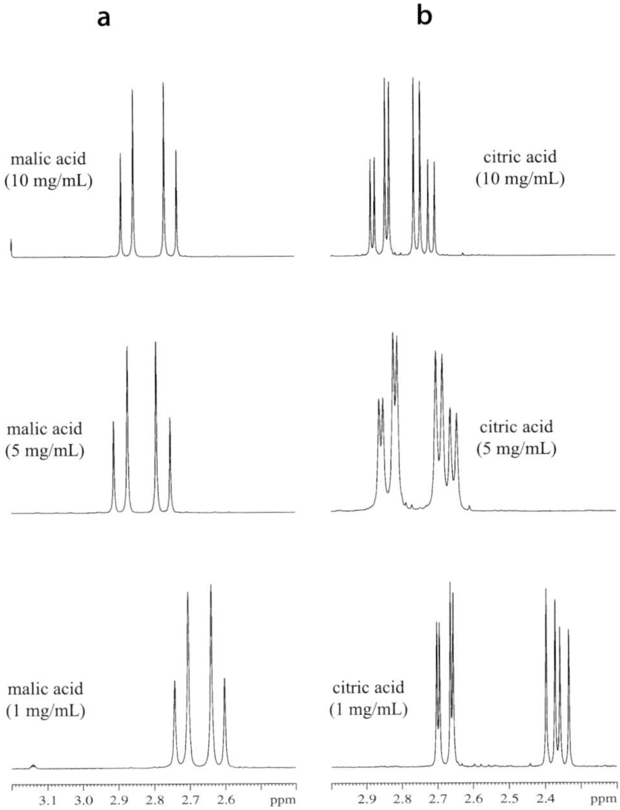

Fig. 3. ^{1}H NMR spectra of: **a** malic acid; **b** citric acid in different concentrations – *top*: 10 mg/mL, *middle*: 5 mg/mL, *bottom*: 1 mg/mL in MeOD-KH_2PO_4 buffer (pH 6.0). Note that the signals of these organic acids were upfield shift in the lower concentration. Protons of citric acid appear at δ 2.7–3.0 (d, J = 17.6 Hz) and δ 2.6–2.8 (d, J = 17.6 Hz), malic acid at δ 2.8–2.6 (dd, J = 16.6 Hz, 4.7 Hz) and δ 2.7–2.3 (dd, J = 16.6 Hz, 6.6 Hz) (unpublished data)

spectra of a healthy plant and phytoplasma-infected *C. roseus* plant. In the chloroform fraction (Fig. 4), most of the signals come from the aliphatic chains of fatty acids and methyl and methylene groups of triterpenoids or steroids. The expanded aromatic region shows the characteristic signal of vindoline: H-9 at δ 6.89 (d, J = 8.2 Hz), H-10 at δ 6.29 (dd, J = 8.5 Hz, 2.3 Hz), H-12 at δ 6.07 (d, J = 2.2 Hz), H-14 at δ 5.85 (ddd, J = 10.2 Hz, 4.9 Hz, 1.7 Hz). Together with these signals, other signals of vindoline such as OCH_3 of C-11 at δ 3.79 (s), OCH_3 of C-22 at δ 3.78 (s), H-18 at δ 0.49 (t, J = 7.4 Hz) could also be observed. The intensities of H-9 signal at 6.89 indicate that in infected plants vindoline content is two times higher than in healthy plants. Figure 5 shows the ^{1}H NMR spectra of the aqueous fraction of the *C. roseus*. Most of the signals in the crowded region at δ 3.0–5.0 come from the carbohydrates present in high amounts in the plant. Besides the signals of carbohydrates and amino acids,

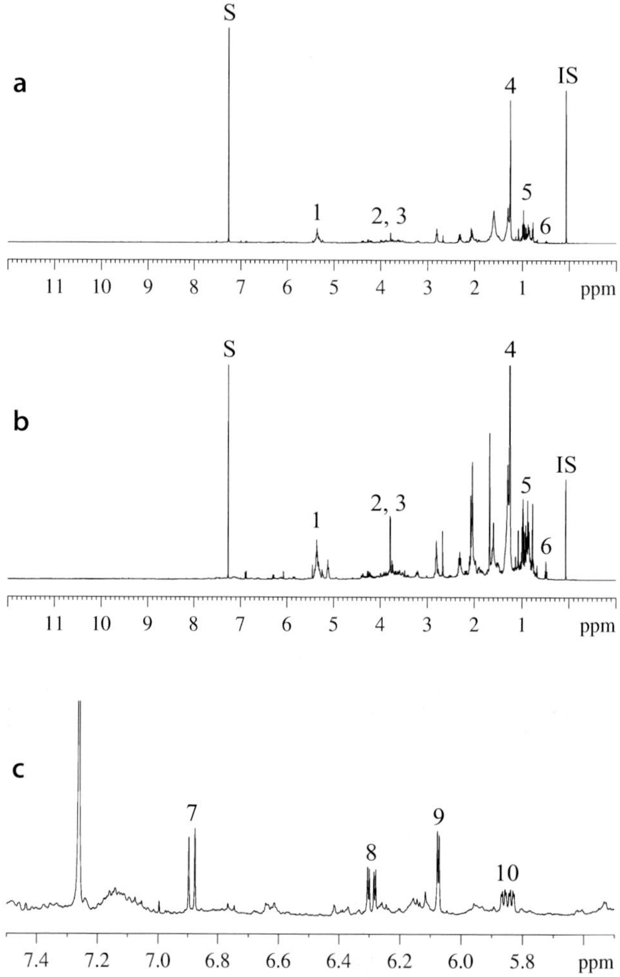

Fig. 4. ^{1}H-NMR spectra of $CHCl_3$ extract of: **a** healthy *C. roseus* leaves; **b** phytoplasma (BIL) infected *C. roseus* leaves; **c** the expansion of δ 5.5–7.5. 1; olefinic signals of fatty components or terpenoids, 2; OCH_3 of C-11 of vindoline, 3; OCH_3 of C-22 of vindoline, 4; long chain CH_2 of fatty material, 5; steroidal or triterpenoidal CH_3, 6; H-18 of vindoline, 7; H-9 of vindoline, 8; H-10 of vindoline, 9; H-12 of vindoline, 10; H-14 of vindoline, S; residual chloroform signal, IS; internal standard (HMDSO). (With kind permission of American Society of Plant Biologists, reproduced from Choi et al. 2004c)

characteristic signals from secologanin and loganic acid, important precursors of the indole alkaloids, can be found, e. g. for the H-3 of secologanin is at δ 7.57 and δ 7.49 and of loganic acid at δ 7.57 and δ 7.49. Other signals from secondary metabolites such as phenolic acids and chlorogenic acid also could be detected in the aqueous extracts. The compounds found in the *C. roseus* plant by ^{1}H NMR are listed in Table 1. In the infected leaves, the contents of

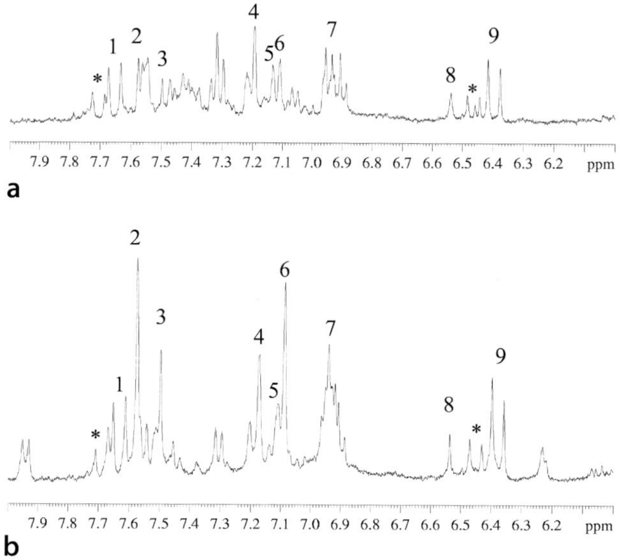

Fig. 5a,b. ^{1}H-NMR spectra of methanol-water fraction of phytoplasma (UDINESE) infected *C. roseus* leaves in the rage of δ 6.0–8.0. 1; H-7′ of chlorogenic acid, 2 and 3; H-3 of secologanin, 4; H-2′ of chlorogenic acid, 5; H-6′ of chlorogenic acid, 6; H-3 of loganic acid, 7; H-5′ of chlorogenic acid and aromatic signals of polyphenols, 8; fumaric acid, 9; H-8′ of chlorogenic acid, *; possible signals of chlorogenic acid derivatives. (With kind permission of American Society of Plant Biologists, reproduced from Choi et al. 2004c)

secologanin, loganic acid, chlorogenic acid and sugar are higher than in the healthy plants.

3.2 J-resolved NMR

NMR-based metabolomic studies typically employ one-dimensional NMR methods to minimize sample acquisition times and therefore maximize throughput. However, the spectral complexity and overlapping signals of one dimensional ^{1}H NMR limits the number of metabolites that can be identified and quantified. Moreover, few databases for ^{1}H NMR spectra of plant metabolites are available if compared to MS spectra. There are some databases related to ^{13}C NMR spectra (e. g. NMRshiftDB). However, ^{13}C NMR spectrometry has limitations in the field of plant metabolomics in the aspect of acquisition time and quantitation. It takes more than 14 h to obtain informative ^{13}C NMR spectra from the same concentration of samples from which ^{1}H NMR spectra are obtained within 10 min. In addition, broad band decoupling adopted to increase the sensitivity of ^{13}C NMR signals cause non reproducible signal increase (up to 200%) by the nuclear Overhauser effect. Therefore other two-dimensional NMR methods should be considered for application to plant

Table 1. ^{1}H Chemical shifts of metabolites of *Catharanthus roseus* leaves detected from NMR

Number	Chemical shifts (ppm) and coupling constants (Hz)	Metabolites
1	1.00 (d, J = 7.0)	H-10 of loganic acid
2	1.33 (d, J = 6.7)	H-4 of threonine
3	1.48 (d, J = 7.4)	H-3 of alanine
4	2.49 (s)	Succinic acid
5	3.56 (s)	Glycine
6	4.22 (d, J = 8.8)	Anomeric proton of fructose (sucrose)
7	4.64 (d, J = 9.5)	Anomeric proton of β-glucose
8	5.24 (d, J = 3.7)	Anomeric proton of α-glucose
9	5.42 (d, J = 3.8)	Anomeric proton of glucose (sucrose)
10	6.39 (d, J = 15.9)	H-8′ of phenylpropanoid
11	6.54 (s)	Fumaric acid
12	6.93 (d, J = 8.5)	H-5′ of chlorogenic acid
13	7.09 (d, J = 1.1)	H-3 of loganic acid
14	7.11 (d, J = 8.5)	H-6′ of chlorogenic acid
15	7.18 (s)	H-2′ of chlorogenic acid
16	7.57 (s)	H-3 of secologanin
17	7.64 (d, J = 15.9)	H-7′ of phenylpropanoid
18	9.65 (s)	Aldehyde proton of secologanin

extracts. Among the 2D NMR methods, the J-resolved technique is an interesting option. It greatly improves the resolution of the ^{1}H NMR spectra within comparably shorter time (25 min) than other 2D NMR techniques and it is easy to build up a database since a projection of the 2D spectrum on the chemical shift axis results in a spectrum in which most of protons are observed as singlet (Viant 2003).

One of the advantages of J-resolved spectra is that it provides spin multiplicities which are sometimes difficult to determine in the 1D ^{1}H NMR due to overlapping of signals. J-resolved spectra separate the chemical shift and spin–spin coupling data onto different axes, F1 for spin–spin coupling and F2 for chemical shifts. Complex aromatic signals in the ^{1}H NMR spectrum of *C. roseus* leaves appeared as less congested signals in the 2D J-resolved spectra (Fig. 6). It was quite difficult to identify indole alkaloids such as catharanthine and vindoline in the plants due to overlapping with other signals in the ^{1}H NMR (e. g. phenolics). However, the resolution of these signals in the J-resolved spectra is dramatically increased. The signals of vindoline H-9 at δ 7.0 (d, J = 2.2 Hz) are clearly separated from other signals. Also the signals from catharanthine H-9 at δ 7.6 (d, J = 7.8 Hz), H-12 at δ 7.2 (d, J = 8.0 Hz), H-10 at δ 7.2 (t, J = 8.0 Hz), H-11 at δ 7.1 (t, J = 8.0 Hz) and H-9 at δ 6.3 (d, J = 2.6 Hz) can be clearly identified.

The enhanced resolution obtained from J-resolved NMR spectra can be applied to monitor minor metabolic change in the plants which might be difficult to detect by 1D ^{1}H NMR spectra (Choi et al., unpublished data).

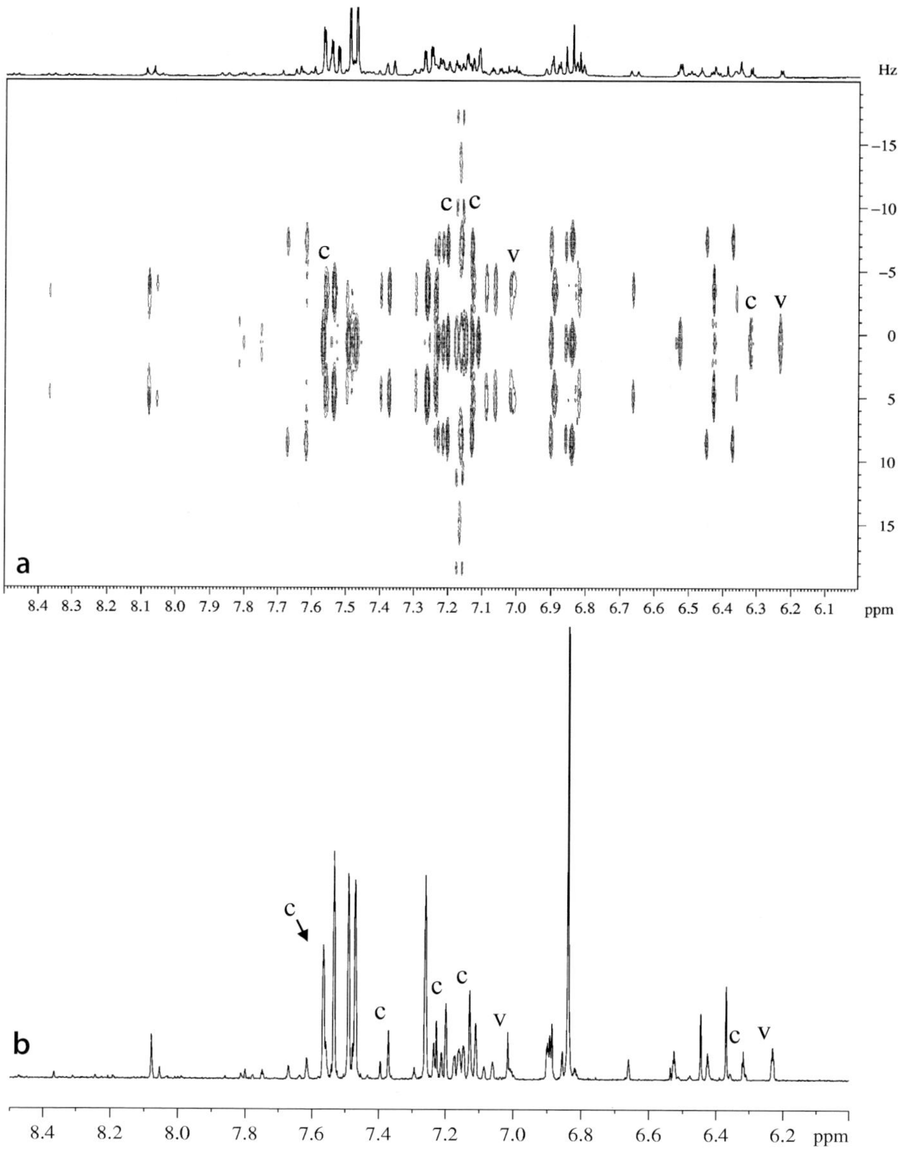

Fig. 6. a J-resolved NMR spectra of MeOD-KH_2PO_4 buffer (pH 6.0) extraction of *C. roseus* leaves. **b** Projection of J-resolved spectra. Vindoline signals (v) of H-9 and H-10, catharanthine signals (c) of H-9, 10, 11, 12 were indicated (unpublished data)

3.3 2D NMR for Structural Confirmation

Although ^{1}H NMR provides a wealth of structural information, extensive overlapping in ^{1}H NMR spectrum often makes it difficult to identify metabolites in plants. Therefore, 2D NMR methods are essential to identify the metabolites. There are a number of 2D NMR methods which can be applied. Homonuclear correlated spectroscopy (COSY) and total correlated spectroscopy (TOCSY) are helpful to obtain information of connectivity and correlation between protons (Braunschweiler and Ernst 1983; Bax and Davis 1985). The most crowded region in the ^{1}H NMR spectra of plant extracts is in the range of δ 3.0 to δ 5.0

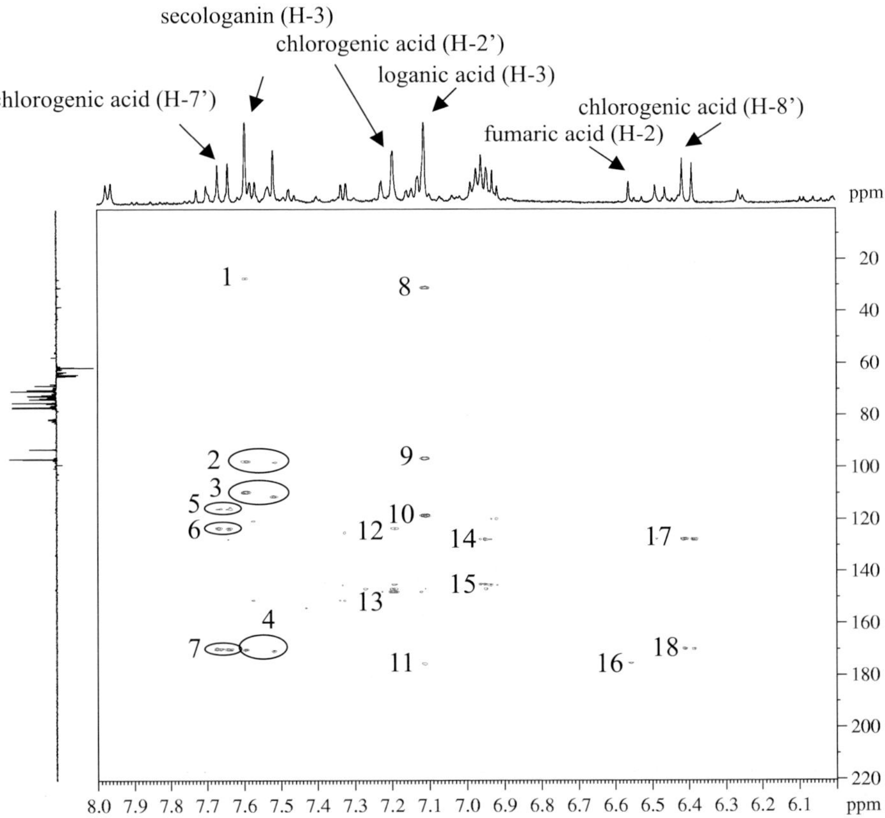

Fig. 7. HMBC spectrum of water fraction of phytoplasma (UDINESE) infected *C. roseus* leaves in the rage of δ 5.5–7.5. *1*; correlation of H-3 and C-5 of secologanin, *2*; correlation of H-3 and C-1 of secologanin, *3*; correlation of H-3 and C-4 of secologanin, *4*; correlation of H-3 and carbonyl group of secologanin, *5*; correlation of H-7′ and C-2′ of chlorogenic acid, *6*; correlation of H-7′ and C-6′ of chlorogenic acid, *7*; correlation of H-7′ and carbonyl group of chlorogenic acid, *8*; correlation of H-3 and C-5 of loganic acid, *9*; correlation of H-3 and C-1 of loganic acid, *10*; correlation of H-3 and C-4 of loganic acid, *11*; correlation of correlation of H-3 and carbonyl group of loganic acid, *12*; correlation of H-2′ and C-1′ of chlorogenic acid, *13*; correlation of H-2′ and C-3′, *14*; correlation of H-2 and C-1 of gallic acid derivatives, *15*; correlation of H-2 and C-3 of gallic acid derivatives, *16*; correlation of H-2 and carbonyl group of fumaric acid, *17*; correlation of H-8′ and C-1′ of chlorogenic acid, *18*; correlation of H-8′ and carbonyl group of chlorogenic acid. (With kind permission of American Society of Plant Biologists, reproduced from Choi et al. 2004c)

where various amino acids and carbohydrates have their signals. The complex signals in this region can be assigned by COSY and TOCSY.

Furthermore, there are several kinds of C-H correlation spectra. Heteronuclear multiple quantum coherence (HMQC) and heteronuclear single quantum coherence (HSQC) spectra give information of direct C-H correlations (J_1). These two-dimensional NMR spectra are very useful for identifying anomeric carbons of carbohydrates (δ 90–δ 110), C-6 and C-8 of flavonoids (δ 95–δ 110) and methyl protons of terpenoids (δ 10–δ 25). For long range correlations (J_2 and J_3) in molecules, heteronuclear multiple bond correlation (HMBC) is applied to confirm structures of plant metabolites. Figure 7 shows an example of the HMBC spectrum to identify the metabolites in the aqueous fraction of *C. roseus* leaves. In case of amino acids, H-2 or H-3 is correlated with the carbonyl group of the amino acid. The proton of alanine at δ 1.48 (H-3d, J, = 4.8 Hz) correlates with the carbon at δ 178.6, glutamic acid at δ 2.14 (m) and δ 2.38 (m) correlates with δ 179.2, and glycine at δ 3.56 (s) correlates with δ 174.7. For iridoids such as loganic acid and secologanin, several correlations between protons and carbons (see figure) confirm the identity of these compounds.

4 Principal Component Analysis

The goal of metabolomic studies is either to characterize an organism or to determine the effect of certain conditions on the organism. It thus requires one to determine first the biological variability of the system followed by determining any significant change. This requires the comparison of a large number of spectra. Thus unbiased or non-targeted analysis is required for these huge data sets. For this purpose multivariate analysis and in particular principal component analysis (PCA) are suited. PCA is an unsupervised clustering method requiring no knowledge of the data set. It reduces the dimensionality of multivariate data while preserving most of the variance within it (Goodacre et al. 2000). All samples are plotted on the coordinates consisting of raw variables (chemical shifts in the case of NMR applications) and a line is constructed based on the best approximation of the data in the least squares sense. Each sample is projected onto this line. The co-ordinate value along the line is a PC1 score. Other PCs can be calculated by the line orthogonal to former PCs (Eriksson et al. 2001). Generally, the separation takes place in the first two components (PC1 and PC2). For the PCA, care must first be taken to choose an appropriate scaling method. The unit variance scaling method uses a reciprocal of standard deviation. It results in normalizing the effect of big and small signals. However, noise in the spectra might have a bigger effect on the result than expected. No scaling is used in PCA combined with NMR spectra because it could preserve the original effect of each variable but the effect of minor metabolites (in particular plant secondary metabolites) is probably neglected. The Pareto scaling method is preferred for application to the analysis of NMR spectra. It gives each variable a variance numerically equal to its initial standard deviation instead of unit variance. Therefore, the Pareto scaling is an intermediate between no scaling and unit variance scaling. The principal components can be displayed graphically as a scores plot. This plot is useful for observing any grouping in the data set. Coefficients by which the original variables must be multiplied to obtain the PC are called loadings. The numerical value of a loading of a given variable on a PC shows how much the variable has in common with that component. Thus, for NMR data, loading plots can be used to detect the metabolites responsible for the separation in the data.

Figure 8 shows an example of PCA of healthy and phytoplasma-infected *C. roseus*. PCA score plots of healthy and infected plants by ten different phytoplasmas show that healthy *C. roseus* leaves are clearly separated from the phytoplasma infected leaves in both chloroform fraction (Fig. 8a) and water fraction (Fig. 8b). Loading plots explain that, in the chloroform fraction, the

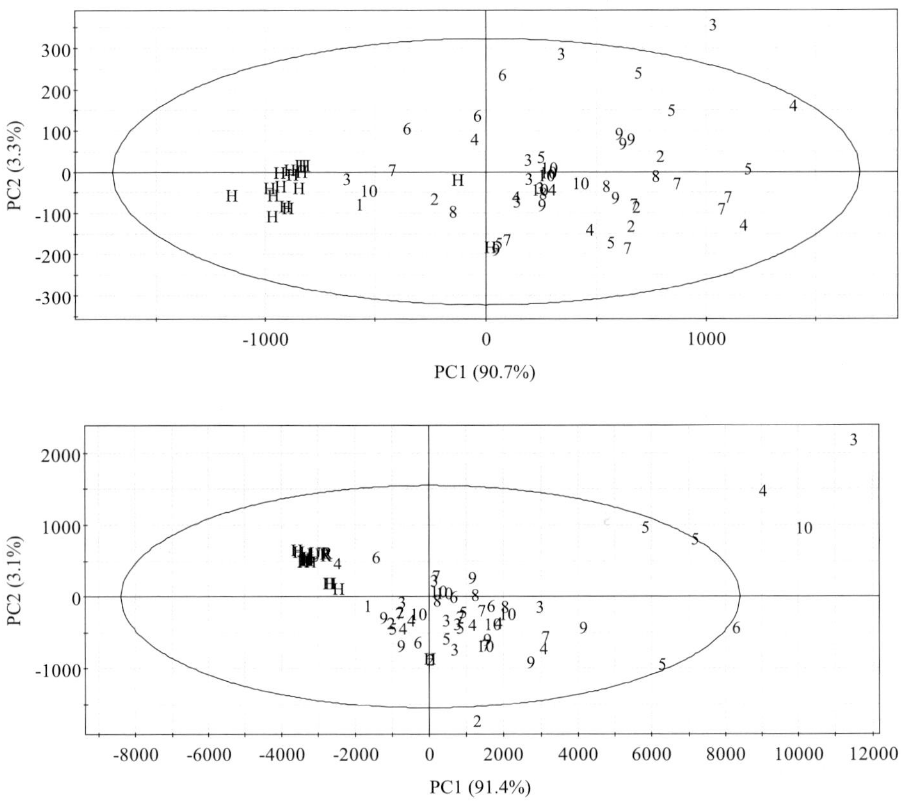

Fig. 8. Score plot of principal component analysis of: **a** $CHCl_3$ extracts; **b** water extracts of healthy and phytoplasma infected *C. roseus* leaves. *1–10*: Infected plants by 10 different phytoplasmas, H; healthy plant. The *ellipse* represents the Hotelling T2 with 95% confidence in score plots. (With kind permission of American Society of Plant Biologists, reproduced from Choi et al. 2004c)

responsible components for separation are: fatty component – the signals at 1.2–1.4 (CH_2), 5.0–5.5 (olefinic CH_2), and indole alkaloids such as vindoline – the signals at 3.79 (OCH_3 of C-11), 3.78 (OCH_3 of C-22), 0.49 (H-18). It indicates that *C. roseus* leaves infected by phytoplasma contain less fatty components and higher vindoline compared to healthy leaves. When the intensity of each signal was compared, it is clear that infected leaves have two to four times increased level of vindoline relative to healthy plants. For the water extract, the score plot shows that healthy leaves are well separated from infected plants by both PC1 and PC2 (Fig. 8b). The healthy leaves have lower PC1 and higher PC2 relative to infected ones. The loading plot of PC1 and PC2 explained that most of infected *C. roseus* leaves have higher amounts of sucrose, chlorogenic acid, loganic acid, secologanin, and polyphenols compared to healthy plants.

By using ^{1}H NMR in combination with PCA, it is clearly shown that the metabolites related to the biosynthesis of terpenoid indole alkaloid (loganin acid, secologanin, vindoline) and phenylpropanoids (chlorogenic acid, polyphenol) are present in higher amounts in the phytoplasma-infected leaves.

5 Concluding Remarks

Several analytical methods may be used for metabolomic profiling of plants; however, ^{1}H NMR spectra offers a wealth of information of metabolites compared to other methods. Decoupled NMR spectra (J-resolved) provide even more information since metabolites can be accurately integrated and it can exclude broad resonances from macromolecules and spin–spin coupling data. In addition, combinations of two dimensional NMR methods are quite helpful to identify the metabolites in plant extract.

To be able to compare all the data generated from NMR from different experiments and different laboratories, a large database is required. However, that requires a high degree of reproducibility, which can be achieved by using a standardized method for sample preparation and data acquisition.

So far, NMR has been successfully used for the metabolomic fingerprinting and profiling of plants and is successfully applied in quality control of among others, food and botanicals. The use of NMR metabolomics in functional genomics will be the challenge for the coming year.

References

Aharoni A, Ric de Vos CH, Verhoeven HA, Maliepaard CA, Kruppa G, Bino R, Goodenowe D (2002) Non-targeted metabolomic profiling using Fourier transform ion cyclotron mass spectrometry (FTMS). OMICS J Intergrative Biol 6:217–234

Bax A, Davis DG (1985) MLEV-17-based two-dimensional homonuclear magnetization transfer spectroscopy. J Magn Reson 65:355–360

Braunschweiler L, Ernst RR (1983) Coherence transfer by isotropic mixing: application to proton correlation spectroscopy. J Magn Reson 53:521–528

Brescia MA, Caldarola V, de Giglio A, Benedetti D, Fanizzi FP, Sacco A (2002) Characterization of the geographical origin of Italian red wine based on traditional and nuclear magnetic resonance spectrometric determinations. Anal Chim Acta 458:117–186

Charlton AJ, Farrington WHH, Brereton P (2002) Application of ^{1}H NMR and multivariate statistics for screening complex mixtures: quality control and authenticity of instant coffee. J Agric Food Chem 50:3098–3103

Choi H-K, Choi YH, Verberne M, Lefeber AWM, Erkelens C, Verpoorte R (2004a) Metabolic fingerprinting of wild type and transgenic tobacco plants by ^{1}H NMR and multivariate analysis technique. Phytochemistry 65:857–864

Choi YH, Kim HK, Hazekamp A, Erkelens C, Lefeber AWM, Verpoorte R (2004b) Metabolomic differentiation of *Cannabis sativa* cultivars using ^{1}H NMR spectroscopy and principal component analysis. J Nat Prod 67:953–957

Choi YH, Tapias EC, Kim HK, Lefeber AWM, Erkelens C, Verhoeven JTJ, Brzin J, Verpoorte R (2004c) Metabolomic discrimination of *Catharanthus roseus* leaves infected by phytoplasma using ^{1}H-NMR spectroscopy and multivariate data analysis. Plant Physiol 135:2398–2410

Choi YH, Sertic S, Kim HK, Wilson EG, Michopoulosa F, Lefeber AWM, Erkelens C, Verpoorte R (2005) Classification of *Ilex* species based on metabolomic fingerprinting using NMR and multivariate data analysis. J Agric Food Chem 53:1237–1245

Duarte I, Barros A, Belton PS, Righelato R, Spraul M, Humpfer E, Gil AM (2002) High-resolution nuclear magnetic resonance spectroscopy and multivariate analysis for the characterization of beer. J Agric Food Chem 50:2475–2481

Eriksson L, Johansson E, Kettaneh-Wold N, Wold S (2001) Multi-and megavariate data analysis. Umetrics Academy, Umeå, Sweden

Fiehn O (2002) Metabolomics-the link between genotypes and phenotypes. Plant Mol Biol 48:155–171

Frédérich M, Choi YH, Angenot L, Harnischfeger G, Lefeber AWM, Verpoorte R (2004) Metabolomic analysis of *Strychnos nux-vomica*, *Strychnos icaja* and *Strychnos ignatii* extracts by ^{1}H nuclear magnetic resonance spectrometry and multivariate analysis techniques. Phytochemisty 65:1993–2001

Goodacre R, Shann B, Gilbert RJ, Timmins EM, McGovern AC, Kell DB, Logan NA (2000) Detection of the dipicolic acid biomarker in *Bacillus* spores using Curie-point pyrolysis mass spectrometry and Fourier transform infrared spectroscopy. Anal Chem 72:119–127

Huhman DV, Sumner LW (2002) Metabolic profiling of saponins in *Medicago sativa* and *Medicago trumcatula* using HPLC coupled to an elcetrospreay ion-trap mass spectrometer. Phytochemisty 59:347–360

Kim HK, Choi YH, Luijendijk TJC, Vera Rocha RA, Verpoorte R (2004) Comparison of secologanin extraction methods and quantitative analysis of secologanin from *Symphoricarpos albus* by using ^{1}H-NMR. Phytochem Anal 15:257–261

Rhee IK, Appels N, Hoete B, Karabatak B, Erkelens C, Stark LM, Flippin LA, Verpoorte R (2004) Isolation of the acetylcholinesterase inhibitor ungeremine from *Nerine bowdenii* by preparative HPLC-coupled on-line to a flow assay system. Biol Pharm Bull 27:1804–1809

Schripsema J, van Beek TA, Verpoorte R, Erkelens C, Perera P, Tibell C (1987) A reinvestigation of the stereochemistry of tubotaiwine using NMR spectroscopy. J Nat Prod 50:89–101

Sumner LW, Mendes P, Dixon RA (2003) Plant metabolomics:large-scale phytochemistry in the functional genomics era. Phytochemistry 62:817–836

Tolstikov VV, Fiehn O (2002) Analysis of highly polar compounds of plant origin: combination of hydrophilic interaction chromatography and electrospray ion trap mass spectroscopy. Anal Biochem 301:298–307

Viant MR (2003) Improved methods for the acquisition and interpretation of NMR metabolomic data. Biochem Biophys Res Commun 310:943–948

Vogels JTWE, Terwel L, Tas AC, van den Berg F, Dukel F, van der Greef J (1996) Detection of adulteration in orange juices by a new screening method using proton NMR spectroscopy in combination with pattern recognition techniques. J Agric Food Chem 44:175–180

III.7 Metabolomics of Plant Secondary Compounds: Profiling of *Catharanthus* Cell Cultures

M. Orešič, H. Rischer, and K.-M. Oksman-Caldentey[1]

1 Introduction

Plants have supplied mankind with potent medication since ancient times and still provide a largely untapped resource for the discovery of novel pharmaceuticals (Verpoorte 1998). Currently only about 10% of higher plants are chemically characterized to some extent. Low molecular weight compounds, usually referred to as 'secondary metabolites', exhibit many biological functions such as stress response (Hirai et al. 2004) but for many the exact function remains unknown. Because of their extremely diverse chemical structures and hence their pharmacophoric properties, these natural products constitute an important addition to compound libraries forming the basis for all drug discovery and development efforts (Hostettman and Terreaux 2000).

Alkaloids are one of the most studied groups of plant secondary metabolites. Currently about 15,000–16,000 different alkaloids are known (Verpoorte 2000) and they can be further classified into several subclasses according to their chemical structures. In contrast to, e. g. phenolic compounds, which are abundant throughout the whole plant kingdom, alkaloids are often restricted to certain plant families or even certain plant species. The reason why alkaloids have been of such wide interest can be explained by their strong physiological properties leading to their use as, e. g. pharmaceuticals or pesticides. Furthermore, the isolation of alkaloids from plant matrices is relatively simple compared to many other plant compounds. This has allowed scientists to measure and isolate very small amounts of various alkaloids using different chromatographical systems (GC, LC) combined later with spectrometry (e. g. MS, NMR) for their structure elucidations.

The importance of plants as a source of new drug molecules can be nicely illustrated by the following figures. During the past 20 years, 28% of new drug entities were either natural products or derived from them as semi-synthetic derivatives and, in addition to that, 24% of the drugs were synthesized after the molecule was first discovered from natural resources (Newman et al. 2003). Modern high throughput screenings (HTS) allow enormous numbers of samples to be tested automatically for biological effects using molecular targets (Cordell 2000). There are three strategies usually applied for the discovery of bioactive compounds from plants and all of them have provided promising substances for further testing. The simplest case is the indiscriminate

[1] VTT Biotechnology, P.O. Box 1500, 02044 VTT, Finland, e-mail: Kirsi-Marja.Oksman@vtt.fi

Biotechnology in Agriculture and Forestry, Vol. 57
Plant Metabolomics (ed. by K. Saito, R.A. Dixon, and L. Willmitzer)

extraction of as many species as possible. Another possibility is the exploitation of ethnobotanical knowledge gathered from indigenous communities, and finally the targeted search for useful compounds within related groups of plants which have already been shown to contain potent metabolite classes. However, no matter how a new lead compound had been discovered, plant derived substances share the same problem: their chemical synthesis is often economically unfeasible. The development of urgently needed alternative production systems such as genetically engineered plant cell cultures are hampered by the incomplete knowledge of biosynthetic pathways leading to the target compounds (Oksman-Caldentey and Inzé 2004).

2 Metabolomics as a Platform to Study Plant Secondary Metabolites

The 'omics' revolution has empowered us with the ability to measure large numbers of biomolecular components in parallel, therefore enabling the systems approach. Metabolites are known to be involved as key regulators of systems homeostasis. As such, level changes of specific groups of metabolites may be descriptive of systems responses to environmental or genetic interventions, and their study may therefore be a powerful tool for characterization of complex phenotypes (Orešič et al. 2004). Among the emerging *omics* technologies, metabolomics has gained the prominence most recently, yet it may also be considered as the oldest of the *omics* approaches. The pioneering research on use of metabolic fingerprinting as a phenotyping tool dates back to the 1970s and 1980s (Jellum 1977; van der Greef et al. 1983; Windig and Meuzelaar 1984). On entering a post-genomic era and with new advances in analytical and informatics technologies, the metabolomics approach is becoming more feasible, making it one of the core components of systems biology research.

The general problem encountered when characterizing the plant metabolome is the highly complex nature and the enormous chemical diversity of the compounds. Additionally, under physiological conditions the metabolites exist across a very broad concentration range, possibly 10 orders of magnitude. Plants produce approximately 200,000 metabolites (Fiehn 2002), many of which play specific roles in allowing adaptation to specific ecological niches. The range of chemical properties sets a challenge to the analytical tools both for profiling multiple metabolites in parallel, and for quantitatively analyzing the selected ones. This has especially become obvious in secondary metabolite analysis, which is far more complex than metabolite profiling of primary metabolites. Metabolites have very different chemical natures, which influence their extractability in various solvents, pH requirements and sensitivity for extraction conditions (e. g. temperature, pressure, time). As a consequence, if applying one general extraction system, it is very likely that many metabolites remain in the plant matrix and cannot be profiled. This holds also if the

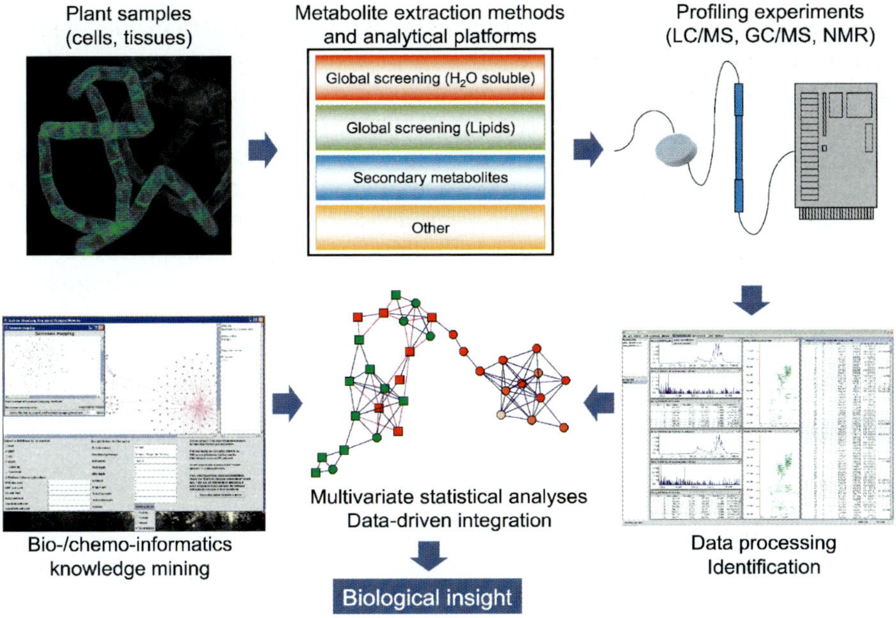

Fig. 1. Schematic representation of an integrated metabolomics platform for studying plant metabolites

specific extractions system is developed for compounds of particular chemical properties; further extracts have to be analysed (Oksman-Caldentey and Saito 2005).

One of the key technological challenges of metabolite profiling is therefore finding the optimal balance between the accuracy and coverage of metabolite measurements. This can be achieved by first dividing the metabolomics platform into multiple methods of varying coverage and specificity (Fig. 1). Advances in instrumentation for metabolite analyses are empowering us with the ability to increase the coverage of metabolites within a single analysis. Most commonly, gas chromatography-mass spectrometry (GC/MS) and liquid chromatography-mass spectrometry (LC/MS) based approaches have been applied in plant metabolomics applications (Fiehn 2002). For example, it is estimated that a single plant can contain 10,000 metabolites and currently by using GC/ToF-MS technology it is possible to detect around 1000 of them from a single sample (Weckwerth et al. 2004). Following analytical measurements, the role of data processing algorithms is to detect the peaks in spectral data (peak detection), match the corresponding peaks across multiple samples (alignment), and correct the peak intensities due to instrumental variability (normalization). These methods enable us to track differentially the metabolite levels in multiple environmental conditions or time points, even if some of the compound identities are not known.

However, when studying plant secondary metabolites and their role in physiological responses to various environmental stress conditions, we are also interested in finding and identifying compounds that are either unknown or not previously analyzed, so there is insufficient data available from profiling experiments alone for accurate identification. The data processing methods outlined above may play an essential role in elucidating the biological role of such compounds, and multivariate approaches combining the profiles of unknown compounds with known metabolites, transcriptional, proteomic and phenotype information may help us direct the process of identifying the most relevant compounds based on their correlations with known compounds and specific biological processes (Fig. 1). This is particularly important since the process of identification can be very difficult and time consuming, and it is unlikely that all peaks found in spectral data can be identified with sufficient confidence.

3 Case Study: Metabolic Profiling of *Catharanthus roseus* Cells

In this section we focus on the question whether a metabolomics approach using LC/MS is feasible for the exploration of secondary metabolites in cell cultures of the medicinal plant *Catharanthus roseus* L. G. Don (Apocynaceae). The intention was strictly to find out how many peaks could be detected unambiguously in only one sample fraction known to contain the most important secondary metabolites from the pathways leading to terpenoid indole alkaloids (TIAs), and to apply the data processing tools which proved useful in other projects.

C. roseus represents an extensively studied object because of the presence of TIAs, several of which are in high demand for pharmaceutical use. Notable is that closely related alkaloids derived from the same pathway may possess completely different pharmacological properties. The clinically used anticancer substances vincristine and vinblastine are still derived from field grown plants and therefore yield high market prices. So far all attempts to produce these phytopharmaceuticals by means of plant cells in bioreactors have economically failed (Moreno et al. 1995). Compared to other compound classes there is a relatively clear picture of the biosynthetic pathway leading to TIAs (Fig. 2). The central metabolite strictosidine is derived from moieties delivered by the shikimate and by the plastidic non-mevalonate (MEP) pathway. From here multiple routes lead to a great diversity of alkaloids of which vindoline and catharanthine constitute the building blocks for the formation of the previously mentioned bisindole alkaloids. In cell cultures low levels of vindoline are generally the limiting factor for bisindole alkaloid accumulation. This has been attributed to transcriptional blockage (Vazquez-Flota et al. 2002) and multicellular compartmentation of the vindoline biosynthesis (St-Pierre et al. 1999). Recently good progress has been made in elucidating partly the transcriptional

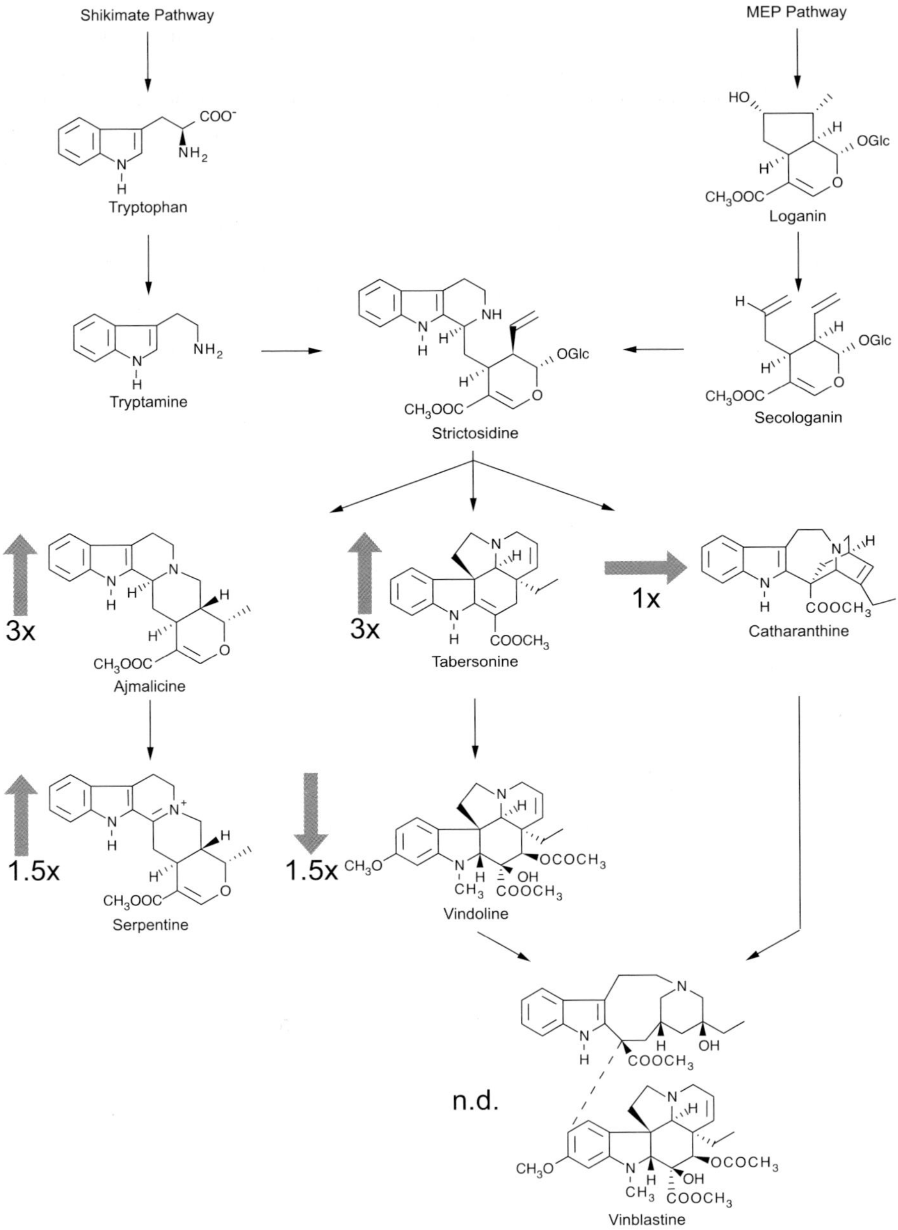

Fig. 2. Scheme of TIA biosynthesis. The up-/downregulation of several identified compounds is indicated at $p < 10^{-4}$

regulation after elicitation (Pauw et al. 2004). Nevertheless, at the metabolite level, the picture is generally very incomplete because, even if a universal extraction method is used, analytics comprise rarely more than 10 metabolites at once (Tikhomiroff and Jolicoeur 2002). This is a clear motivation for the attempt to use more comprehensive metabolomic approaches for the profiling of secondary metabolites in the *Catharanthus* system as described here.

3.1 Results

The principal of elicitation was used to induce a differential response on the secondary metabolite level as a stress response (Reymond and Farmer 1998). We applied jasmonates since they act as signalling molecules by activating gene expression of biosynthesis genes in a coordinated way (Memelink et al. 2001).

We profiled 20 samples in total, 10 control strains and 10 elicited strains. The replicates are the same strain in parallel cultures corresponding to the same time point. They can thus be considered as biological replicates. Total ion chromatograms already reveal clear differences at the retention time of approximately 21 min, as can be seen from Fig. 3 comparing two representative samples. However, from chromatography alone we cannot determine differences at the level of individual compound peaks, which can be found after processing LC/MS profile data.

Following data processing as explained in Sect. 4.3, we found 4190 peaks, and analyzed the data using Principal Components Analysis (Jackson 1991)

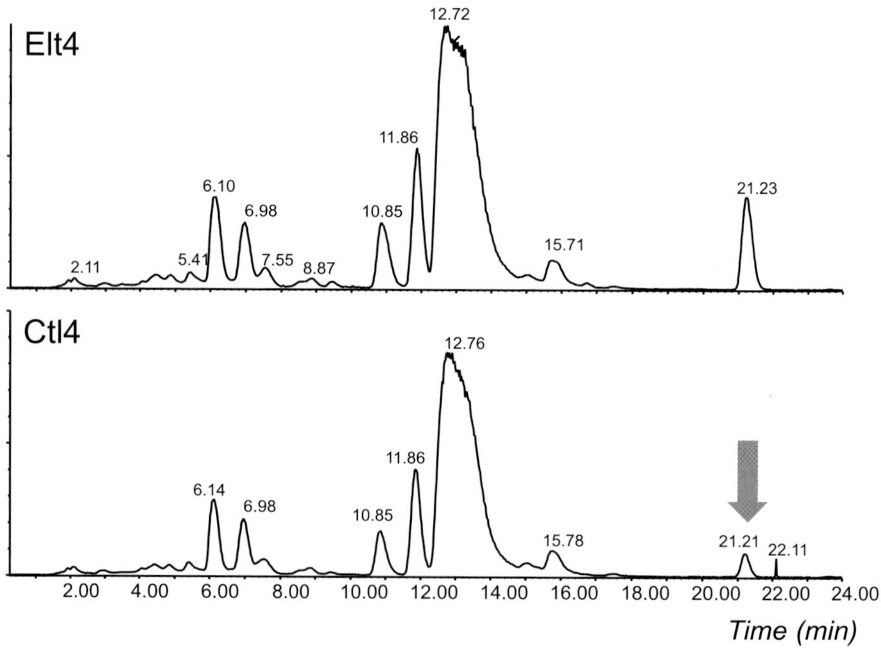

Fig. 3. Total ion chromatograms of two representative LC/MS *Catharanthus roseus* profiles. Elt4 = elicited cells, Ctl4 = control cells

using mean centering transformation of data. PCA is a linear dimensionality reduction method which finds the orthogonal axes of maximum variance in data, e. g. the principal components. Figure 4a shows that excellent separation between the control and elicited groups can be found already with the first principal component, which accounts for 83% of total variance. In order to identify the variables contributing most to the differences between the two groups of samples, we investigated the loadings for the first principal component. The loading plot in Fig. 4b shows that there are three major compounds descriptive of the elicited cultures: ajmalicine, tabersonine, and a compound we were unable to identify from the internal reference database containing spectral compound information.

Extracting and interpreting the results for these compounds from the peak list serves as a nice control whether the data make biological sense. Figure 2 shows a few examples of relevance to the known pathways. As expected, vinblastine was not detected in the samples, which is also explainable by the observed significant downregulation of vindoline, only present in trace amounts, by a factor of 1.5. Upstream tabersonine on the other hand differentially accumulated threefold. It has been known for several years that the late stages in the vindoline pathway are strictly regulated in a development-specific, tissue-specific and light-dependant manner (St-Pierre and de Luca 1995) and are therefore partly not functional in undifferentiated cells. Catharanthine levels were unaltered at the observed time point 24 h after elicitation, whereas the ajmalicine concentration increased threefold and serpentine 1.5-fold. These results are again very much in line with reports in the literature using a targeted approach (Lee-Parsons et al. 2004).

We have also found several differentially regulated compounds which are not part of the shikimate or MEP pathways. Based on results of PCA and univariate statistical analyses, we are currently pursuing targeted MS/MS analyses on selected peaks that we were unable to identify from the internal reference database.

Overall, we can conclude that the chosen differentially responding system, i. e. elicited vs non-elicited *Catharanthus* cell cultures, is very suitable for a metabolomic approach. Even using only one extracted fraction and one chromatographic condition, more than 4000 peaks were unambiguously detected. The PCA employed proved sufficient for reducing the complexity of the data and afforded the selection of potential compounds exhibiting the greatest modulation. Peak identification was possible for a number of compounds involved in the TIA pathway by comparison of spectral characteristics and retention times of references in a database and their behaviour is sound in the biological sense.

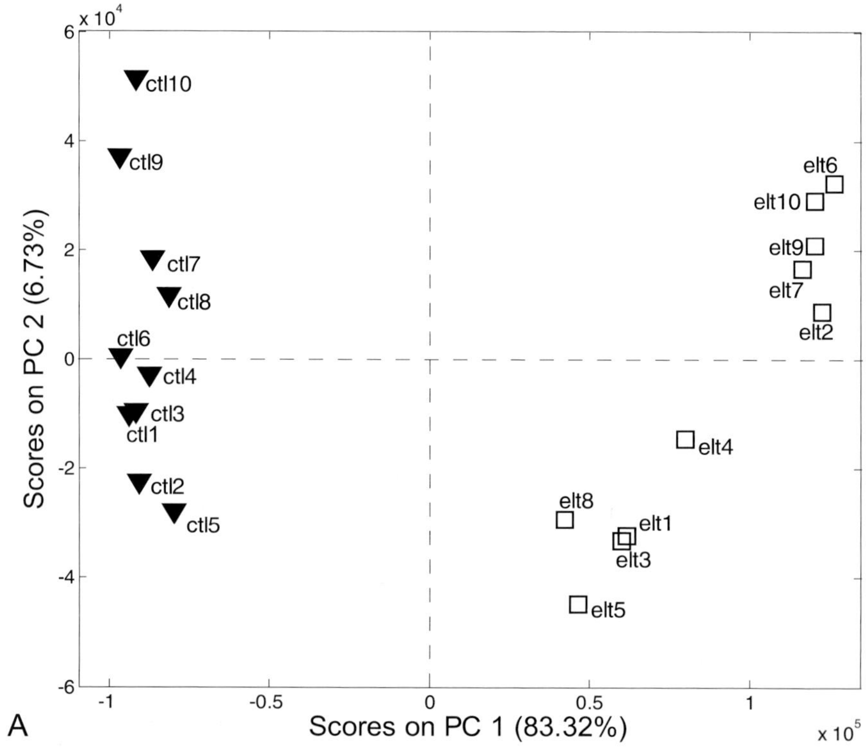

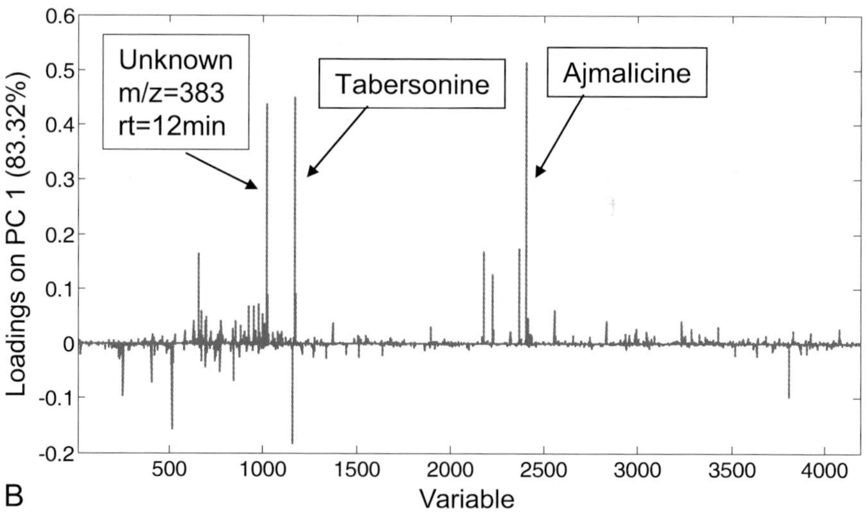

Fig. 4. A Principal Components Analysis (PCA) shows differences between the elicited and control samples. **B** The factor analysis reveals three compounds as major contributors to the clustering of elicited samples

4 Protocol

4.1 Plant Material and Sample Preparation

Catharanthus roseus cell suspensions were grown in liquid Gamborg B5 medium (Gamborg et al. 1968) containing 20 g/L sucrose and 1.86 mg/L NAA in an incubator shaker at 26 °C, continuous light and 130 rpm. Elicitations were started at day six after inoculating 2 g fresh weight of cells in 25 mL medium contained in 100-mL Erlenmeyer flasks by addition of methyl jasmonate dissolved in DMSO at a final concentration of 50 μmol/L or DMSO alone as a control. Cells were harvested by vacuum filtration after 24 h. Extraction followed the modified protocol of Whitmer et al. (2002). Prior to extraction, 50 mg of lyophilized cells were spiked with vincamine as internal standard and extracted with 15 mL ethanol in an ultrasonic bath for 10 min. Following centrifugation at 5000 rpm for 10 min, the solvent was decanted and evaporated to dryness. Dry samples were stored at −20 °C until analysis. Then the samples were redissolved in a 1:1 mixture of acetonitrile and 10 mmol/L ammonium acetate pH 10 and 25 μL of the solution were injected into the HPLC after centrifugation.

4.2 HPLC/ESI/MS

HPLC separation was performed using a Waters HT-Alliance 2795 system and was monitored with a Micromass Quattro Micro triple quadrupole mass spectrometer equipped with an electrospray source. The ion source was operated at capillary voltage 3.20 kV and cone voltage 45 V. Source and desolvation temperatures were 130 °C and 290 °C, respectively. Desolvation gas flow was 900 L/h and cone gas flow 30 L/h. The scan mode function was applied to record the protonated molecular ions. An aliquot of 25 μL of sample were loaded onto a reverse-phase C18 column (Xterra MS C18, 4.6 × 150 mm, 5 μm, Waters) at 35 °C. The sample was eluted within 30 min using isocratic conditions of 10 mmol/L ammonium acetate at pH 10 and acetonitrile (55:45) applying a flow of 1 mL/min and a split of 0.2 mL/min reaching the mass spectrometer.

4.3 Processing of LC/MS Data

Raw data from the MS instrument were converted to NetCDF format using the DataBridge application from MassLynx software. We used our MZmine software package to process the data (Katajamaa and Orešič 2005). While the software, which contains several options at each stage of LC/MS data processing, will be discussed in detail elsewhere, we summarize below the methods and their parameters used for this chapter.

Each scan was filtered using the mean window filter with a 0.3 mass unit window, followed by peak detection in the m/z dimension using the recursive threshold method. The m/z peaks were then compared across different retention time scans. A peak was connected with another peak in the neighbouring scan if the m/z difference was below a threshold, set to 0.1 mass units. Only peaks connected by more than 8 scans and less than 100 were retained in our analysis. For each peak we recorded the m/z and retention time at the position of the maximum height across connected m/z peaks, as well as peak height in this position as the measure of intensity. We then performed alignment across different samples by creating a master list of all peaks and finding the best matched peaks from each sample. The peaks were matched across different samples if they met specific similarity criteria, in our case 0.15 mass units in m/z and 15 scans in retention time. Following this process, we created a data matrix of intensities, where as row indices each peak was described by m/z and retention time value, and as column indices were the samples.

The normalization method is based on multiplicative error model normalization first applied to gene expression data (Hartemink et al. 2001). The method has since been applied to LC/MS data as well (Orešič et al. 2004). Log-values of intensities (with index i for peaks, and j for samples) were modelled by a linear model:

$$y_{ij} = \mu_i + s_j + \varepsilon_{ij} \tag{1}$$

where the peak effects are described by μ_i, the sample-specific effect by s_j, and the error by ε_{ij}. We assume the error is normally distributed with zero mean and the variance σ_i, i. e. we do permit the variance to be different for each peak. The optimal parameters of the model were calculated using the maximum likelihood method, with the difference that the model is applied to the whole dataset, not only to the internal standards. Each sample was then scaled by the factor $\exp(-s_j)$.

5 Perspectives

The recent advances in functional genomics offer unprecedented opportunities to use the biochemical capacity of plants to produce and to design novel compounds. At the same time the integrations of metabolomics into other 'omics' (transcriptomics and proteomics) have brought us closer to understanding different levels towards systems biology. However, before we can metabolically engineer, e. g. medicinal plants or their cell cultures to produce secondary metabolites of high-value, we need to profile the changes in the whole metabolome. This cannot be done only with the current tools of targeted metabolite analysis which gives us very limited information how the whole metabolic machinery works. Therefore sophisticated tools are needed in metabolite profiling to connect the functions of individual genes or in combination in a system.

The method utilized for metabolite profile analysis in this chapter, i. e. PCA, is a common dimensionality reduction method which has been applied to analyses of metabolite fingerprint data since the 1970s. PCA effectively enables study of the objects (i. e. biological samples) with a reduced number of variables. PCA is a powerful first-pass method for analysis of profile data, which can detect major patterns or trends in data. Several related methods such as Sammon's mapping (Sammon 1969), self organizing maps (Kohonen 2001), or independent component analysis (Hyvärinen and Oja 2000) may also be applied for the same purpose. However, when the changes between phenotypes of interest are minor, i. e. not contributing significantly to overall variation in data, and are possibly limited to specific sub-groups of compounds, PCA and similar methods may fail detecting such changes.

An intuitive method to study subtle patterns of changes at the level of individual compound peaks is correlation networks. With this method we estimate the level of co-regulation between the pairs of compounds by calculating the correlation coefficients across their profiles. This can be based on linear methods such as Pearson correlation (Kose et al. 2001) or other estimates of similarity. In the domain of plant secondary metabolites where the pathways are largely unknown, it is also important to understand the co-regulation between the metabolite and other levels, such as transcript, protein, and genome. The correlation network analysis can be extended to such integrated data, with additional precautions in regards to normalization and similarity measures (Griffin et al. 2004; Orešič et al. 2004). With appropriate experiment design, such approaches may also facilitate discovery of novel pathways. In this respect peak identification plays a crucial role, too, as only this piece of information allows for matching new data with already known reactions in biochemical pathways.

Acknowledgements. This research has been funded by the National Technology Agency of Finland (Tekes) programme "NeoBio" to KMOC and supported by a Marie Curie Fellowship of the European Community programme 'Quality of Life' under contract number QLK4-2002-51547 granted to HR. We thank J. Rikkinen for technical assistance regarding tissue culture work, T. Seppänen-Laakso for analytical help, M. Katajamaa for the *masso* software development, and R. Verpoorte, Division of Pharmacognosy, Leiden University for the *Catharanthus roseus* cell line.

References

Cordell GA (2000) Biodiversity and drug discovery – a symbiotic relationship. Phytochemistry 55:463–480

Fiehn O (2002) Metabolomics – the link between genotyopes and phenotypes. Plant Mol Biol 48:155–171

Gamborg OL, Miller RA, Ojima K (1968) Nutrient requirements of suspension cultures of soybean root cells. Exp Cell Res 50:151–158

Griffin JL, Bonney SA, Mann C, Hebbachi AM, Gibbons GF, Nicholson JK, Shoulders CC, Scott J (2004) An integrated reverse functional genomic and metabolic approach to understanding orotic acid-induced fatty liver. Physiol Genomics 17:140–149

Hartemink AJ, Gifford DK, Jaakkola TS, Young RA (2001) Maximum likelihood estimation of optimal scaling factors for expression array normalization. SPIE BiOS, San Jose, California

Hirai MY, Yano M, Goodenowe DB, Kanaya S, Kimura T, Awazuhara M, Arita M, Fujiwara T, Saito K (2004) Integration of transcriptomics and metabolomics for understanding of global responses to nutritional stresses in *Arabidopsis thaliana*. Proc Natl Acad Sci USA 101:10205–10210

Hostettman K, Terreaux C (2000) Search for new lead compounds from higher plants. Chimia 54:652–657

Hyvärinen A, Oja E (2000) Independent component analysis: algorithms and applications. Neural Comput 5:411–430

Jackson JE (1991) User's guide to principal components. Wiley, New York, NY

Jellum E (1977) Profiling of human body fluids in healthy and disease states using gas chromatography and mass spectrometry, with special reference to organic acids. J Chromatogr 143:427–462

Katajamaa M, Orešič M (2005) Processing methods for differential analysis of LC/MS profile data. BMC Bioinformatics 6:179.1–179.12

Kohonen T (2001) Self organizing maps. Springer, Berlin Heidelberg New York

Kose F, Weckwerth W, Linke T, Fiehn O (2001) Visualizing plant metabolomic correlation network using clique-metabolite matrices. Bioinformatics 17:1198–1208

Lee-Parsons CWT, Ertürk S, Tengtrakool J (2004) Enhancement of ajmalicine production in *Catharanthus roseus* cell cultures with methyl jasmonate is dependent on timing and dosage of elicitation. Biotechnol Lett 26:1595–1599

Memelink J, Verpoorte R, Kijne JW (2001) ORCAnization of jasmonate – responsive gene expression in alkaloid metabolism. Trends Plant Sci 6:212–219

Moreno PRH, van der Heijden R, Verpoorte R (1995) Cell and tissue-cultures of *Catharanthus-roseus* – a literature survey. 2. Updating from 1988 to 1993. Plant Cell Tissue Organ Cult 42:1–25

Newman DJ, Cragg GM, Snader KM (2003) Natural products as sources of new drugs over the period 1981–2002. J Nat Prod 66:1022–1037

Oksman-Caldentey K-M, Inzé D (2004) Plant cell factories in the post-genomic era: new ways to produce designer secondary metabolites. Trends Plant Sci 9:433–440

Oksman-Caldentey KM, Saito K (2005) Integrating genomics and metabolomics for engineering plant metabolic pathways. Curr Opin Biotechnol 16:174–179

Orešič M, Clish CB, Davidov EJ, Verheij E, Vogels JTWE, Havekes LM, Neumann E, Adourian A, Naylor S, van der Greef J, Plasterer T (2004) Phenotype characterization using integrated gene transcript, protein and metabolite profiling. Appl Bioinformatics 3:205–217

Pauw B, Hilliou FAO, Martin VS, Chatel G, de Wolf CJF, Champion A, Pre M, van Duijn B, Kijne JW, van der Fits L, Memelink J (2004) Zinc finger proteins act as transcriptional repressors of alkaloid biosynthesis genes in *Catharanthus roseus*. J Biol Chem 279:52940–52948

Reymond P, Farmer EE (1998) Jasmonate and salicylate as global signals for defense gene expression. Curr Opin Plant Biol 1:404–411

Sammon JW Jr (1969) A nonlinear mapping for data structure analysis. IEEE Trans Comp C-18:401–409

St-Pierre B, de Luca V (1995) A cytochrome-P-450 monooxygenase catalyzes the first step in the conversion of tabersonine to vindoline in *Catharanthus roseus*. Plant Physiol 109:131–139

St-Pierre B, Vazquez-Flota FA, De Luca V (1999) Multicellular compartmentation of *Catharanthus roseus* alkaloid biosynthesis predicts intercellular translocation of a pathway intermediate. Plant Cell 11:887–900

Tikhomiroff C, Jolicoeur M (2002) Screening of *Catharanthus roseus* secondary metabolites by high-performance liquid chromatography. J Chromatogr A 955:87–93

Van der Greef J, Tas AC, Bouwman J, Ten Noever de Brauw MC, Schreurs WHP (1983) Evaluation of field-desorption and fast atom-bombardment mass spectrometric profiles by pattern recognition techniques. Anal Chim Acta 150:45–52

Vazquez-Flota F, de Luca V, Carrillo-Pech M, Canto-Flick A, Miranda-Ham MD (2002) Vindoline biosynthesis is transcriptionally blocked in *Catharanthus roseus* cell suspension cultures. Mol Biotechnol 22:1–8

Verpoorte R (1998) Exploration of nature's chemodiversity: the role of secondary metabolites as leads in drug development. Drug Discov Today 3:232–238

Verpoorte R (2000) Secondary metabolites. In: Verpoorte R, Alfermann AW (eds) Metabolic engineering of plant secondary metabolism. Kluwer Academic Publ, Dordrecht, pp 1–29

Weckwerth W, Loureiro ME, Wenzel K, Fiehn O (2004) Differential metabolic networks unravel the effects of silent plant phenotypes. Proc Natl Acad Sci USA 101:7809–7814

Whitmer S, van der Heijden R, Verpoorte R (2002) Effect of precursor feeding on alkaloid accumulation by a strictosidine synthase over-expressing transgenic cell line S1 of *Catharanthus roseus*. Plant Cell Tissue Organ Cult 69:85–93

Windig W, Meuzelaar HL (1984) Nonsupervised numerical component extraction from pyrolysis mass spectra of complex mixtures. Anal Chem 56:2297–303

III.8 The *Taxus* Metabolome and the Elucidation of the Taxol®* Biosynthetic Pathway in Cell Suspension Cultures

R.E.B. Ketchum and R.B. Croteau[1]

1 Introduction

Plants of the genus *Taxus* (yews) produce a class of natural products known as taxane diterpenoids or taxoids characterized by the unusual taxane (pentamethyl $[9.3.1.0]^{3,8}$ tricyclopentadecane) skeleton. To date, nearly 400 taxoids have been isolated and characterized (Baloglu and Kingston 1999; Itokawa 2003). The most economically and pharmaceutically important of these compounds is the anticancer drug paclitaxel, known commercially as Taxol® (**1**; Fig. 1). All taxoids are derived from the same parent diterpene olefin, taxa-4,11-diene (**2**; Fig.1) or products of its rearrangement. The biochemical pathway that originates with taxa-4,11-diene and culminates in the production of taxol likely branches from the central pathway leading to the formation of the more abundant taxines present in *Taxus* tissue (Ketchum et al. 2003). Departures from this pathway, the result of variations in the pattern of cytochrome P450 oxygenations and subsequent acylations, lead to the formation of a diverse assortment of taxoids, of which taxol is most often a minor component (Fig. 2).

It is estimated that the biosynthesis of taxol from the universal diterpenoid precursor geranylgeranyl diphosphate involves 19 distinct enzymatic steps, with a similar number of relevant intermediates out of the approximately 400 taxoid metabolites characterized to date (Hezari and Croteau 1997; Croteau et al. 2005). An understanding of the enzymatic reactions that lead to taxol requires definition of the comparatively few intermediates that are directly involved in taxol biosynthesis.

A detailed metabolic profiling of taxoids produced by *Taxus* cell suspension cultures, via both constitutive and induced pathways, is essential for the identification of intermediates directly involved in taxol biosynthesis. Equally important is the identification of metabolites that are intermediates of parallel or divergent taxoid pathways. Guided manipulation of the genes that encode these pathway steps, either by up- or down-regulation, also requires quantification of the equilibrium levels of these taxol intermediates, as well as of those

[1] Institute of Biological Chemistry, Washington State University, Pullman, WA 99164, USA, e-mail: rketchum@wsu.edu

* Taxol® is a registered trademark of the Bristol-Meyers-Squibb company. The approved generic term for the drug is paclitaxel. Due to historic precedent and the abundance of taxoids with names derived from "taxol," we will use the more familiar term "taxol" when referring to this compound.

Biotechnology in Agriculture and Forestry, Vol. 57
Plant Metabolomics (ed. by K. Saito, R.A. Dixon, and L. Willmitzer)

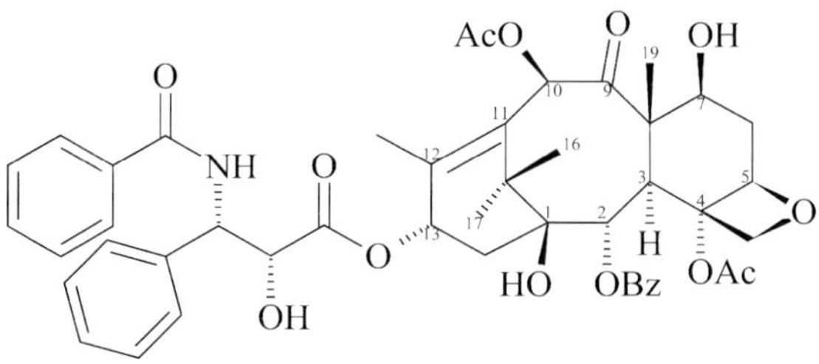

1 Paclitaxel (Taxol)

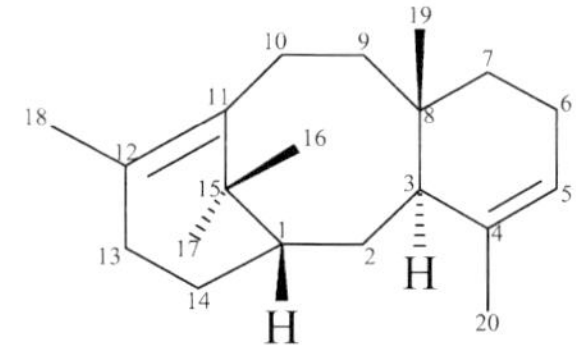

2 Taxa-4,11-diene

Fig. 1. Taxol (paclitaxel; **1**) is the final product of an 18-step biosynthetic pathway that begins with the parent diterpene olefin taxa-4,11-diene (**2**)

taxoids not involved in taxol formation, and of the flux through the various pathway branches during the normal growth cycle and following elicitation. By metabolic engineering of taxoid metabolism in cell culture, it should be possible to increase the production yields of taxol well over the 0.01 to 0.08% dry weight concentration range now reported (Ketchum et al. 1999c).

Plant cell suspension cultures of *Taxus*, that are inducible by methyl jasmonate and are capable of producing significant amounts of taxol, are an essential tool for investigating taxol biosynthesis (Mirjalili and Linden 1996; Yukimune et al. 1996; Ketchum et al. 1999a, 2003). Such cultures provide metabolic intermediates, enzymes, and genes that allow biochemical and molecular elucidation of the taxol pathway. In previous reports we have utilized methyl jasmonate as an inducer of taxol biosynthesis, and have presented comparative information on the distribution of taxoids produced both with and without methyl jasmonate elicitation (Ketchum et al. 2003). Whereas methyl jasmonate is an excellent elicitor of taxol production, it is also well known as an inducer of other defense-related genes (Creelman and Mullett 1997; Miller et al. 2005). Methyl jasmonate can increase the in vitro production of taxol by over 20-fold relative to uninduced cultures (Ketchum et al. 1999a); however, it also increases the production of taxoids that are not involved in taxol formation, such as the 14β-hydroxy taxoids (Ketchum et al. 2003). To simplify the profiling of the taxoids produced by cell suspension cultures, and to prevent the induction of genes and enzymes not directly involved in taxol biosynthesis, only constitutive (uninduced) taxoid metabolism is described in this report.

► **Fig. 2.** Biosynthesis of taxol from primary metabolism illustrating proposed intermediates and "off-pathway" metabolites. Abbreviations: MEP – methylerythritol phosphate; IPP – isopentenyl diphosphate; DMAPP – dimethylallyl diphosphate; IPPI – isopentenyl diphosphate isomerase; GGPPS – geranylgeranyldiphosphate synthase; TS – taxadiene synthase. [a]This compound has not been confirmed to be an intermediate in taxol biosynthesis. [b]This hypothetical epoxide intermediate has not been found

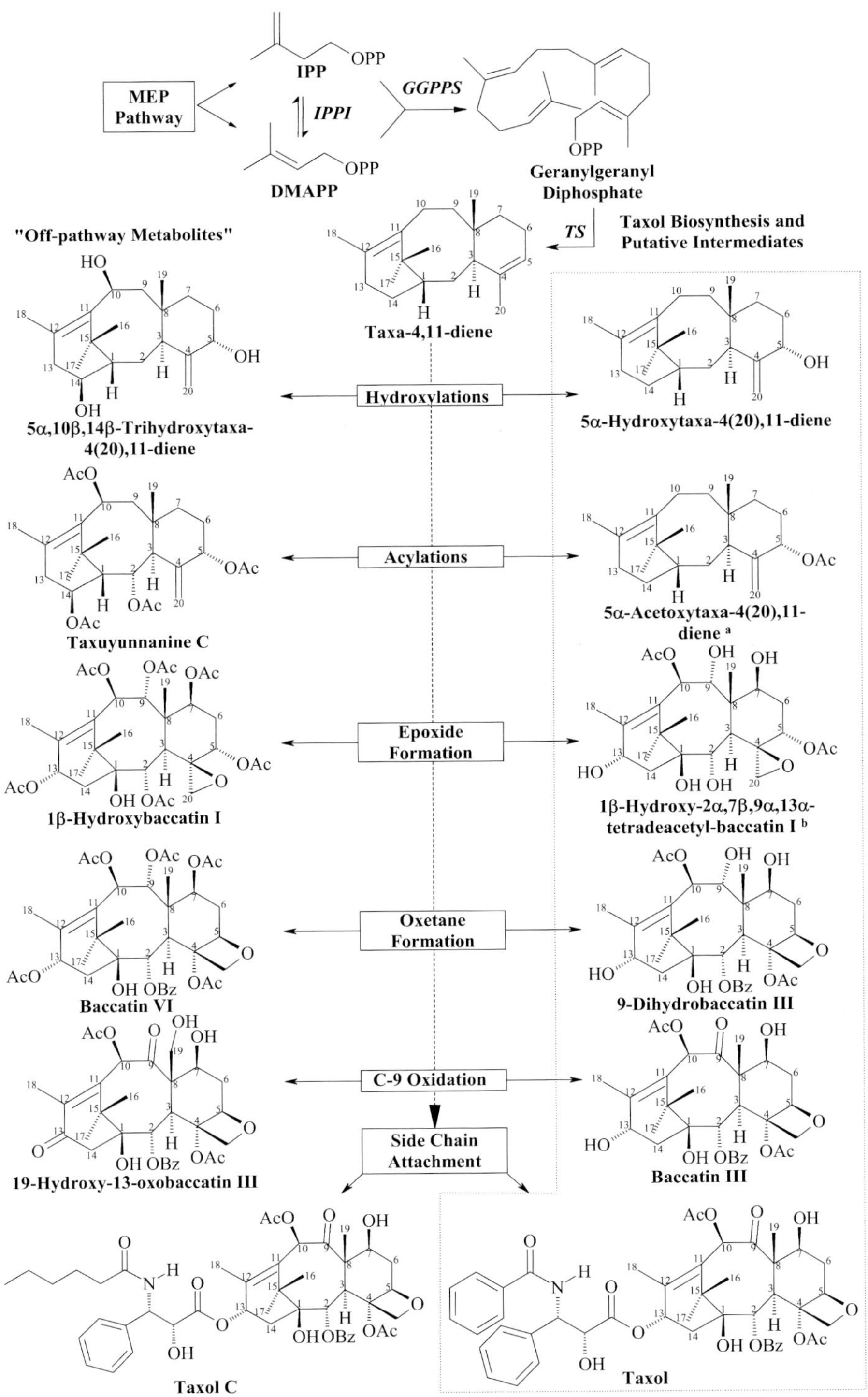

MEP Pathway
IPP
OPP
IPPI
DMAPP
GGPPS
Geranylgeranyl Diphosphate
TS
Taxol Biosynthesis and Putative Intermediates
"Off-pathway Metabolites"
Taxa-4,11-diene
Hydroxylations
5α,10β,14β-Trihydroxytaxa-4(20),11-diene
5α-Hydroxytaxa-4(20),11-diene
Acylations
Taxuyunnanine C
5α-Acetoxytaxa-4(20),11-diene [a]
Epoxide Formation
1β-Hydroxybaccatin I
1β-Hydroxy-2α,7β,9α,13α-tetradeacetyl-baccatin I [b]
Oxetane Formation
Baccatin VI
9-Dihydrobaccatin III
C-9 Oxidation
19-Hydroxy-13-oxobaccatin III
Baccatin III
Side Chain Attachment
Taxol C
Taxol

2 Results and Discussion

The diterpene olefin taxa-4,11-diene is the first committed intermediate of taxoid biosynthesis (Koepp et al. 1995), and is formed via the cyclization of geranylgeranyl diphosphate by taxadiene synthase (Hezari et al. 1995). To elucidate the sequence of subsequent enzymatic steps, and to determine which steps are potential targets for genetic manipulation, it is necessary to characterize all of the taxoids produced from this early intermediate by *Taxus* cell cultures.

2.1 Mass Spectral Fragmentation

Coupling mass spectroscopy to existing HPLC-UV methods of analysis of taxoids produced by *Taxus* cell suspension cultures has added the additional information of mass and fragmentation pattern of unknown compounds to UV absorbance and retention time measurements. Atmospheric pressure chemical ionization (APCI) as the source of ionization, rather than electrospray (ES), was employed because of the greater sensitivity of APCI, the diagnostic fragmentations obtained, and the ability to use existing LC methods without modification (Ketchum et al. 2003).

Using APCI under typical LC conditions, NH_4^+ adducts ([M+18]) to the parent ion are often observed. These ions are readily differentiated from fragment ions that result from loss of H_2O [M-18], as these are typically accompanied by an ion at [M-17] due to the [M + H^+] species. For example, the spectrum of baccatin III (MW = 586; Fig. 3a) exhibits the protonated parent ion [M + H^+] = 587 and also the parent ion as the ammonium adduct [M+NH_4^+] = 604, the difference between the two ions being the apparent [M-17] fragmentation. In addition to the ammonium adducts, the loss of hydroxyls and side-chain substituents from the taxoid core is common and diagnostic. Typical fragment ions observed are the result of loss of hydroxyl groups as water [M-18], acetate groups as acetic acid [M-60], or stepwise as ketene [M-42] and water, and benzoyloxy groups as benzoic acid [M-122].

In addition to the typical loss of the groups mentioned above, the oxetane function contributes to the formation of a stable and diagnostic fragment ion. Thus, the loss of an acetoxy group from baccatin III yields a characteristically abundant ion at [*m*/*z*] = 527 (Fig. 3a); this and all other taxoids examined with an oxetane ring and a ketone function at C-9 have this characteristic ion. In addition to the ion at [*m*/*z*] = 527, an ion at [*m*/*z*] = 509, corresponding to the additional loss of water [M-18], is generally observed with many oxetane-containing taxoids, and the [*m*/*z*] = 509 ion is often more abundant than the [*m*/*z*] = 527 ion. The presence of these two ions, at the same retention time in the extracted ion chromatogram, has indicated the presence of an oxetane-bearing taxoid in every case examined to date. A similar fragmentation pattern is observed for 9α-dihydrobaccatin-type taxoids, containing a hydroxyl group instead of the ketone function at C-9 (Ketchum et al. 1999b). For these taxoid types there is usually a major ion at $[m/z] = 511$, resulting from the presence of the two additional hydrogens, as well as a detectable ion at $[m/z] = 529$. The presence of these two ions at the same retention time is most often an indication of the presence of a 9α-dihydrobaccatin-type taxoid structure.

The molecular ion is usually significantly reduced or absent in the spectrum of taxadiene polyols and often in the spectrum of taxadiene polyacetates (see Fig. 3b–d). The loss of a hydroxyl or acetoxy group often results in the same fragmentation pattern in a molecule that contains the same number of these substitutions. This phenomenon is observed, for example, in comparing the fragmentation pattern of the taxadiene tetraacetate taxuyunnanine c (Fig. 3c) to that of the taxadiene tetraol derived from taxusin (Fig. 3d). Both compounds fragment to a core structure with $[m/z]$ of 265.

The diagnostic losses from the taxadiene core of substituent esters allows for the identification of taxoid classes based on the fragmentation observed. Using this characteristic feature of fragment loss of ester side chains from the taxadiene core, it is possible to diagnose fragment ions of unknown oxygenated taxoids (Table 1). By scanning a mass spectrum for such predicted fragments, it is possible to target unknown taxoids for isolation and more detailed structure elucidation.

Table 1. Predicted ions from APCI fragmentation of hydroxylated taxadienes

			Ion resulting from loss of H_2O						
Number of hydroxy groups	Mass	$[M+H^+]$	$[M+H^+ -H_2O]$	$[-H_2O]$	$[-H_2O]$	$[-H_2O]$	$[-H_2O]$	$[-H_2O]$	$[-H_2O]$
0	272[a]	273							
1	288	289	271						
2	304	305	287	269					
3	320	321	303	285	267				
4	336	337	319	301	283	265			
5	352	353	335	317	299	281	263		
6	368	369	351	333	315	297	279	261	
7	384	385	367	349	331	313	295	277	259

[a] Calculated mass of taxa-4,11-diene

The spectra of taxadiene polyols are generally easier to interpret than those of the corresponding polyacetates. One interesting and complicating feature of the fragmentation of such compounds is that the same ions at $[m/z]$ 265, 283, and 301 are observed in the spectra of both tetraacetoxy and tetrahydroxy taxoids, presumably resulting from loss of ketene ($O{=}C{=}CH_2$, $[m/z] = 42$) from the acetoxy groups, followed by loss of water ($[m/z] = 18$). This can make interpretation of the complex fragmentation patterns more difficult but, for practical purposes the combined loss of both groups can be viewed as resulting in elimination of the complete acetoxy group as acetic acid ($[m/z] = 60$; Fig. 3c).

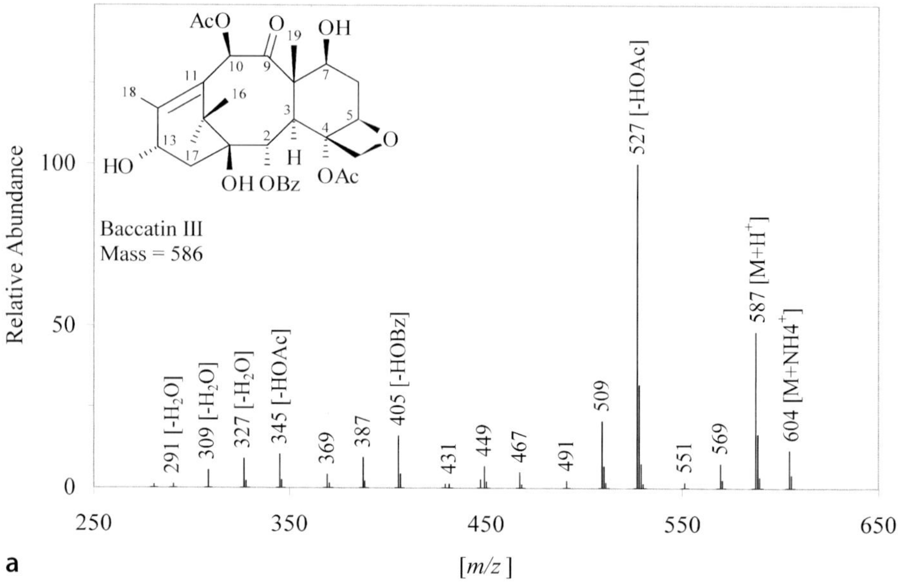

a

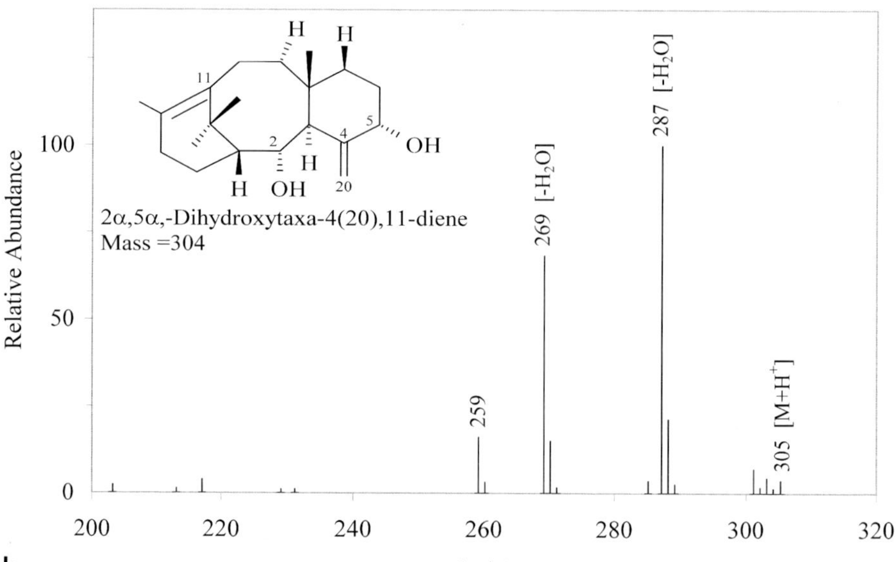

b

Fig. 3. Spectra of representative taxoids: **a** baccatin III; **b** 2α, 5α-dihydroxytaxa-4(20),11-diene; **c** taxuyunnanine C (2α, 5α, 10β, 14β-tetraacetoxytaxa-4(20),11-diene); **d** 5α, 9α, 10β, 13α-tetrahydroxytaxa-4(20),11-diene

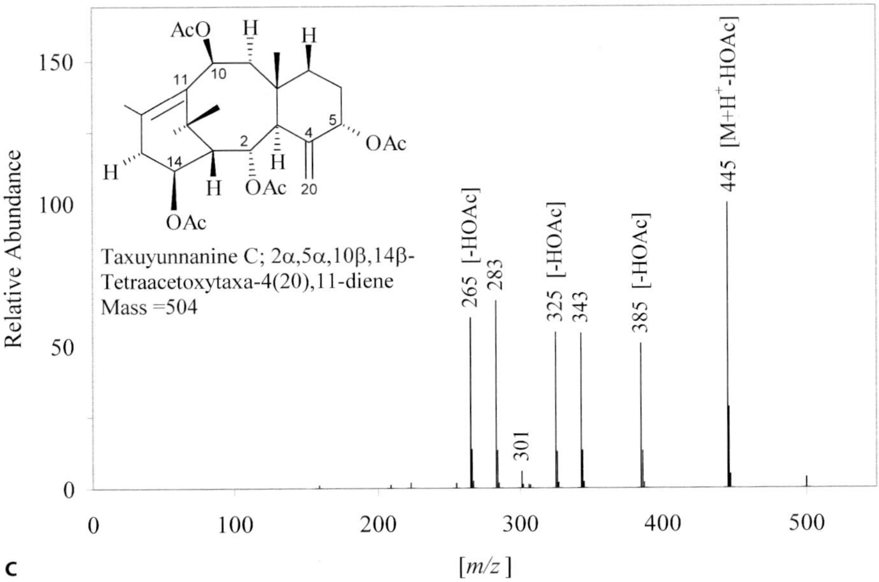

c

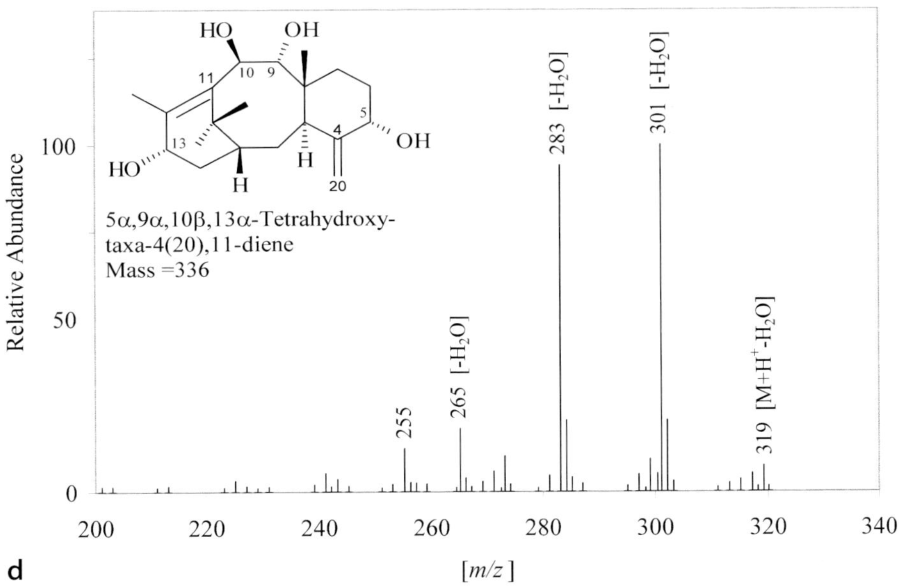

d

Fig. 3. (continued)

2.2 Taxoids from Plant Cell Culture

In previous work, the identifications of 25 taxoids produced by *Taxus cuspidata* and *Taxus x media* suspension cultures were described (Ketchum et al. 2003). To this original list of cell culture metabolites, 13 additional taxoids have been newly identified (Table 2).

Table 2 lists the most abundant taxoids identified in the present cell suspension cultures. Of these taxoids, taxa-4,11-diene, 5α-hydroxytaxa-4(20),11-diene, taxol, 10-deacetylbaccatin III, baccatin III, and possibly 5α, 9α, 10β, 13α-tetraahydroxytaxa-4(20),11-diene are the only metabolites that we have identified in plant cell cultures that could be intermediates in the biosynthesis of taxol.

Although it is conceivable that intermediates of taxoid biosynthesis undergo reversible acylations and deacylation (perhaps for the purpose of targeting, trafficking, or flux control), such processes seem unlikely. Thus, "inappropriate" acylations, for example at C-9 of baccatin VI or at C-13 of baccatin I derivatives (Fig. 2 and Table 2), would appear to block pathway progression to taxol. Consequently, most of the identified taxoids that accumulate to significant levels in cell cultures are not likely to be intermediates in the biosynthesis of taxol because these intermediates are acylated at incorrect positions on the taxane core.

2.2.1 *Taxoids in a Typical Methanol Extract of Fresh Cells*

The amount of taxol in the cells used for methanol extraction was 42.6 µg/g fresh weight; thus, the entire 1-L culture (cells and medium) contained 11.9 mg of taxol, with the cells contributing about 39% of the total. This distribution is typical of cell suspension cultures that have not been elicited with methyl jasmonate and are harvested two weeks after subculture, i. e., about 30–35% of the taxol is contained in the cells while about 65–70% of the taxol is released into the medium.

The total ion mass chromatogram of the "taxoid region" (15–50 min) of the methanol extract is illustrated in Fig. 4, with the inset showing the entire chromatogram over the period data were collected (5–55 min). The 31 compounds identified as taxoids are listed in Table 3. There are seven compounds whose fragmentation patterns indicate that they are oxetane-containing or polyoxygenated taxoids, but the spectra of which were too ambiguous to permit identification; these compounds will require isolation and structure elucidation by NMR spectroscopy.

Taxol is the second most abundant taxoid identified, next to taxol D. However, taxol represents only 6.8% of the total taxoids based on the integrated chromatogram. If the three components, 10-deacetylbaccatin III, baccatin III, and 5α, 9α, 10β, 13α-tetrahydroxytaxa-4(20),11-diene are considered to be intermediates on the pathway to taxol, then the remaining 27 taxoids (comprising

Table 2. Taxane diterpenoids (taxoids) isolated from suspension cell cultures of *Taxus*

Taxoids with an oxetane ring and C-13 phenylisoserine side chain

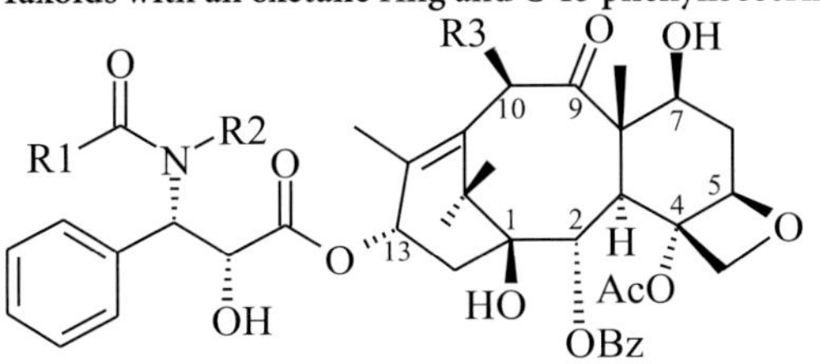

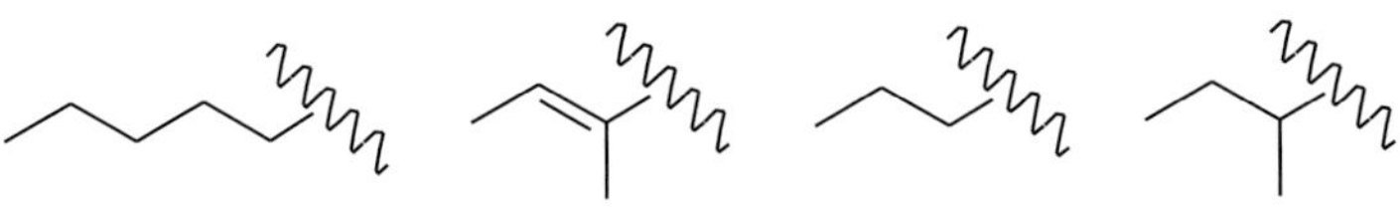

a **b** **c** **d**

Taxoid[a]	MW	R1	R2	R3
10β-Deacetyltaxuyunnanine A[b]	805	a	H	OH
10β-Deacetyltaxol	811	Ph	H	OH
Taxol D	819	c	H	OAc
Cephalomannine (Taxol B)	831	b	H	OAc
N-Debenzoyl-*N*-(2-methylbutyryl)taxol[b]	833	d	H	OAc
Taxol C	847	a	H	OAc
Taxol (paclitaxel)	853	Ph	H	OAc
N-Methyltaxol C[b]	861	a	CH_3	OAc

Taxoids with an oxetane ring

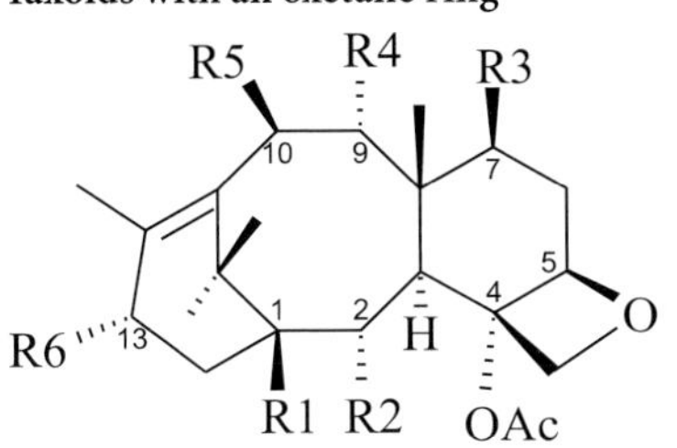

Taxoid	MW	R1	R2	R3	R4	R5	R6
10β-Deacetylbaccatin III	544	OH	OBz	OH	=O	OH	OH
Baccatin III	586	OH	OBz	OH	=O	OAc	OH
9α-Dihydrobaccatin III	588	OH	OBz	OH	OH	OAc	OH
9α-Dihydro-13α-acetylbaccatin III	630	OH	OBz	OH	OH	OAc	OAc
Baccatin IV	652	OH	OAc	OAc	OAc	OAc	OAc
1β-Dehydroxybaccatin VI	698	H	OBz	OAc	OAc	OAc	OAc
Baccatin VI	714	OH	OBz	OAc	OAc	OAc	OAc

Table 2. continued

Taxoids with a C-4(20) epoxide

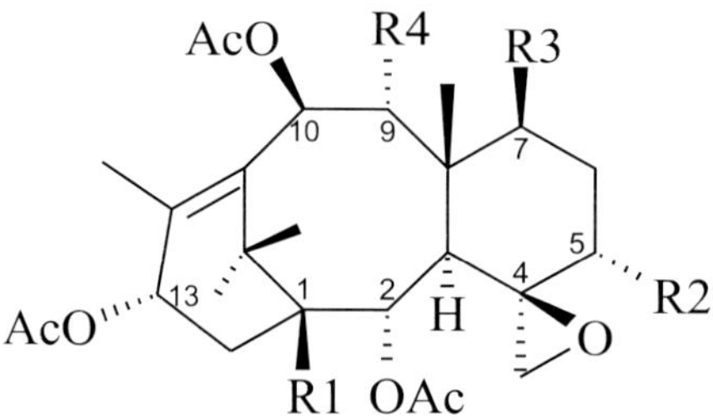

Taxoid	MW	R1	R2	R3	R4
1β-Hydroxy-7β,9α-deacetylbaccatin I	586	OH	OAc	OH	OH
1β-Hydroxy-5α-deacetylbaccatin I	610	OH	OH	OAc	OAc
Baccatin I	636	H	OAc	OAc	OAc
1β-Hydroxybaccatin I	652	OH	OAc	OAc	OAc

Taxoids with a C-4(20) double bond

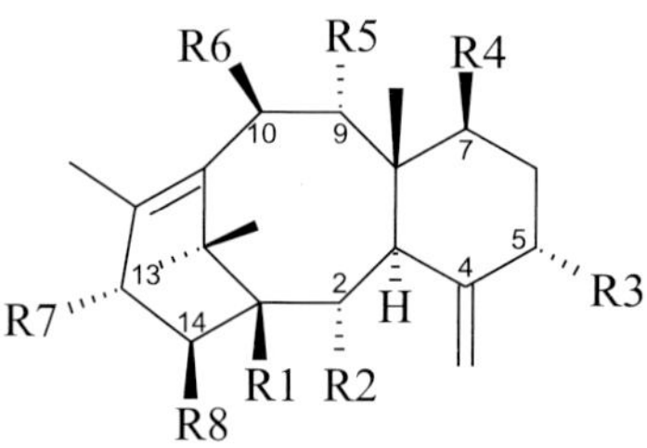

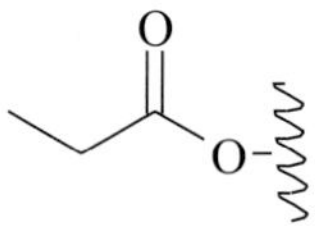

a

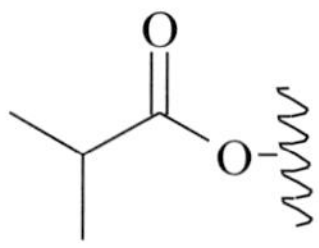

b

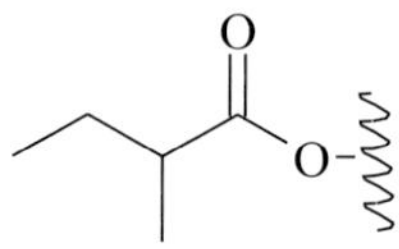

c

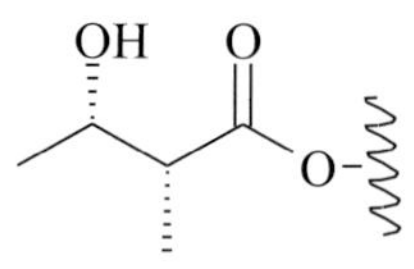

d

Taxoid	MW	R1	R2	R3	R4	R5	R6	R7	R8
Taxa-4,11-diene[c]	272	H	H	H	H	H	H	H	H
5α-Hydroxytaxa-4(20),11-diene[d]	288	H	H	OH	H	H	H	H	H
5α,10β, 14β-Trihydroxytaxa-4(20),11-diene	320	H	H	OH	H	H	OH	H	OH
5α-Acetoxytaxa-4(20),11-diene[d]	330	H	H	OAc	H	H	H	H	H
5α, 9α, 10β, 13α-Tetrahydroxytaxa-4(20), 11-diene	336	H	H	OH	H	OH	OH	OH	H
2α,5α,10β-Triacetoxytaxa-4(20),11-diene[e]	446	H	OAc	OAc	H	H	OAc	H	H
2α,10β, 14β-Triacetoxy-5α-hydroxytaxa-4(20),11-diene	462	H	OAc	OH	H	H	OAc	H	OAc

Table 2. continued

Taxoid	MW	R1	R2	R3	R4	R5	R6	R7	R8
2α,5α,10β, 14β-Tetraacetoxytaxa-4(20), 11-diene; Taxuyunnanine C	504	H	OAc	OAc	H	H	OAc	H	OAc
5α, 9α, 10β, 13α-Tetraacetoxytaxa-4(20), 11-diene; Taxusin	504	H	H	OAc	H	OAc	OAc	OAc	H
2α,5α,10β-Triacetoxy-14β-propionyloxytaxa-4(20),11-diene	518	H	OAc	OAc	H	H	OAc	H	a
2α,5α,10β-Triacetoxy-14β-isobutyryloxytaxa-4(20),11-diene	532	H	OAc	OAc	H	H	OAc	H	b
2α,5α,10β-Triacetoxy-14β-(2-methyl)butyryloxytaxa-4(20),11-diene	546	H	OAc	OAc	H	H	OAc	H	c
2α,5α,9α,10β,14β-Pentaacetoxytaxa-4(20),11-diene	562	H	OAc	OAc	H	OAc	OAc	H	OAc
5α,7β,9α,10β,13α-Pentaacetoxytaxa-4(20),11-diene	562	H	H	OAc	OAc	OAc	OAc	OAc	H
2α,5α,10β-Triacetoxy-14β-(3-hydroxy-2-methyl)butyryloxytaxa-4(20),11-diene	562	H	OAc	OAc	H	H	OAc	H	d
5α-Hydroxy-2α,7β,9α,10β,13α-pentaacetoxytax-4(20),11-diene	578	H	OAc	OH	OAc	OAc	OAc	OAc	H
2α,5α,7β,9α,10β,13α-Hexaacetoxytaxa-4(20),11-diene[f]	620	H	OAc	OAc	OAc	OAc	OAc	OAc	H
5α,7β,9α,10β,13α-Pentaacetoxy-2α-benzoyloxytaxa-4(20),11-diene[e]	682	H	OBz	OAc	OAc	OAc	OAc	OAc	H

11(15→1)-abeo-Taxoids with an oxetane ring

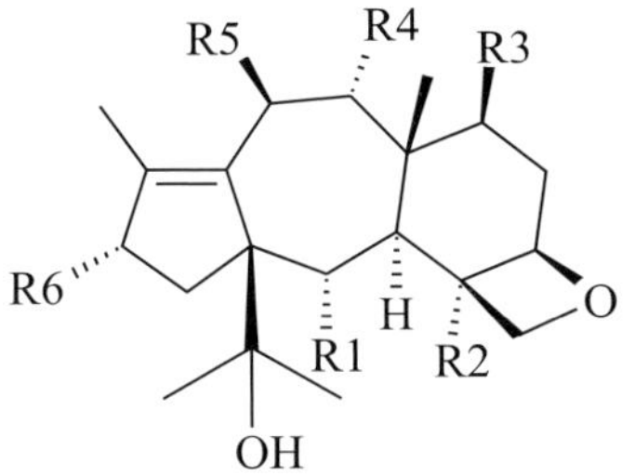

Taxoid	MW	R1	R2	R3	R4	R5	R6
4α,7β-Diacetoxy-2α,9α-dibenzoyloxy-5β,20-epoxy-10β,13α,15-trihydroxy-11(15→1)-*abeo*-taxene[b]	692	OBz	OAc	OAc	OBz	OH	OH

[a] The organization of this table and all compounds are described and cited in Baloglu and Kingston (1999) unless otherwise noted

[b] Identified by mass spectroscopy based on molecular ion and fragmentation pattern, but not yet confirmed by NMR

[c] Koepp et al. (1995)

[d] Hefner et al. (1996)

[e] Ketchum et al. (2003)

[f] Kingston et al. (1993)

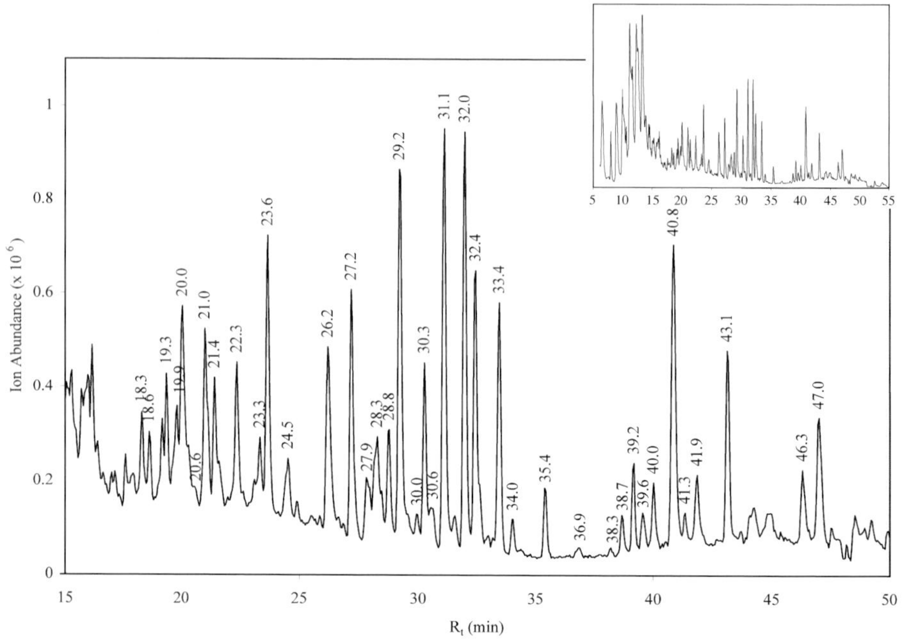

Fig. 4. Total ion chromatogram of the methanol extract of *Taxus x media* cv. Hicksii cells. The taxoid region (from 15 to 50 min) is shown. *Inset* shows the total ion chromatogram over the entire sampling period (5 to 55 min); note the large amount of polar material eluting before 15 min. Peak identities are listed by retention time in Table 3

about 84% of the integrated peak area) may be considered as side-route or dead-end products. Thus, only about 6% of the taxa-4,11-diene produced is converted to taxol. We previously reported that taxadiene synthase, the committed enzyme responsible for the cyclization of geranylgeranyl diphosphate to tax-4,11-diene (Fig. 2), is a slow step in the biosynthesis of taxol (Hezari et al. 1997). This enzyme is, nevertheless, sufficiently active to promote the formation of high levels of other taxoids (85% of the mix) in cell culture. Genetic redirection of the pathway toward taxol, and away from these alternate routes to other taxoids, may be the most efficient way of increasing taxol production yields.

2.2.2 *Early Taxol Biosynthetic Intermediates*

Low taxol-producing, 14-day old *Taxus x media* cv. Hicksii cell cultures were used for pentane extraction and the search for early taxol biosynthetic intermediates, based on the assumption that such cultures may be blocked in late steps and so accumulate early precursors. At harvest, these cells produced 5.1 μg/g fresh weight of taxol, a contribution of 0.55 mg taxol to the entire culture. Thus, the entire 1-L culture contained 1.80 mg taxol with the cells contributing ~31% to the total.

Table 3. Identification of components at the indicated retention time in Fig. 4. The percentage of total taxoids is the proportion of peak area relative to that of the total area of confirmed taxoids

R_t (min)	**Taxoid**	**% of total taxoids**
18.3	Unknown oxetane taxoid	3.19
18.6	Unknown oxetane taxoid	2.19
19.3	Unknown hexaoxy taxoid	2.80
19.9	Unknown hexaoxy taxoid	3.34
20.0	10β-Deacetylbaccatin III	5.12
20.6	5α,9α,10β,13α-Tetrahydroxy-taxa-4(20),11-diene	1.37
21.0	1β-Hydroxy-7β,9α-deacetylbaccatin I	5.09
21.4	Unknown taxoid	2.71
22.3	9α-Dihydrobaccatin III	4.26
23.3	Unknown taxoid	2.45
23.6	1β-Hydroxy-5α-deacetylbaccatin I	6.51
24.5	Baccatin III	3.16
26.2	9α-Dihydro-13α-acetylbaccatin III	4.73
27.2	Baccatin IV	4.74
27.9	5α-Hydroxy-2α,7β,9α,10β,13α-pentaacetoxy-4(20),11-diene	1.96
28.3	10β-Deacetyltaxol	2.77
28.8	1β-Hydroxybaccatin I	2.53
29.2	Taxol D	7.49
30.0	4α,7β-Diacetoxy-2α,9α-dibenzoyl-5α,20-epoxy-10β,13α,15-trihydroxy-11(15>1)-*abeo*-taxene	0.92
30.3	Cephalomannine	2.87
30.6	*N*-Debenzoyl-*N*-(2-methylbutyryl)taxol	1.34
31.1	**Taxol**	**6.80**
32.0	Taxol C	6.47
32.4	Baccatin I	5.30
33.4	2α,5α,10β-Triacetoxy-14β-(3-hydroxy-2-methyl)butyryloxy-taxa-4,11-diene	4.10
34.0	5α,7β,9α,10β,13α-Pentaacetoxy-taxa-4,11-diene	0.79
35.4	2α,5α,10β,14β-Tetraacetoxy-taxa-4,11-diene	1.36
36.9	2α,5α,10β-Triacetoxy-14β-propionyloxy-taxa-4,11-diene	0.49
38.3	2α,5α,10β-Triacetoxy-14β-isobutyryloxy-taxa-4,11-diene	0.41
38.7	Unknown taxoid (MW = 590)	1.10
39.2	2α,5α,10β-Triacetoxy-14β-(2-methyl)butyryloxy-taxa-4,11-diene	1.63
39.6	Unknown phytosterol	
40.0	Unknown non-taxoid	
40.8	Unknown non-taxoid	
41.3	Unknown	
41.9	Unknown non-taxoid	
43.1	Unknown non-taxoid	
46.3	Campesterol	
47.0	β–Sitosterol	

The total ion chromatogram of the concentrated pentane fraction that eluted from a silica column, following drying, powdering, and pentane extraction of these cells, is shown in Fig. 5a. From 1.51 g of cells, 7.2 μg of taxa-4,11-diene and 0.5 μg of taxa-4(20),11-diene were obtained, equivalent to a concentration of 4.8 μg/g dry weight of taxa-4,11-diene and 0.33 μg/g dry weight of taxa-4(20),11-diene (Fig. 5a,c). The taxa-4(20),11-diene isomer is present in the extract at 6.9% of the amount of the taxa-4,11-diene isomer, consistent with previous reports obtained with the product distribution of taxadiene synthase (Williams et al. 2000). The concentration of taxa-4,11-diene in dried cells from these actively growing cultures is 3700 times the concentration previously reported in *T. brevifolia* bark, 1 mg/750 kg (Koepp et al. 1995)! It is worth noting that both taxadiene isomers are converted to 5α-hydroxytaxa-4(20),11-diene at comparable rates in the subsequent hydroxylation step (Jennewein et al. 2004), which likely accounts for the 93:7 proportion of residual olefins observed. Given that taxadiene accounts for only 0.3% of the total taxoids produced, it is not surprising that this important intermediate has been difficult to detect in the milieu of taxoids found in extracts of *Taxus* cells.

As indicated above, taxa-4,11-diene is the first committed intermediate in taxol biosynthesis and is converted to 5α-hydroxytaxa-4(20),11-diene in a subsequent cytochrome P450-mediated hydroxylation reaction (Hefner et al. 1996). Recent work has demonstrated incorporation of radiolabeled 5α-hydroxytaxa-4(20),11-diene into taxol and other highly functionalized taxoids, supporting cyclization and 5α-hydroxylation as the first two steps of the taxol biosynthetic pathway (unpublished data). Analysis of the hexane:ethyl ether (9:1) wash of the silica column yielded an estimated concentration of 1.47 μg/g dry weight of 5α-hydroxytaxa-4(20),11-diene, only about 20% of the amount of taxa-4,11-diene and just over 1% of the concentration of taxol found in the cells (Fig. 5b,d). Similar to the concentration of taxa-4,11-diene, the concentration of 5α-hydroxytaxa-4(20),11-diene in dried cells from actively growing cultures is nearly 1000 times the concentration previously reported in *T. brevifolia* bark, 5–10 μg/kg (Hefner et al. 1996). That both taxa-4,11-diene and 5α-hydroxytaxa-4(20),11-diene occur at such low levels suggests that both of these intermediates are rapidly turned over in subsequent pathway steps. Due to the difficulty in identifying these two early intermediates in cell cultures, and because of their extremely low abundance, these are the only two pathway intermediates that have been confirmed by incorporation into taxol in *Taxus* cell feeding studies (unpublished data).

Attempts to incorporate other radiolabeled putative intermediates into taxol in *Taxus* cell feeding studies have been unsuccessful, thus far, possibly due to uptake limitations. However, the success of the focused searches for taxa-4,11-diene and 5α-hydroxytaxa-4(20),11-diene in cell extracts, coupled to the metabolic implication for rapid turnover of these very low abundance taxol precursors, has encouraged similar directed approaches to identify other predicted early pathway intermediates. These efforts, now focused on low abun-

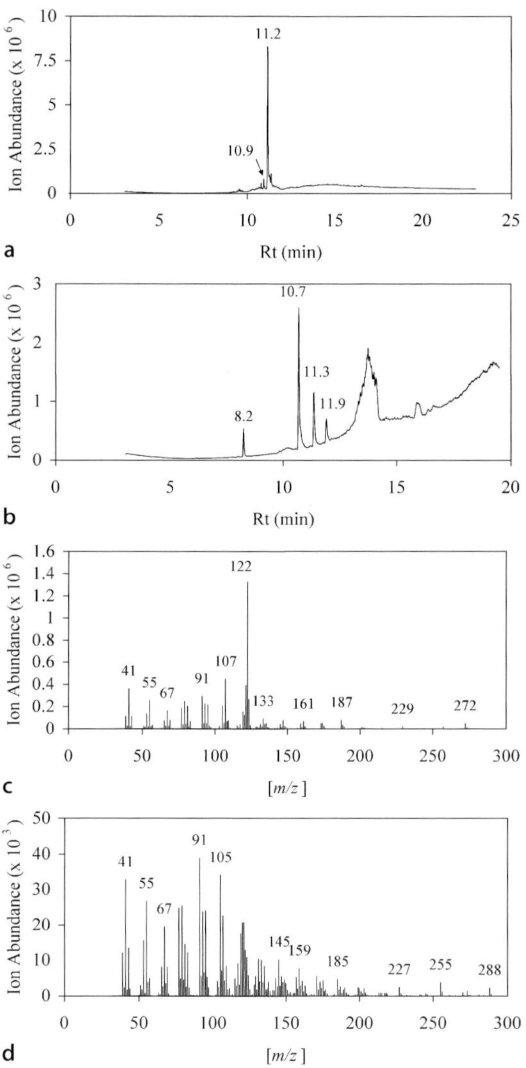

Fig. 5a–d. Identification of early intermediates of the taxol biosynthetic pathway by GC/MS analysis of a pentane extract from *Taxus x media* cv. Hicksii cells: **a** total ion chromatogram (TIC) of pentane wash of silica column showing taxa-4(20),11-diene (10.9 min) and taxa-4,11-diene (11.2 min); **b** TIC of pentane:ethyl ether (9:1) wash from same silica column shows solvent contaminant (8.2 min), solvent contaminant (10.7 min), phytol (11.3 min), and 5α-hydroxytaxa-4(20),11-diene (11.9 min); **c** mass spectrum of taxa-4,11-diene; **d** mass spectrum of 5α-hydroxytaxa-4(20),11-diene

dance taxoids of moderate to high polarity, are certain to add to the list of *Taxus* cell culture metabolites and to contribute to a more detailed understanding of taxol biosynthesis.

3 Protocol

3.1 Plant Cell Cultures

Cell suspension cultures of *Taxus x media* cv. Hicksii were established from callus cultures initiated from embryos excised from freshly collected seeds in September, 2000. Plant cell cultures were initiated, maintained, and screened for taxol production as previously described (Ketchum and Gibson 1996; Hezari et al. 1997; Ketchum et al. 1999b, 2003).

3.2 Cell Extraction

Two different methods were used for the extraction of taxoids. For extraction of taxoids from fresh tissue, cells were grown for two weeks in liquid TM19 medium (Ketchum et al. 2003). The cells were from a 1-L culture of a *Taxus x media* cv. Hicksii cell line that was producing a moderate level of taxol (7.3 mg/L). The cultures were harvested by removing the medium, rinsing the cells in deionized water, and briefly drying on a 60-μm nylon filter in a Büchner funnel with vacuum. The air dried cells (0.52 g) were placed in a Spin-X (Corning) 0.45-μm nylon centrifuge filter. Excess water (0.19 mL) was removed by centrifuging the cells for 5 min at 16,000 g. Cells were extracted on the filter with 300 μL methanol, then centrifuged at 16,000 g for 5 min; methanol extraction was repeated twice more. The 900 μl combined methanol extract was dried in a centrifugal vacuum evaporator. The dried residue was suspended in 50 μL methanol, filtered through a 0.2 μm nylon filter, and a 10-μL aliquot was analyzed by HPLC.

For extraction of the less polar taxoids, cells from 11 separate cultures were combined, briefly air-dried, frozen in liquid nitrogen, lyophilized, and ground to a fine powder in a coffee grinder. At harvest, these cultures produced 1.25±0.2 (S.E., n = 11) mg/L taxol in the cell-free medium and contained 108±6.5 (S.E., n = 11) g/L fresh weight cells. Dried cells (1.51 g) were extracted with 7 ml pentane in a sealed glass test tube in an ultrasonic bath for 30 min. The extraction was repeated three more times with 5-mL aliquots of pentane, and the combined extracts were dried over anhydrous $MgSO_4$, filtered and evaporated under a stream of air. The dried residue (16 mg) was dissolved in 500 μL of pentane and loaded onto a pentane-rinsed silica gel column fashioned from a Pasteur pipette containing approximately 400 mg Silica Gel 60 (Baker), 200–400 mesh, and approximately 100 mg anhydrous $MgSO_4$. The column was rinsed with pentane until just prior to the elution of the first pigmented compounds, approximately 5 ml. This pentane extract was evaporated to yield 1.4 mg of residue. The column was then rinsed with 5 ml pentane:ethyl ether (9:1). This extract was evaporated to yield 3.2 mg of residue. Each dried extract was dissolved in 100 μL pentane. Extracts were centrifuged at 16,000 g for 1 min to pellet insoluble material, and the upper 90 μL was transferred to a clean glass insert. A 1-μL aliquot from each extract was analyzed by GC-MS.

3.3 HPLC Analyses

HPLC/MSD instrumentation consisted of an Agilent Series 1100 HPLC with diode array and G1946A mass detector, with Chemstation Software Rev. 8.03. Extracts were eluted from a Discovery HS-F5 250 × 4.6 mm column (Supelco), 5 μm particle size, with guard column, 5–100% CH_3CN over 50 min, 100% CH_3CN for 5 min, and re-equilibration at 100% CH_3CN for 10 min.

Mass detection of taxoids was by atmospheric pressure chemical ionization (APCI) in the positive ion mode. Drying gas was N_2 at 60 psi, 5 L/min, 350 °C. The vaporizer was set to 400 °C, fragmentor to 60 V, capillary to 3000 V, and corona current to 8 μA (Ketchum et al. 2003).

3.4 GC-MS Analyses

GC-MS analyses were conducted on an Agilent Series 6890 GC system with Series 6890 mass selective detector. A 1-μL aliquot of each pentane extract was analyzed by cool on column injection on a Restek RTX-5MS column, 30 m, 0.25 mm ID, 0.25 μm film thickness. Chromatography was accomplished using He at 0.7 mL/min (30 cm/s) as the carrier gas. Initial temperature was 40 °C, followed by programmed gradient to 300 °C at 20 °C/min, with a hold for 10 min at 300 °C.

Identification and quantitation of taxoids was accomplished by comparison to authentic standards of the retention time, peak area, mass fragmentation pattern, and UV absorbance.

4 Conclusion

The metabolome of an organism strictly refers to all metabolites produced during its life. The *Taxus* metabolome, as briefly described in this chapter, relates to that part of *Taxus* secondary metabolism that begins with the formation of taxa-4,11-diene and ends with the production of a taxane diterpenoid, such as the pharmaceutically important drug taxol. The metabolic pathways that originate from taxa-4,11-diene and result in the formation of the over 400 taxoids so far described form a complex network through which taxol biosynthesis is woven. While less than 40 of these taxoids (comprising about 85 mass %) have been characterized in our plant cell cultures, most are not involved in taxol biosynthesis. Identification of the taxoids that are produced in our *Taxus* plant cell culture system and understanding their relationship to other intermediates in the taxol biosynthetic pathway provide clues to the order of the synthesis of intermediates and branch points in this complex metabolic grid. This knowledge is critical for the intelligent targeting of genes for future metabolic engineering of plant cell cultures for increased taxol production.

Acknowledgements. This investigation was supported by U.S. National Institutes of Health grant CA-55254, and by McIntire-Stennis Project 0967 from the Washington State University Agricultural Research Center.

References

Baloglu E, Kingston DGI (1999) The taxane diterpenoids. Phytochemistry 62:1448–1472

Creelman RA, Mullett JE (1997) Biosynthesis and action of jasmonates in plants. Annu Rev Plant Physiol Plant Mol Biol 48:355–381

Croteau R, Ketchum REB, Long RM, Kaspera R, Wildung MR (2006) Taxol biosynthesis and molecular genetics. Phytochem Rev 4 (in press)

Hefner J, Rubenstein SM, Ketchum REB, Gibson DM, Williams RM, Croteau R (1996) Cytochrome P450-catalyzed hydroxylation of taxa-4(5),11(12)-diene to taxa-4(20),11(12)-dien-5α-ol: the first oxygenation step in taxol biosynthesis. Chem Biol 3:479–489

Hezari M, Croteau R (1997) Taxol biosynthesis: an update. Planta Med 63:291–295

Hezari M, Lewis NG, Croteau R (1995) Purification and characterization of taxa-4(5),11(12)-diene synthase from Pacific yew (*Taxus brevifolia*) that catalyzes the first step of taxol biosynthesis. Arch Biochem Biophys 322:437–444

Hezari M, Ketchum RE, Gibson DM, Croteau R (1997) Taxol production and taxadiene synthase activity in *Taxus canadensis* cell suspension cultures. Arch Biochem Biophys 337:185–190

Itokawa H (2003) Taxoids occurring in the genus *Taxus*. In: Itokawa H, Lee H-K (eds) *Taxus:* The Genus *Taxus*, Taylor and Francis, London, pp 35–78

Jennewein S, Long RM, Williams RM, Croteau R (2004) Cytochrome P450 taxadiene 5α-hydroxylase a mechanistically unusual monooxygenase catalyzing the first oxygenation step of Taxol biosynthesis. Chem Biol 11:379–387

Ketchum REB, Gibson DM (1996) Paclitaxel production in cell suspension cultures of *Taxus*. Plant Cell Tissue Organ Cult 46:9–16

Ketchum REB, Gibson DM, Croteau RB, Shuler ML (1999a) The kinetics of taxoid accumulation in cell suspension cultures of *Taxus* following elicitation with methyl jasmonate. Biotechnol Bioeng 62:97–105

Ketchum REB, Tandon M, Gibson DM, Begley T, Shuler ML (1999b) Isolation of labeled 9-dihydrobaccatin III and related taxoids from cell cultures of *Taxus canadensis* elicited with methyl jasmonate. J Nat Prod 62:1395–1398

Ketchum REB, Luong JV, Gibson DM (1999c) Efficient extraction of paclitaxel and related taxoids from leaf tissue of *Taxus* using a potable solvent system. J Liq Chromatogr Relat Technol 22:1715–1732

Ketchum REB, Rithner CD, Qiu D, Williams RM, Croteau RB (2003) Taxus metabolomics: methyl jasmonate preferentially induces production of taxoids oxygenated at C-13 in *Taxus x media* cell cultures. Phytochemistry 62:901–909

Kingston DGI, Molinero AA, Rimoldi JM (1993) The taxane diterpenoids. Prog Chem Org Nat Prod 61:1–206

Koepp AE, Hezari M, Zajicek J, Stofer Vogel B, LaFever RE, Lewis NG, Croteau R (1995) Cyclization of geranylgeranyl diphosphate to taxa-4(5),11(12)-diene is the first committed step of taxol biosynthesis in Pacific yew. J Biol Chem 270:8686–8690

Miller B, Madilao LL, Ralph S, Bohlmann J (2005) Insect-induced conifer defense. White pine weevil and methyl jasmonate induce traumatic resinosis, de novo formed volatile emissions, and accumulation of terpenoid synthase and putative octadecanoid pathway transcripts in sitka spruce. Plant Physiol 137:369–382

Mirjalili N, Linden JC (1996) Methyl jasmonate induced production of taxol in suspension cultures of *Taxus cuspidata:* ethylene interaction and induction models. Biotechnol Prog 12:110–118
Williams DC, Wildung MR, Jin AQ, Dalal D, Oliver JS, Coates RM, Croteau R (2000) Heterologous expression and characterization of a "pseudomature" form of taxadiene synthase involved in paclitaxel (Taxol) biosynthesis and evaluation of a potential intermediate and inhibitors of the multistep diterpene cyclization reaction. Arch Biochem Biophys 379:137–146
Yukimune Y, Tabata H, Higashi H, Hara Y (1996) Methyl jasmonate-induced overproduction of paclitaxel and baccatin III in *Taxus* cell suspension cultures. Nat Biotechnol 14:1129–1132

III.9 The Use of Non-targeted Metabolomics in Plant Science

T. Daskalchuk, P. Ahiahonu, D. Heath, and Y. Yamazaki[1]

1 Introduction

The emergence of the "omic" technologies has greatly expedited the amount of information available to biologists (Edwards and Batley 2004). The information wave launched with genomics has been followed by transcriptomics, proteomics and metabolomics, all in an attempt to assign functionality to genes and genomes (Oliver et al. 1998). One advantage of the "omics" technologies is their ability to view global changes in a system as a result of some perturbation. Metabolomics, the comprehensive analysis of the whole metabolome under a particular experimental condition, ultimately defines a chemical phenotype of an organism at some point in time (Goodacre et al. 2004). This chemical phenotype, the metabolome, is the global pool of all metabolites, and is the functional result of a genome in a particular environment (Tweeddale et al. 1998). Thus, changes in either the genome or environment will be ultimately manifested in the metabolome as the end result of both gene and protein expression.

A number of analytical techniques are currently available for studying the metabolome of living tissues and organs. The ultimate goal of plant metabolomics, the ability to detect and quantify every metabolite in a plant extract reliably is unlikely to be attained by any single analytical method available at present. Due to the large diversity of chemical and physical properties of plant metabolites, different analytical methods must be combined to achieve this since each of these tools available to the researcher today has its unique advantages as well as limitations. Targeted as well as non-targeted approaches to metabolic profiling are possible depending on the objectives of any particular study with each approach demanding a certain amount of sample preparation. Currently, the most universal, sensitive and versatile detection method applicable to metabolite detection is mass spectrometry (Tolstikov and Fiehn 2002), though the use of nuclear magnetic resonance (NMR) techniques is rapidly gaining ground in the field. The platforms currently available for plant metabolomics research include:

1. Gas Chromatography-Mass spectrometry (GC/MS)
2. Liquid Chromatography-Mass spectrometry (LC/MS)

[1] Phenomenome Discoveries Inc., 204–407 Downey Road, Saskatoon, Saskatchewan, Canada S7N 4L8, e-mail: info@phenomenome.com

Biotechnology in Agriculture and Forestry, Vol. 57
Plant Metabolomics (ed. by K. Saito, R.A. Dixon, and L. Willmitzer)

3. Liquid Chromatography-Ultra violet photodiode array (HPLC/UV)
4. Capillary Electrophoresis-Mass spectrometry (CE/MS)
5. Fourier Transform Ion Cyclotron Resonance-Mass spectrometry (FTICR-MS)
6. Nuclear Magnetic Resonance Spectroscopy (NMR)

Gas chromatography coupled to mass spectrometry (GC/MS) has been used extensively for over half a century to analyze successfully small volatile non-polar organic molecules including de novo identification of small plant metabolites. Derivatization of polar non-volatile metabolites with amino, carboxylic acid, alcohol and phenolic functionalities followed by GC/MS enables the detection of these types of metabolites. However, large and thermally labile compounds such as sugar nucleosides, large oligosaccharides and peptides cannot be detected by GC/MS due to their limited volatility. Liquid chromatography-mass spectrometry (LC/MS) and liquid chromatography-ultra violet photodiode array (HPLC/UV) were developed to complement GC/MS in the analyses of these compounds. The front end chromatographic separation offered by GC/MS and LC/MS allows the mass analyzer to hold and detect ions of different metabolites separately. For both techniques, as column diameter decreases, column resolving power increases, thus improving their capabilities. Also, as the separation technique is downscaled, the amount of material necessary for analysis is reduced. GC/MS and LC/MS provide unique retention time data as well as mass spectral data with suggested molecular formulae depending on the resolution and sensitivity of the mass analyzer. Identification of metabolites is possible by matching data collected with commercially available libraries of standards. For targeted metabolic profiling, GC/MS, HPLC/UV photodiode array and LC/MS remain the methods of choice for quantitative and qualitative analyses (Hall et al. 2002). Capillary electrophoresis (CE) coupled to mass spectrometry (CE/MS) has been used in proteomics research for separation and identification of proteins and peptides but is now being introduced gradually to metabolomics, especially in the sequencing and identification of amino acids. However, GC/MS, HPLC photodiode array, LC/MS and CE/MS technologies cannot be relied on for the identification of novel biomarkers in biological samples.

Fourier Transform Ion Cyclotron Resonance mass spectrometry (FT-ICR-MS) is rapidly becoming the mass analyzer of choice when it comes to non-targeted metabolic profiling of complex mixtures of biological origin. In an FT-ICR (Fourier Transform Ion Cyclotron Resonance) mass spectrometer, ions are held in the analyzer cell by a combination of a static magnetic field and a coincident electrical field generated by potentials applied to the walls of the metal cell (Busch 2002). Ions attain a coherent cyclotron orbit with frequency proportional to mass and are detected by monitoring the alternating electrical current generated in detector plates by their regular orbits. A Fourier transformation converts the monitored frequency to ion mass. The high mass resolving power, sensitivity, and mass accuracy ($\leq$ 1 ppm) that FT-ICR mass spectrometry provides make it ideal for the study of complex mixtures without front end chromatographic separation and purification since the components are simultaneously resolved and identified as to elemental composition. The potential limitation this technology has is differentiation between identical molecular mass isomers and low reproducibility due to ion-suppression effect. These then would not enhance the complete structure elucidation of novel biomarkers discovered in samples.

Nuclear magnetic resonance spectroscopy (NMR) is a powerful and theoretically complex analytical technique used in the determination of the structure of unknown organic compounds and more recently biomacromolecules. It also provides comparative analysis of numerous samples. Therefore, NMR is quickly becoming a key technology in plant metabolomics with the use of stable isotope labeling and advanced hetero-nuclear NMR methodologies. Since the NMR-based approach has an advantage in comparison with different samples, spectral subtraction between different mutants or stimuli enable quantification of increased or decreased metabolites among those samples. The limitation of NMR technology is the lack of front end chromatographic separation and purification of the components of the complex mixture to enable complete structure elucidation of novel biomarkers. This is resolved by coupling the LC system to NMR (LC/NMR) though this provides lower sensitivity. With improvements in sensitivity, the use of LC/NMR is likely to grow.

In the light of these advantages and limitations of the various platforms available to the plant metabolomics researcher, FT-ICR mass spectrometry would be the preferred initial analytical technique for discovery of novel biomarkers and metabolites in complex mixtures of biological origin. Their isolation and structure determination would then be achieved with LC-NMR technology.

A number of examples in the literature show how FT-ICR MS has been applied in the plant metabolomics field. These include studies on metabolic changes in strawberry fruit development (Aharoni et al. 2002), exploration of whole cellular processes at levels of transcriptome and metabolome under sulfur deficiency related stresses in *Arabidopsis* (Hirai et al. 2004), and metabolic analysis of medicinal diversity in *Scutellaria baicalensis* (Georgi) genotypes (Murch et al. 2004). Current research in plant metabolomics conducted at Phenomenome Discoveries Inc., are discussed below.

2 Fundamental Investigations into Plant Metabolomics

2.1 Metabolomic Analysis of Cold Acclimation in *Arabidopsis*

The model plant *Arabidopsis* is able to acclimate in non-freezing temperatures (4–10 °C) by modifying its gene expression and biochemistry (Thomashow 1999; Stitt and Hurry 2002). For example, soluble sugars, proline, and other osmoprotectants are known to be increased in order to confer freezing tolerance

(Stitt and Hurry 2002). A non-targeted metabolomic method, FT-ICR MS, was utilized to study changes in the *Arabidopsis* leaf metabolome after treating plants with cold acclimating conditions. *Arabidopsis* plants were grown at a non-acclimating temperature of 23 °C for 27 days, and then were transferred to cold acclimation conditions of 4 °C for a further 49 days. Existing leaves were sampled at 0 days immediately before shifting the plants to lower temperature, and afterwards, shifted leaves were sampled at days 1, 7, 28, 49 (Gray and Heath 2005). New leaf formation occurs slowly at 4 °C, so samples of new leaves that developed de novo at cold temperature were taken at days 28 and 49 only.

Inner rosette leaves, referred to as '*developed*' leaves, that originally developed entirely in 4 °C conditions were compared to the outer rosette leaves, referred to as '*shifted*' leaves, which originally formed at 23 °C and were later shifted to 4 °C. The leaves that develop only at 4 °C have the highest freezing tolerance, while leaves that are shifted to 4 °C after a growth phase at 23 °C show variable freezing tolerance (Strand et al. 1997).

By comparing the intensity of each common detected metabolite mass tag from *shifted* leaves (days 0, 1, 7, 28, 49 after shifting to 4 °C) to the average from *developed* leaves (days 28, 49 of development at 4 °C), two distinct metabolic profile changes were observed that describe the requirement for leaves to develop at 4 °C in order to achieve increased freezing tolerance (Fig. 1). The first profile consists of metabolites that were found in *developed* leaves at either high or low levels, and over the 49 day acclimation time course, *shifted* leaves were able to adjust the concentration of the metabolite to the same level, within a twofold cutoff. Thus, these are "temperature modulated" metabolites because all leaves, given enough time, were able to adjust specific metabolites to the same relative concentrations. The second profile consisted of metabolites that were found in *developed* leaves at either high or low intensity, and over the entire 49 day cold temperature time course, *shifted* leaves were unable to adjust the metabolite concentrations to the same levels as in *developed* leaves. These are "developmentally modulated" metabolites, because their levels in freezing-tolerant leaves can only be attained when a leaf has developed solely at cold acclimating temperatures.

Looking at a set of known metabolites would likely have missed these results that a global non-targeted metabolomic approach found. There is a set of metabolites that a plant leaf can adjust in relation to cold temperature stress independent of leaf development, and a separate set that can only be modulated when a leaf has undergone some degree of development entirely in acclimating temperatures.

2.2 Compositional Analysis of Flax Seeds (*Linum usitatissimum* L)

The oil from linseed flax is predominantly used in industrial applications such as paints, oils and varnishes due to the high levels of linolenic (C18:3) and linoleic (C18:2) acids. Because of the high levels of these unsaturated fatty

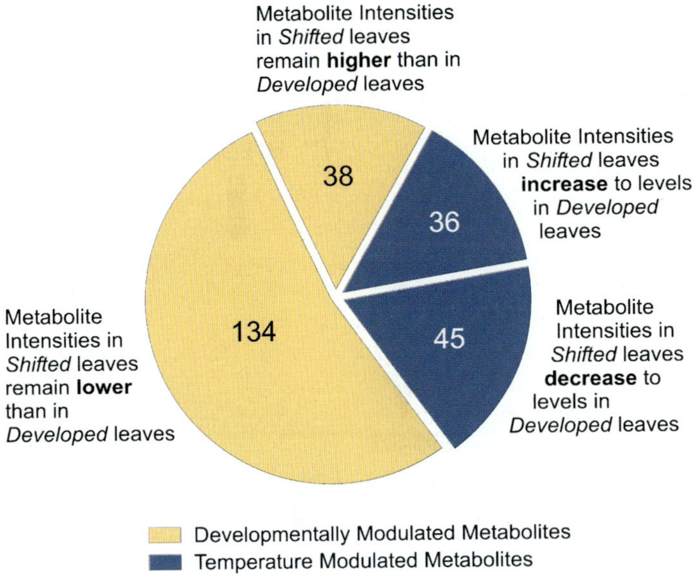

Fig. 1a. Developmental vs temperature-only modulation of the *Arabidopsis* metabolome during cold acclimation. The number of metabolites following four trends is shown over the 49 day time course at 4 °C where the intensities of individual metabolites are the same in developed and shifted leaves within a twofold cutoff, or have a difference greater than twofold. Here, '*Developed*' refers to the average intensity of metabolites from *developed* leaves harvested 28 days and 49 days after the temperature change from 23 °C to 4 °C. '*Shifted*' leaves refer to pre-existing leaves harvested immediately before this temperature shift (0 days), or 49 days after

acids, linseed oil is easily oxidized, and unstable at high temperatures, making it unsuitable as a cooking oil. To be more competitive with the vegetable oil market, much work has been done to alter the oil composition of linseed flax, the outcome being the development of solin varieties. As a result, linolenic acid levels have been reduced from over 50% to less than 5%, and linoleic acid levels have been increased from 20% to 70% (Rowland 1991).

Seeds from ten different flax lines (*Linum usitatissimum* L) were analyzed by FT-ICR MS to evaluate differences in seed composition (Fig. 2). An advantage of FT-ICR MS to analyze seed composition is the ability to evaluate a wide range of both known and unknown or unexpected metabolites simultaneously in all samples analyzed. In this particular study, most methods currently utilized in plant research for genetic alterations or manipulations were included, and consisted of somaclonal variation, chemical mutagenesis (ethyl methanesulphonate), plant breeding, and genetic engineering vis-à-vis *Agrobacterium* mediated transformations. A total of 2606 spectral peaks, each representing a unique m/z (mass-to-charge) were identified among the ten different flax lines, with approximately 1100–1200 independent m/z detected in each flax sample analyzed. Global analysis of all ten cultivars with both principal

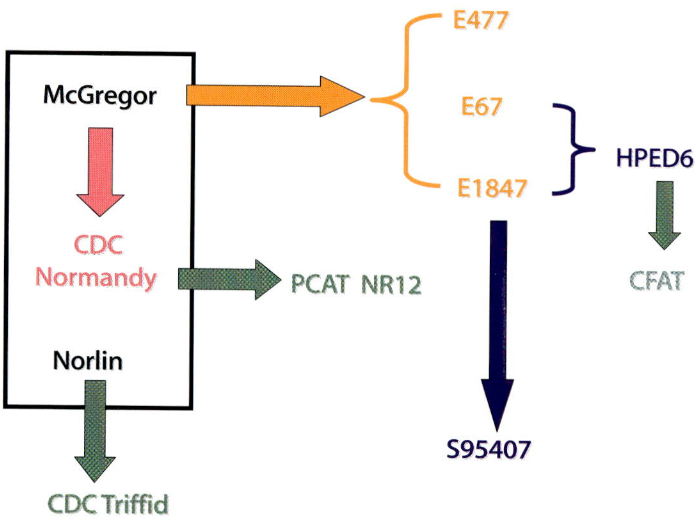

Fig. 2. Schematic diagram of the ten flax varieties analyzed by FT-ICR MS, and how they relate to each other. The registered conventional varieties are *boxed in black*. CDC Normandy is a somaclonal variant of McGregor. The introduction of transgenes is indicated with *green arrows*, and the resultant transgenic line is indicated in *green*. EMS mutagenesis is indicated in *orange*, while advanced breeding lines are indicated in *blue*. All flax varieties are unregistered except the conventional varieties McGregor, CDC Normandy, and Norlin, which are registered and commercially grown, and contain high levels of linolenic acid. E67, an EMS McGregor mutant, was selected for its increased linoleic and decreased linolenic acid content. E67 was also found to contain elevated palmitic acid levels. E477, another EMS McGregor mutant, was selected for its altered mucilage (fibre) content. S95407, a solin variety with decreased linolenic acid content, was developed from another EMS McGregor mutant (E17847). HPED6 was derived from an original cross of E67 and E1747, and contained increased palmitic and linoleic acid and decreased linolenic acid content. The transgenic variety CDC Triffid was created as a herbicide resistant "Norlin" variety (McHughen et al. 1997) by the introduction of the *Arabidopsis thaliana* CSR-1 gene (Haughn et al. 1988) encoding a sulfonylurea resistant acetolactate synthase (ALS) under control of the 35S CaMV promoter. The transgenic line PCAT NR12 was created to increase the oleic acid and decrease the linolenic acid content in the seed oil through the introduction of the soybean phosphatidyl choline acyl transferase gene AAPT1 (Dewey et al. 1994) under the control of the *Arabidopsis thaliana* oleosin promoter. The transgenic line CFAT was created to increase the short and medium chain fatty acid levels in the seed by the introduction of the *Cuphea wrightii Cw* FATB1 (Lenoard et al. 1997) gene under control of the 35S CaMV promoter into HPED6. All transgenic lines were generated by *Agrobacterium tumefaciens* mediated transformations

component analysis (PCA) and hierarchical clustering (HCA) were able to separate the linseed-like from the solin-like varieties base on the 2606 independent m/z detected (Fig. 3). These results suggested differences in the metabolomes all of the flax lines analyzed, with the greatest metabolome changes occurring between the linseed and solin-like varieties which separated out along the first principal component.

Relative changes in the fatty acid content of each flax variety were determined by FT-ICR (Fig. 4a). As expected, linolenic acid levels were lower, while

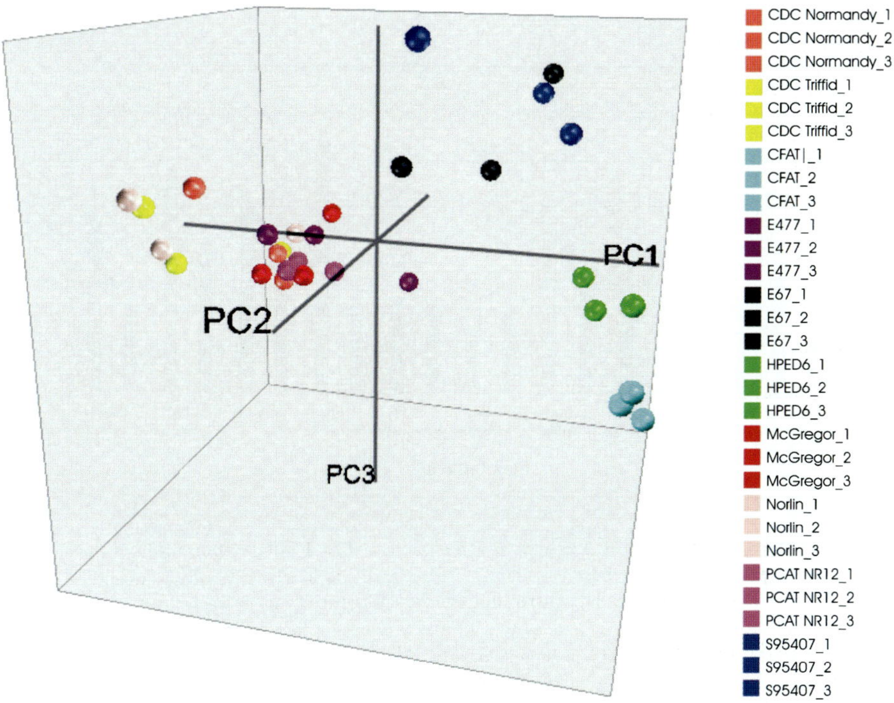

Fig. 3. Principal component analysis of the seed metabolomes from ten flax varieties analyzed by FT-ICR MS. The linseed-like varieties separate from the solin-like varieties along principal component 1

linoleic acid levels were higher in the solin varieties. Furthermore, levels of stearic acid (C18:0) which were not selectively altered remained fairly constant between the linseed and solin varieties. An increase in palmitic acid (C16:0) was observed in the CFAT line, which was genetically engineered to increase the short chain fatty acid content. In conjunction to being able to monitor fatty acid levels, the levels of mono-, di- and triacylglycerides containing linolenic acid could be monitored simultaneously between the different flax varieties. For example, as the levels of linolenic acid decreased and linoleic acid increased in the solin flax, so did putative triacylglycerides which contained either fatty acid (Fig. 4b). Furthermore, relative levels of other known compounds such as the beneficial secoisolariciresinol diglucoside (SDG), and the detrimental cyanogenic glucosides linustatin, neolinustatin and linamarin could be analyzed simultaneously between the flax varieties.

Although many of the above analyses can be done by targeted methods, which could also quantitate the observed differences in both fatty acids and triacylglycerides, FT-ICR MS allows for the global non-biased observation of many classes of metabolites (polar, non-polar etc.) at once. Thus, the researcher

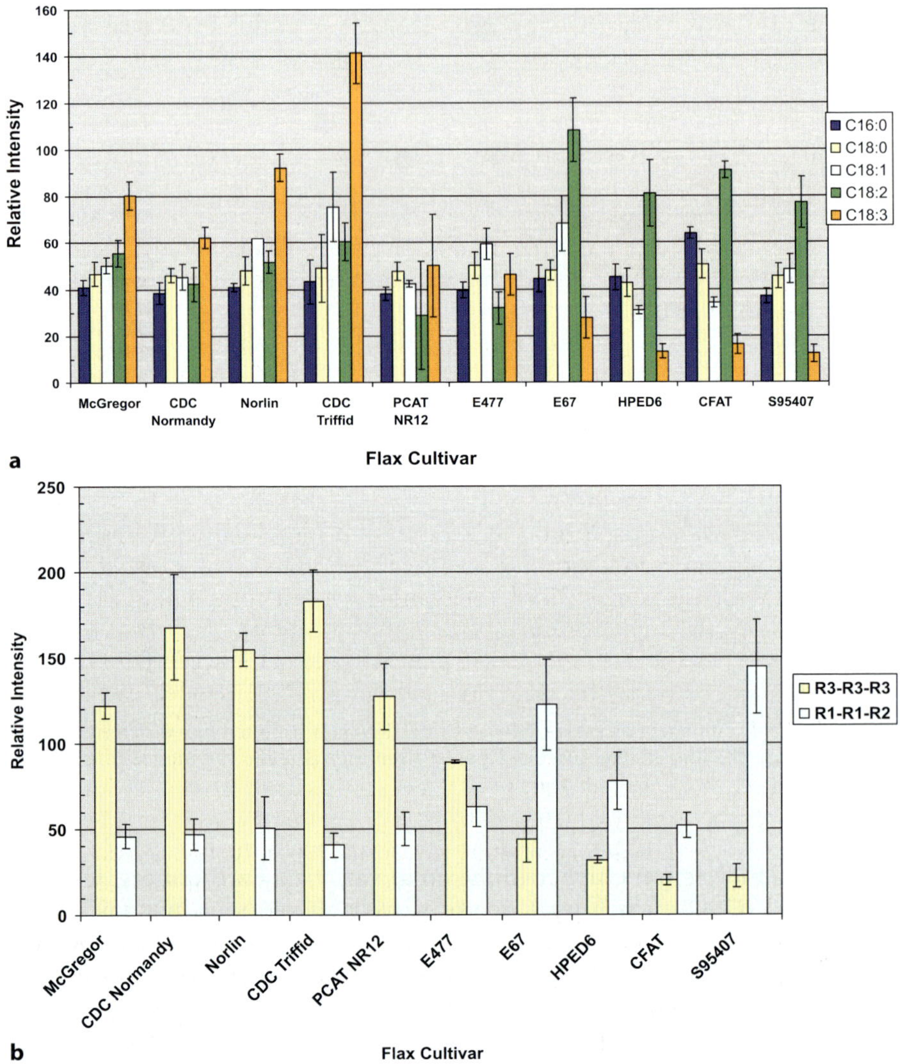

Fig. 4. a Relative abundance of fatty acids in the ten flax lines. Overall, linolenic acid (C18:3) decreased while linoleic acid (C18:2) increased in the solin-like varieties relative to the linseed-like varieties. **b** Relative abundance of two putative triacylglycerides in the ten flax lines. A putative triacylglyceride containing only linolenic acid (R3-R3-R3) was observed to decrease in the solin-like varieties, mirroring the observed decrease in the C18:3 fatty acid. The putative triacylglyceride (R1-R1-R2) containing both oleic acid (R1) and linoleic acid (R2) was observed to increase in the solin-like varieties mirroring the increase in C18:2 fatty acid in solin-like varieties

can note relevant changes with FT-ICR MS, and then validate the changes with known and appropriate targeted methods.

2.3 Flavonoid Identification in Wheat (*Triticum aestivum* L.)

Metabolomics can be useful in identifying novel metabolites and candidate genes. The identification of a metabolite or groups of metabolites associated with a particular trait or phenotype can be used to deduce possible genes and/or metabolic pathways involved.

Kernel color due to pigmentation has become an important discriminator among hard-white wheats (*Triticum aestivum* L.) in Canada. White-seeded hard-white wheat differs from red-seeded hard-white wheat predominantly in the color of the bran, with no red-pigmentation being present in the bran of the white-seeded varieties. Although the red-pigment associated with the red-seeded varieties has been shown to be a highly heritable trait (Cooper and Sorrells 1984), no specific genes have been identified to be responsible for the red-pigmentation. Work by several different groups suggested that the red-pigmentation was most likely a flavonoid-related compound(s) (Lamkin and Miller 1980; Matus-Cadíz, personal communication).

Flavonoids are major compounds in plants, found in the seeds, fruits, leaves, stems and flowers (Iwashina 2000; Winkel-Shirley 2001; Hodek et al. 2002). However, the various combinations of hydroxyl and methoxyl groups on the basic flavonoid structure (C6-C3-C6) yield many different possible flavonoid configurations (Hodek et al. 2002), with several thousand flavonoids identified to date in plants (Iwashina 2000; Pietta 2000). Furthermore, many flavonoids exist as glycosides (Hodek et al. 2002; Iwashina 2000), further adding to the structural complexity. Simply deducing that the pigment in the red-seeded wheat varieties was likely a flavonoid still posed a daunting task in identifying which flavonoid(s) out of the possible thousands was responsible for the pigmentation in the red-seeded hard white wheat varieties. The use of FT-ICR MS was an ideal platform to attempt to initially identify pigmentation molecules.

FT-ICR MS analysis of the whole grain from 18 independent wheat varieties consisting of red-pigmented and non-pigmented samples showed compositional differences, but no flavonoid-like compounds were observed. However, FT-ICR MS analysis of the bran from one of the red-seeded and two of white-seeded wheat varieties indicated possible differences in flavonoid-like compounds (Fig. 5). The accurate masses generated by FT-ICR MS suggested these flavonoid compounds were not previously characterized in wheat. Three of the putative flavonoids most likely to be responsible for the pigmentation, appeared to be chemically related to each other as the flavonoid itself, its glycosylated form, and an acetylated form of the glycoside. Interestingly, one metabolite, thought to be an acetylated-glycosylated flavonoid-like molecule, increased in the non-pigmented bran sample. Examination of the KEGG flavonoid biosynthesis pathway suggested that this metabolite could be the result of a block

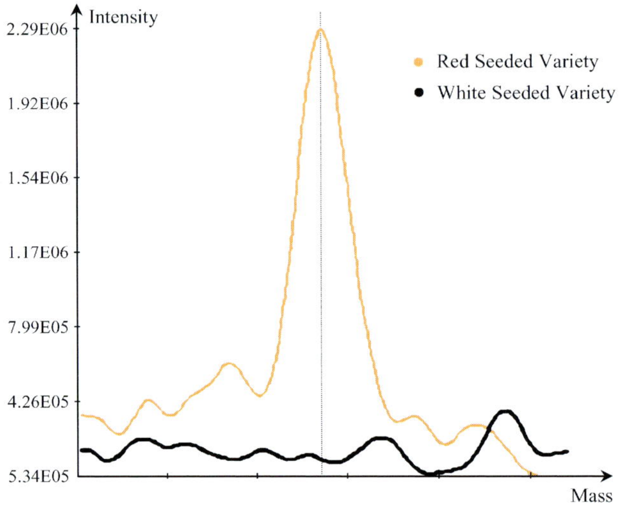

Fig. 5. Actual spectral data of a metabolite thought to be a flavonoid involved in the pigmentation of the wheat bran in red-seeded varieties. The metabolite was found to be present in the red-seeded variety, but was absent in the white-seeded variety

in the wheat pigmentation pathway, and was likely a build-up product in the non-pigmented varieties. Thus, knowing both where the block in the pathway is located, and where the pigmentation molecules are located along the biosynthetic pathway, researchers can speculate which set of genes may be responsible for either pigmentation or lack of pigmentation.

Although not conclusive proof that the pigment molecules have been identified, the use of FT-ICR MS has narrowed the focus, and allowed a starting point to formulate hypotheses around the pigmentation in wheat bran. Further experiments are required either to validate or to reject these hypotheses.

2.4 Effect of Glutamate Dehydrogenase on the Metabolic Profiling of Transgenic *Nicotiana tabacum*

Glutamate dehydrogenase (GDH) is a broadly distributed enzyme that atalyzes a reversible oxidative deamination of glutamate to α-ketoglutamate and ammonia in the presence of the coenzyme (NADH or NADPH). Roots and leaves from both wild-type and *gdhA* transgenic tobacco encoding a NADPH-dependent GDH were analyzed using FT-ICR MS in a non-targeted manner. Approximately 2000 peaks were detected within a single sample. The fold change value of each mass peak observed was calculated as the ratio of signal intensity in normal tobacco to that in transgenic tobacco. Figure 6a shows the number of metabolites which are significantly (Student's *t*-test; $p < 0.01$) increased or decreased in *gdhA* tobacco compared to control tobacco. In the root samples, a total of 210 metabolites are elevated and 86 metabolites are decreased in *gdhA* tobacco. In contrast, the leaves exhibit a total of 179 metabolites that are increased and 197 metabolites that are decreased in *gdhA* tobacco. The ratios of metabolites altered in abundant GDH activity are almost the same in root (18%) and leaf (20%) samples.

Successive additions of sugar moieties to metabolites were observed in the *gdhA* tobacco. For example, quercetin (theoretical mass of 302.0427 Da) (Da = Dalton), a plant pigment, is a unique flavonoid that was highly increased in the leaves of *gdhA* tobacco. Also, a related metabolite, quercetin 3-*O*-glucoside (theoretical mass of 464.0955 Da), a glycoside of quercetin, was increased. As shown in Fig 6b, a total of 72 increased metabolites in roots were glucuronidated ($p < 0.05$; in *gdhA* tobacco compared with control tobacco). In contrast, only 21 glucuronidated metabolites in leaves were increased in *gdhA* transgenic tobacco. Glucose and sucrose levels were significantly elevated both in roots (1.6-, 3.5-fold, respectively) and in leaves (1.6-, 2.3-fold, respectively).

Figure 7A shows principal component analysis (PCA) of *gdhA* tobacco and control tobacco metabolome data. Regardless of genotype, FT-ICR MS metabolic fingerprints of leaves and roots differed significantly (in Fig. 7A the proportion of principal component (PC) 1 is 83%). Figure 7B shows the Phenomenome Profiler™ screen shot which lists the top 20 metabolites (ten most positive and ten most negative loadings) for PC1. Each row represents the signal to noise ratio of each metabolite and each column represents the metabolite profile of each sample (replicate analysis, $n = 5$). Metabolites having a high positive loading score were mainly derived from the root extract, and metabolites having a high negative loading score were primarily derived from the leaf extract. Qualitative differences (metabolites only detected in one organ) approached 40% (819 metabolites) of the total number of metabolites observed.

This study used FT-ICR MS metabolic profiling to associate phenotypes with biochemical changes resulting from endogenous effects of glutamate synthesis in transgenic plants. The GDH plants were a suitable test for FT-ICR MS because they exhibit cell composition alterations that result from a specific biochemical change in a well-characterized pathway targeting the cellular glutamate pools. From a plant metabolic engineering perspective, GDH could be useful for inducing increases or decreases in the yield of a large number of chemical compounds. In particular this attribute may be useful as the pharmaceutical industry discovers new plant-derived compounds of therapeutic value.

The work presented here demonstrates that metabolite analysis by FT-ICR MS provides a useful tool for the analysis of cryptic phenotypes in transgenic plants. The acquisition of data from extracts that have not undergone derivatization enables the analysis of the relationships between various metabolites and the determination of equivalence, or lack thereof, between samples. The sensitivity and resolution of FT-ICR MS provides a useful method for cataloguing chemical diversity; within existing technological limits differences may be measured between samples whose mass is as low as 50 mg for fresh tissue and

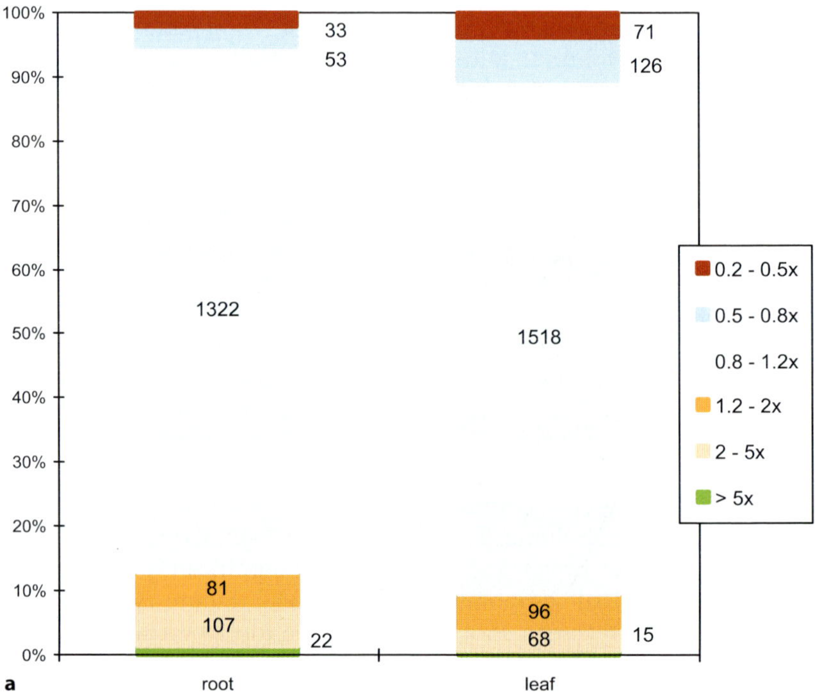
100%
90%
80%
70%
60%
50%
40%
30%
20%
10%
0%
33
53
1322
81
107
22
71
126
1518
96
68
15
root
leaf
0.2 - 0.5x
0.5 - 0.8x
0.8 - 1.2x
1.2 - 2x
2 - 5x
> 5x

a

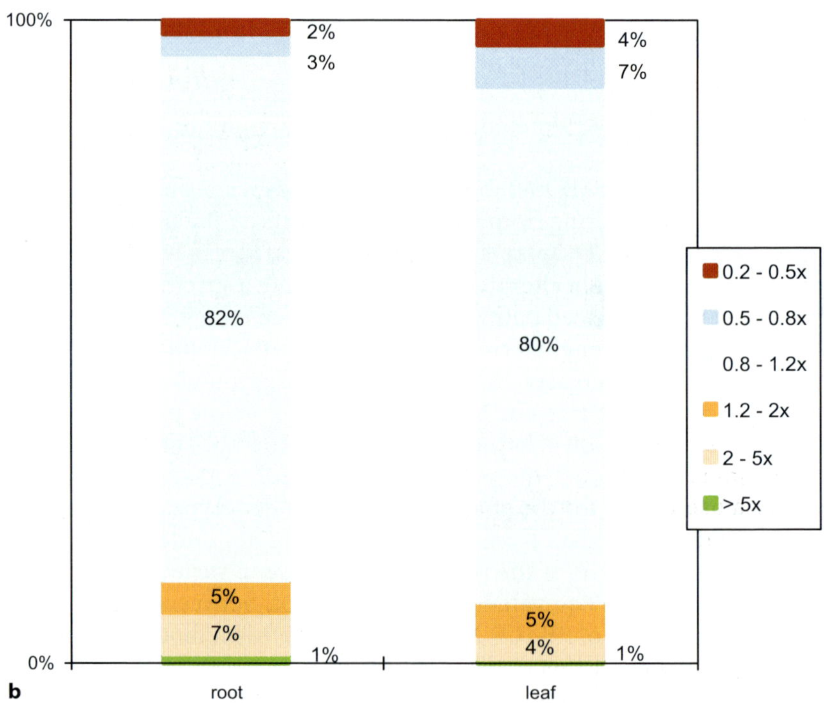
100%
0%
2%
3%
82%
5%
7%
1%
4%
7%
80%
5%
4%
1%
root
leaf
0.2 - 0.5x
0.5 - 0.8x
0.8 - 1.2x
1.2 - 2x
2 - 5x
> 5x

b

◀ **Fig. 6. a** Number of metabolites detected in this experiment and indicated as fold changes in *gdhA* tobacco compared with control tobacco. **b** Number of metabolites related via a glucuronidated reaction and indicated as fold changes in *gdhA* tobacco compared with control tobacco

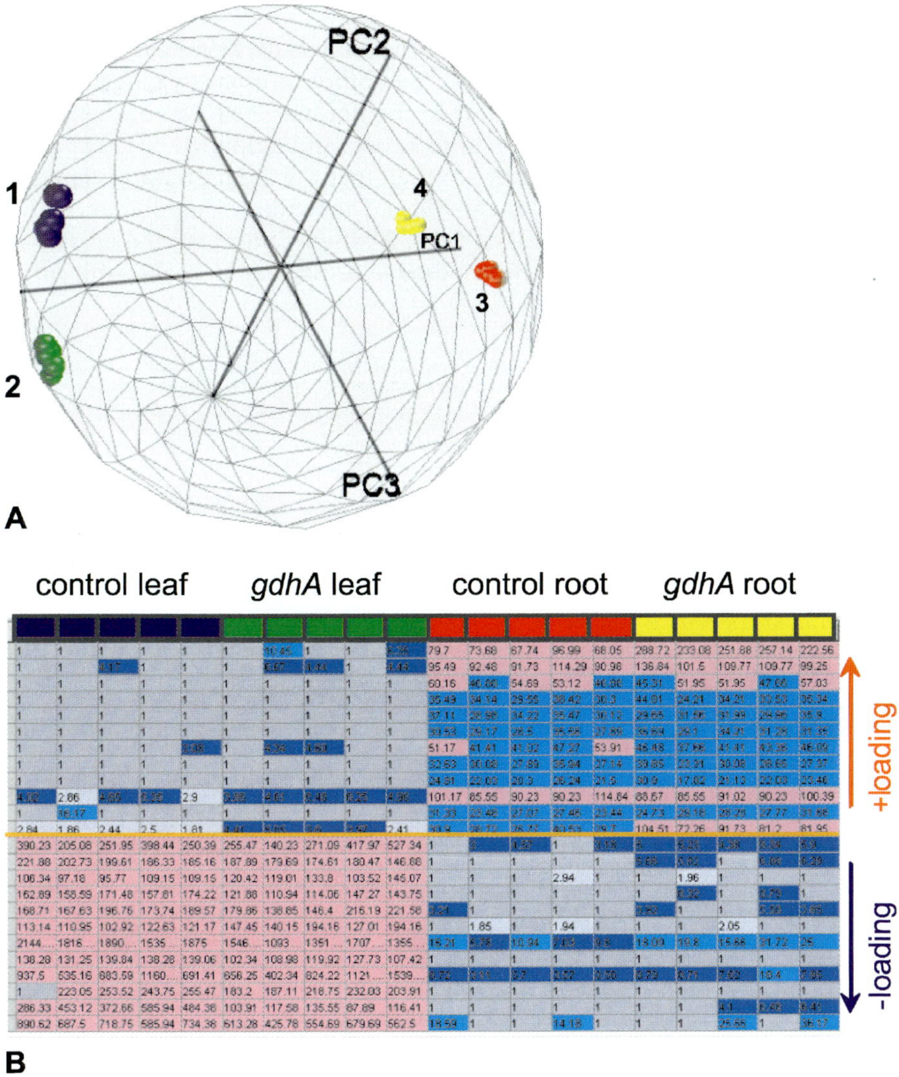

Fig. 7. A Metabolomic PCA analysis of the samples. Each spheroid represents a single sample. Spheroid clusters 1 to 4 are control leaf, *gdhA* leaf, control root, and *gdhA* root respectively. Proportions of the first, second, and third components are 83%, 5%, and 4%. **B** The Phenomenome Profiler™ screenshot of the top 20 metabolites for principal component 1

even lower for dried tissue. Therefore, the occurrence of novel compounds in individuals, populations, species and genera may be catalogued with relative ease.

3 Conclusion

FT-ICR MS has many applications within plant sciences. The major advantage for utilizing FT-ICR MS technology is the ability to monitor global system changes in a non-targeted manner. It is the non-targeted approach that allows the visualization of changes of both known and unknown or unexpected metabolites, allowing the researcher to then focus or target their research to a specific metabolite or group of metabolites. The examples highlighted in this chapter have shown, with the use of FT-ICR MS in a non-targeted approach, discoveries that may not have been possible or as easy to make with other metabolomic platforms.

Acknowledgements. Gordon Gray and Ron Wilen Department of Plant Sciences, University of Saskatchewan; Gordon Rowland, Pierre Hucl and Maria Matus-Cadìz Crop Development Centre, Department of Plant Sciences, University of Saskatchewan; David A. Lightfoot, Department of Plant and Soil Science, Southern Illinois University.

References

Aharoni A, Ric De Vos CH, Verhoeven HA, Malipaard CA, Kruppa G, Bino R, Goodenowe DB (2002) Nontargeted metabolome analysis by use of Fourier Transform Ion Cyclotron Mass Spectrometry. Omics 6:217–234

Busch KI (2002) A glossary of mass spectrometry. Mass Spectrometry 17(6S):29

Cooper DC, Sorrells ME (1984) Selection for white kernel color in the progeny of red/white wheat crosses. Euphytica 33:227–232

Dewey RE, Wilson RF, Novitzky WP Goode JH (1994) The AAPT1 gene of soybean complements a cholinephosphotransferase-deficient mutant of yeast. Plant Cell 6:1495–1507

Edwards D, Batley J (2004) Plant bioinformatics: from genome to phenome. Trends Biotechnol 22:232–237

Goodacre R, Vaidyanathan S, Dunn WB, Harrigan GG, Kell DB (2004) Metabolomics by numbers: aquiring and understanding global metabolite data. Trend Biotechnol 22:245–252

Gray GR, Heath D (2005) A global reorganization of the metabolome in *Arabidopsis* during cold acclimation is revealed by metabolic fingerprinting. Physiol Plant (accepted)

Hall R, Beale M, Fiehn O, Hardy N, Sumner L, Bino R (2002) Plant metabolomics: the missing link in functional genomics strategies. Plant Cell 14:1437–1440

Haughn GW, Smith J, Mazur B, Somerville C (1988) Transformation with a mutant Arabidopsis acetolactate synthase gene renders tobacco resistant to sulfonylurea herbicides. Mol Gen Genet 211:266–271

Hirai MY, Yano M, Goodenowe DB, Kanaya S, Kimura T, Awazuhara M, Arita M, Fujiwara T, Saito K (2004) Integration of transcriptomics and metabolomics for understanding of global responses to nutritional stresses in *Arabidopsis thaliana*. Proc Natl Acad Sci USA 101:10205–10210

Hodek P, Trefil P, Stiborová M (2002) Flavonoids-potent and versatile biologically active compounds interacting with cytochromes P450. Chem Biol Interact 139:1–21

Iwashina T (2000) The structure and distribution of flavonoids in plants. J Plant Res 113:287–299

Lamkin WM, Miller BS (1980) Note on the uses of sodium hydroxide to distinguish red wheat from white common, club and durum cultivars. Cereal Chem 57:293–294

Lenoard JM, Slabaugh MB, Knapp SJ (1997) *Cuphea wrightii* thioesterases have unexpected boad specificities on saturated fatty acids. Plant Mol Biol 34:669–679

McHughen A, Rowland GG, Holm FA, Bhatty RS, Kenaschuk EO (1997) CDC Triffid transgenic flax. Can J Plant Sci 77:641–643

Murch SJ, Rupasinghe HP, Goodenowe DB, Saxena PK (2004) A metabolomic analysis of medicinal diversity in Huang-qin (*Scutellaria baicalensis* Georgi) genotypes: discovery of novel compounds. Plant Cell Rep 23:419–425

Oliver SG, Winson MK, Kell DB, Baganz F (1998) Systematic functional analysis of the yeast geneome. Trends Biotechnol 16:373–378

Pietta P-G (2000) Flavonoids as antioxidants. J Nat Prod 63:1035–1042

Rowland GG (1991) An EMS-induced low-linolenic-acid mutant in McGregor flax (*Linum usitatissimum*). Can J Plant Sci 71:393–396

Stitt M, Hurry V (2002) A plant for all seasons: alterations in photosynthetic carbon metabolism during cold acclimation in Arabidopsis. Curr Opin Plant Biol 5:199–206

Strand A, Hurry V, Gustafsson P, Gardestrom P (1997) Development of *Arabidopsis thaliana* leaves at low temperature releases the suppression of photosynthesis and photosynthetic gene expression despite the accumulation of soluble carbohydrates. Plant J 12:605–614

Thomashow MF (1999) Plant cold acclimation: freezing tolerance genes and regulatory mechanisms. Annu Rev Plant Physiol Plant Mol Biol 50:571–599

Tolstikov VV, Fiehn O (2002) Analysis of highly polar compounds of plant origin: combination of hydrophilic interaction chromatography and electrospray ion trap mass spectrometry. Anal Biochem 301:298–307

Tweeddale H, Notley-McRobb L, Ferenci T (1998) Effect of slow growth on metabolism of *Escherichia coli*, as revealed by global metabolite pool ("metabolome") analysis. J Bacteriol 180:5109–5116

Winkel-Shirley B (2001) Flavonoid biosynthesis. A colorful model for genetics, biochemistry, cell biology and biotechnology. Plant Physiol 126:485–493

III.10 Plant Metabolite Profiling for Industrial Applications

R.N. Trethewey[1]

1 Introduction

The industrial application of plant metabolite profiling is less than a decade old. However, the growth of interest and application of this technology has been rapid due to a range of new commercial opportunities from functional genomics to the development of improved crop varieties through conventional breeding or genetic modification. In particular the expectation that future generations of plant biotechnology products will be based upon metabolic characteristics (for example improved yield or enhanced compositions of nutritionally important compounds) has provided a huge impetus to this field. Indeed plant scientists have been the pioneers of many of the contemporary technologies of metabolite profiling. In this chapter I will review the requirements for the technology of metabolite profiling in an industrial setting, discuss the challenges of achieving a high throughput, and illustrate applications in crop protection, plant biotechnology and nutrition.

2 The Metabolome

Before diving into the technology of metabolite profiling, it is worth pausing to consider what it is that is to be measured. Based on the nomenclature of genomics, the term "metabolome" has recently emerged to represent the total metabolite complement of a cell, tissue or organism. No one really knows what the extent of the metabolome is (Bino et al. 2004), nor where to draw the line between a metabolite and a small molecule or a slightly larger molecule! Some scientists include DNA, RNA and protein in the definition of the metabolome, others prefer to limit it to compounds with a molecular mass of less than an arbitrary number such as 1000. In addition, metabolites of similar or different chemical nature can occur *in vivo* as conjugates which adds to the confusion. Natural product chemistry is a field with a long and extensive tradition and there have been estimates that more than 100,000 secondary metabolites have been identified and that this represents less than 10% of the total amount present in the plant kingdom (Wink 1988). Some scientists who take on an

[1] metanomics GmbH and metanomics Health GmbH, Tegeler Weg 33, 10589 Berlin, Germany, e-mail: richard.trethewey@metanomics.de

Biotechnology in Agriculture and Forestry, Vol. 57
Plant Metabolomics (ed. by K. Saito, R.A. Dixon, and L. Willmitzer)

advocate role for the field of plant metabolomics often emphasize very large numbers in order to draw attention to the challenge (and themselves). Definitions of the global metabolome are of limited value, as ultimately it is necessary to understand the metabolome in the species being studied. Estimates for an individual plant species, or humans, vary in the 3–25,000 range and a likely figure for an *Arabidopsis* leaf is 5000.

In the case of industrial applications, it is important to precisely define which metabolites need to be measured and adapt methods to give an optimal quantification of these key compounds. For commercial projects, it is often the case that key compounds are available and can be purchased as standards to allow calibration and optimization of methods. In the case of plant biotechnology products, primary metabolism is at the forefront of interest, for example, amino acids, fatty acids, organic acids, sugars and vitamins. In some cases, specific secondary metabolites such as constituents of wine, or known growth regulators also play a role. It is therefore sufficient for most commercial applications to generate profiles covering hundreds of metabolites, a fraction of the overall metabolome.

One of the characteristics of contemporary plant metabolomics is that the majority of metabolites that are determined in a typical profile are unknown (Bino et al. 2004; Kopka et al. 2005). Often the unknown analytes deliver reliable measurements, perhaps having a characteristic mass spectrum signal and retention time behavior in a chromatographic system. Such unknowns have the potential to be biomarkers, or can contribute to precise classification of metabolite profiles. Thus many companies assign a significant value to the "known unknowns" and these are often determined in parallel with the key known metabolites.

3 Profiling Technologies

There are currently no methods that are even close to delivering a complete quantification of the metabolome. NMR based spectroscopy has historically been used for plant metabolite profiling (Ratcliffe and Shachar-Hill 2005) and some industrial applications have been reported (see Sect. 5.1). However, this technology is limited by sensitivity and the range of metabolites covered is low. Thus mass spectrometry (MS) techniques have emerged in the last few years as the method of choice for metabolite profiling. There are however a bewildering range of different MS technologies each with a different list of advantages and disadvantages. These have been expertly reviewed by Sumner et al. (2003) and I will not go into this diversity in more depth in this chapter.

For industrial applications, there are three key requirements which significantly influence the choice of system: robust quantification, economic reliability, and high throughput operation. The first requirement excludes direct-injection MS approaches, where the quantification can be subject to ion-suppression effects due to unpredictable variability of matrix derived compounds (Annesley 2003; Matuszewski et al. 2003). Many biologists have been seduced by the apparent speed and ease of direct injection, but there is little acceptance for metabolite profiling without chromatography amongst analytical chemists. To date, the remaining two requirements have been unequivocally met only by MS coupled to gas chromatography (GC) or liquid chromatography (LC). However, newer technologies such as capillary electrophoresis coupled to MS will no doubt reach a similar maturity in the near future (Sato et al. 2004).

3.1 GC-MS Based Metabolite Profiling

GC-MS based profiling has a long and extensive history. The origins can be found in clinical research in the 1960s. Indeed one of the first definitions in the field was GC specific: "Metabolic profiles are multi-component GC analyses that define or describe metabolic patterns for a group of metabolically or analytically related metabolites" (Horning and Horning 1971). This early paper described the analysis of around 20 analytes (steroids, acids and drug metabolites) in human and rat urine samples. Clinical applications were pursued consistently by a small number of groups through the 1970s and 1980s (Niwa 1986). However, the field was limited by the separation capabilities of GC columns, the expense of MS couplings and computational limitations. These barriers slowly crumbled in the 1990s as improvements in both the engineering of GC-MS systems and in the affordability of computing power led to the development of robust bench-top systems.

The pioneer of GC and GC-MS based metabolite profiling in plants was the group of Sauter working at the German chemical company, BASF AG (Sauter et al. 1991) (Sect. 5.1). The authors were able to resolve around 200 peaks with a high degree of reproducibility and could determine the structure of around 70 compounds. The method that they developed involved the silylation of plant extracts before GC analysis. However, this has the disadvantage that a large number of the peaks are multiple isomers of sugars, which are abundant metabolites in plants. Thus in the late 1990s the Max Planck Institute for Molecular Plant Physiology adapted the method in two ways. First, the extracts were separated into polar and non-polar components and, second, a second derivatisation step was introduced to reduce isomer complexity in the polar fraction. Roessner et al. (2000) published this new method and showed that in potato tubers it allowed the quantitative determination of more than 150 tuber metabolites including sugars, sugar alcohols, dimeric and trimeric saccharides, amines, amino acids and organic acids.

In the meantime, bench top GC-MS systems have evolved to be highly robust and very precise. At metanomics we are able to operate over 40 systems in routine practice with an "up-time" availability for each system of over 95% (Fig. 1). The costs of GC-MS have also fallen significantly over the years and

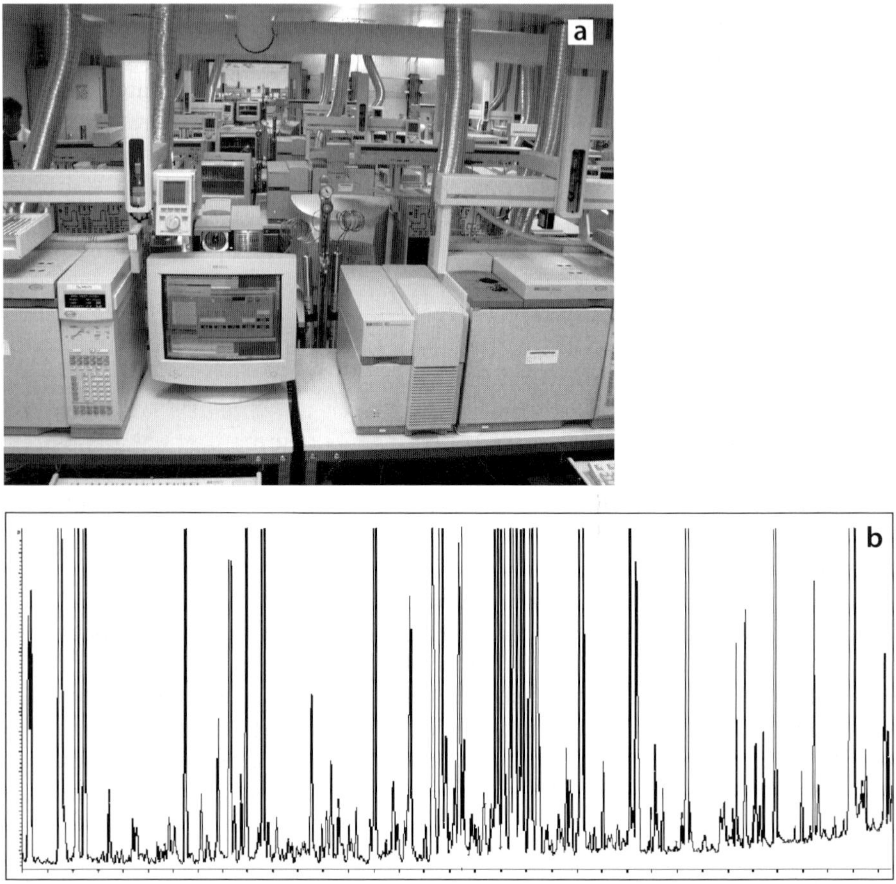

Fig. 1. a High Throughput GC-MS at metanomics. More than 40 systems are in operation. **b** A typical GC-MS chromatogram of polar compounds in *Arabidopsis* leaves generated in the laboratory shown in **a**. The figure illustrates the large number of analytes which can be found in a single analysis of a plant extract

this represents an ideal technology for a research group to enter into the field of metabolite profiling. Indeed it remains mysterious why this technology is not even more widespread than it currently is.

Despite the maturity of GC-MS technology it is reassuring that innovation continues (Santos and Galceran 2003). A coupling between GC and a time-of-flight (TOF) mass spectrometer was commercially introduced in the late 1990s and is now available from several instrument providers. TOF has the decisive advantage that is scans faster than conventional quadrupole technology allowing either improved deconvolution of peaks in complex mixtures, or shorter run times. In addition TOF provides a higher accuracy than quadrupole technology in the determination of mass-to-charge ratios. There are currently no publications on the application of this technology to industrial questions, and only a few reports from the academic community (e. g. Taylor et al. 2002; Fiehn 2003). However, it is to be anticipated that GC-TOF will ultimately replace quadrupoles as the workhorse of industrial metabolite profiling. Even more innovative has been the recent development of two-dimensional GC-MS (Dallüge et al. 2003). This technology offers the possibility to perform a second dimension of GC separation enabling isomers to be separated which otherwise would be submerged into a single peak.

3.2 LC-MS Based Metabolite Profiling

Whilst GC-MS fulfills many of the technical requirements of a system suited for industrial metabolite profiling, it has one key limitation – the range of metabolites that can be analyzed. Obviously, only compounds that can enter the gas phase can be studied and the use of chemical derivatisation protocols to widen the range of volatile compounds has the drawback of introducing more processing steps and sources of error. Thus the potential for LC-MS as a profiling technology has attracted much attention (Wilson et al. 2005a). It took both innovation in the development of LC-MS couplings and successive iterative improvement in LC separation technologies to bring us to the point where LC-MS metabolite profiling is now feasible (Niessen 2003). Indeed, the ionization techniques available for LC-MS such as atmospheric pressure ionization (API), atmospheric pressure chemical ionization (APCI), atmospheric pressure photoionization (APPI) and electrospray (ESI), have been demonstrated to be capable of generating ions from labile analytes (Hayen and Karst 2003).

LC-MS application development has been driven by the pharmaceutical industry, attracted by the sensitivity and precision of the technology for studying drugs or xenobiotics and their metabolic products. The latter application is often termed metabolite profiling which can cause some confusion. Most publications on LC-MS metabolite profiling in plants have targeted specific substance classes (e. g. Lange et al. 2001; Huhman and Sumner 2003) and it is only recently that there have been reports of wider plant metabolite profiling with LC-MS (Tolstikov et al. 2003). Of particular note was a recent paper that reported that 2000 different signals could be profiled in *Arabidopsis* leaves and roots when using capillary LC coupled to TOF-MS (van Roepenack-Lahaye et al. 2004). Some of these signals could be assigned to particular metabolites or classes.

The experience of metanomics has shown that LC-MS methods for wide metabolite profiling can be developed and implemented in high throughput. In addition, a range of other industry players offer LC-MS profiling technologies as part of their company sales material. Thus it seems that LC-MS is establishing itself as a core profiling technology in plant industrial profiling and in due time may come to supercede GC-MS. Considerable further development in technologies can be expected in the future. In particular, the combination of LC with cutting-edge mass spectrometric techniques such as Fourier-transform ion cyclotron resonance mass spectrometry (FT-ICR-MS, also FTMS) (Zhang et al. 2005) or linear ion traps (Schwartz et al. 2002) may offer the potential to widen the scope of metabolites being profiled whilst simultaneously supporting the process of structural elucidation of unknown metabolites in the profile. In addition, improvements in liquid chromatography, so called ultra performance LC (UPLC), offer the possibility of enhanced separation in complex mixtures (Wilson et al. 2005b).

4 High Throughput Metabolite Profiling

Many applications in the industrial sphere require high throughput metabolite profiling. For example, in the case of genomics projects large sample numbers have to be processed in order to provide genome scale coverage whilst in the area of breeding it is often the case that samples have to be analyzed in short time periods in order for selection decisions to be taken from one season to the next. Further, high throughput capabilities allow the parallel operation of a range of different analytical methods in order to maximize the width of the profiling information that is generated.

In this section, some of the challenges and requirements of high throughput industrial operation will be introduced based upon the experience of metanomics with GC-MS and LC-MS systems. An overview of the technical and information workflow in high throughput operation at metanomics is provided in Fig. 2. Given that there are currently no "off-the-shelf" solutions for high throughput operation, each step of the analytical process has to be thoroughly optimized for ease of operation and potential automization. Further, upscaling to high throughput operation should only be considered for methods where a full validation process has been successfully concluded. There are currently no commonly agreed standards for the validation of metabolite profiling methods (although there are various initiatives crystallizing to tackle this question, e. g. www.smrsgroup.org; www.metabolomicssociety.org) and extreme care needs to be taken in defining what is sufficient degree of validation.

The analytical process starts with the experimental design, generation of material and sampling. In the case of the experimental design the importance of control material cannot be underestimated in high throughput operation. At metanomics around 30% of the analyses that are performed are on control material and this enables the quality of the process to be monitored and maintained across different systems and over long time periods. Further, great care has to be taken to ensure that sufficient control material is generated to enable the metabolite profiles to be adequately interpreted and ensure that the experimentation leads to a decisive outcome. In order to design the experiments competently, it is necessary to have some idea of the variability that is to be expected, for the metabolites of critical importance, so that statistical power can be estimated and optimized. This may necessitate conducting feasibility and pilot experimentation before upscaling to the final experimental design.

The production of the plant material and the sampling process are key steps that influence the overall variability of the process. In some cases, for example in functional genomics projects, it is desirable to minimize the variability in the process so that the effects of a particular gene can be seen as clearly as possible. This can involve a significant investment in infrastructure for the precise controlled growth of plants and the development of standard operating procedures (SOPs) for the rapid harvesting and snap-freezing of plant tissues. In other cases, for example in breeding work with field grown plants, it may be important to understand the variability associated with particular metabolites and here steps should be undertaken to ensure that a wide diversity of material is sampled.

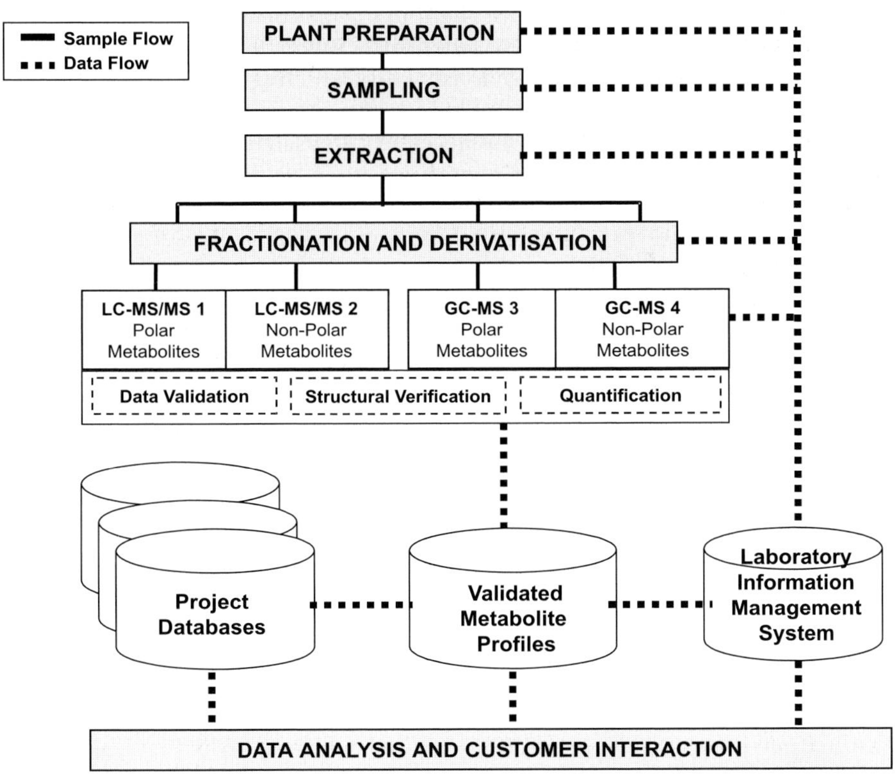

Fig. 2. Schematic representation of sample and data workflow for high throughput metabolite profiling at metanomics

There are a range of extraction methods available in the literature for metabolite profiling. However, in the case of high throughput operation it is of crucial importance to minimize the complexity of the extraction and ensure that manual operations occur as efficiently as possible. Suitable extraction techniques include ball mills, ultrasound-assisted extraction, or extraction systems which use pressurized solvents. All extraction procedures in metabolite profiling represent a compromise between the quality and the width of analysis and the choice of which method is utilized is ultimately dependent on the key metabolites that need to be included in the profiles.

The advantages of GC-MS and LC-MS have already been introduced. For an industrial application, great care needs to be taken in the actual choice of machines and instrument provider. A high degree of reliability coupled with the availability of good maintenance and trouble shooting support is mandatory. Further, if the projects being undertaken are long term and there is a need for data to be compared over months or years then temporal stability becomes a further requirement. In daily operation it is crucial to follow SOPs that monitor the "health" of each system and enable maintenance activities to be triggered once a system starts to drift outside of key performance indicators.

In the case of metanomics, the throughput that has been implemented means that around 400,000 peaks have to be identified, integrated and validated in the chromatograms per day. This is obviously a task which can only be realized with a substantial degree of automation. There is currently no software available on the market place that covers all the necessary steps. Therefore, any company seeking to run high throughput operations must be prepared to undertake a considerable degree of software development. At metanomics software has been developed and implemented which automatically checks all peaks for compliance with up to 50 pre-defined rules. Further it supports the technical personnel to perform a manual appraisal of peaks which do not pass the validation.

The files generated by GC-MS and LC-MS are large and metanomics has had to implement computing systems capable of dealing with around 10 TB per year distributed across 20 million files. This experimental data needs to be fully integrated with sample tracking data, and for this a Laboratory Information Management System (LIMS) is absolutely required. It is normally a multi-year process to successfully implement a LIMS for a process as complex as metabolic profiling.

The challenges of high throughput operation are considerable, but they can be overcome. Metabolite profiling is actually more suited to high throughput than proteomics or transcriptomics. Interestingly, despite this fact, there has been more sustained investment and effort in proteomics/transcriptomics worldwide than in metabolomics. It is therefore to be expected that the high throughput potential of metabolite profiling will lead to it playing an ever increasing role in the upcoming era of system biology.

5 Industrial Applications

The industrial applications of plant metabolite profiling can be divided into four broad areas: agrochemical development, functional genomics, crop improvement and nutrition.

5.1 Agrochemical Development

As discussed in Sect. 3.1, the pioneer of plant metabolite profiling was Sauter, working at BASF AG. His group published a landmark paper in the American Chemical Society Symposium Series (Sauter et al. 1991), describing metabolic profiling of plants as a new diagnostic technique for mode of action studies in herbicide research. This paper describes the analytical procedures and illustrates their application in an experiment where the response profiles of barley plants treated with four chemically unrelated herbicides were determined. Around 200 peaks of known and unknown substances could be followed and response profiles were generated through the comparison between treated and untreated material. The authors found that all four herbicides generated different characteristic responses for known and unknown peaks. Further, by interpreting the known components of the profiles the authors were able to show that if they had just discovered the chemistries they would very rapidly be able to gain an insight into their modes of action. They concluded that the technique was well suited for fingerprinting and classifying compounds according to mode of action. They predicted that classification would become more powerful as more and more different response patterns were registered in the form of a "library". Indeed, this approach became a routine tool in the herbicide research at BASF over the last 20 years (Sauter, personal communication).

Researchers at the company American Cyanamid were able to achieve similar results using a completely different technological approach (Ott et al. 2003). Using H-1 NMR and artificial neural networks this group was able to classify herbicides and bioactive compounds rapidly according to the signals generated from crude plant extracts. The authors tested 19 of the most relevant mode of actions in corn plants and were able to build an expert system that was able to recognize whether a new compound could be classified to a known mode of action. Whilst a powerful approach, this methodology suffers from the general drawback of NMR based studies that the information at the metabolite or pathway level is low, thus restricting the ability to interpret the modes of action.

5.2 Plant Functional Genomics

The industrial application of metabolite profiling to functional genomics has been pioneered by metanomics. Underlying the approach taken by metanomics are metabolite profiling methods for *Arabidopsis* which were first published by Fiehn et al. (2000) working at the Max-Planck-Institute for Molecular Plant

Physiology. In this groundbreaking study, the authors used GC-MS to analyse some 326 different compounds in *Arabidopsis* leaves, of which roughly half were known metabolites. Profiles were generated for two ecotypes (Columbia and C24) along with a biochemical and a morphological mutant. The authors used principle component analysis to explore the overall metabolic phenotypes and concluded that the ecotypes were more divergent than the mutants. In addition there were more extensive changes observed in the biochemical mutant than in the morphological mutant. Overall the work showed for the first time that metabolite profiling could be a very valuable tool for the analysis of genotypes and the association of genes with metabolic functions.

The approach taken by metanomics to functional genomics is therefore to create a large genetic diversity in *Arabidopsis* at a genome scale and subject this to broad based, high throughput metabolic profiling. Plant lines are identified where the levels of key metabolites have changed and the respective gene whose activity was altered is pinpointed. Thus a functional link between gene and metabolite is generated and this information can be used in the development of crops with improved traits. Indeed, the second and third generation products of plant biotechnology are expected to be crops where there is a direct benefit for the consumer such as promoting health or enhancing nutrition. Such beneficial effects will often be due to altered metabolite compositions.

Key to the conduct of efficient genomic disovery projects are the availability of good plant populations with precise genetic modifications. Both a gain-of-function and a loss-of-function strategy have been followed by metanomics. In the loss-of-function, or knockout, approach, a large *Arabidopsis* T-DNA population (Azprioz-Leehan et al. 1997) has been established where insertional mutagenesis has been used to disrupt the endogenous genes. For the gain-of-function approach, a facility has been implemented which is able to clone genes into plant expression vectors at the rate of 200 per week. The overexpression *Arabidopsis* populations generated to date include those where each one of the yeast and *E.coli* genes have been individually overexpressed.

The results from these programs show that they are effective and efficient at linking genes to particular metabolic functions (Fernie et al. 2004). At a statistical level, metanomics has observed that a significant alteration in any one metabolite can be induced by altering the activity of 0.1–1.0% of the genes in a genome. Only a minority of the genes identified at metanomics have previously been associated with the observed alterations in metabolism. Thus plant functional genomics is a field currently wide open for new discoveries of a profound nature.

5.3 Crop Improvement

There are two principle ways in which improved crops will be developed in the future: via targeted breeding or through genetic engineering (Carrari et al. 2003). Metabolite profiling can contribute to the success of both strategies.

As the genetic engineering of crop plants is a lengthy and expensive procedure it is as important to be as precise as possible in selection and hypothesis validation, and as early as possible in the trait development process. In the case of traits developed through metabolic engineering, there are now many examples from academic research that illustrate how metabolite profiling can play a role in the characterisation of transgenic plants (Trethewey 2004). It is therefore likely that the large plant biotechnology companies will rapidly adopt this technology.

Examples of the application of metabolite profiling to breeding are at an earlier stage. The priniciple is very simple in that advancement decisions in the breeding process could be driven through metabolite profiling data in addition to the data that is normally generated, e. g. phenotype and yield. Such approaches would become very powerful if coupled to genetic marker analysis. An early study that illustrates this potential has recently been published on wild tomato (Schauer et al. 2005).

Applications could also extend well beyond crop breeding. Schaneberg et al. (2003) have published a study of one of the oldest medicinal herbs of Traditional Chinese Medicine: *Ephedra sinica*, commonly known as Ma Huang. Using a simple HPLC metabolite profiling characterisation this group was able to distinguish between Ephedra species originating from Eurasia, North America or South America. This illustrates the potential of metabolite profiling to contribute to herb selection procedures. Further profiling could support routine quality assurance procedures within the dietary supplement industry.

5.4 Nutrition

As the nutritional and medical sciences are impacted by genomics, there is an intensifying discussion that metabolite profiling may be an important approach to drive a deeper understanding of the relationship between genotype, nutrition and health (German et al. 2004). A new term has emerged for this combination of disciplines: nutrigenomics. That dietary habits can influence the progression of degenerative diseases such as cancer, cardiovascular disease and diabetes is now well established (Davis and Milner 2004). However, the question of which dietary components are important remains largely unresolved, and this is exactly where wide metabolite profiling might play a key role in the future. Careful application of such methodologies to particular cohorts of patients, perhaps in conjunction with dietary interventions, will lead to an increased understanding of the metabolic basis of such diseases and can lead to the development of new strategies for the preservation of health. Interestingly, metabolite profiling can be applied directly to both plant and animal food components and also to body fluids (e. g. urine and blood) from human participants in the studies. Direct links between metabolites that are ingested and their occurrence, distribution and effects on human subjects might be identified.

6 Outlook

Industrial applications of plant metabolite profiling are in their infancy. It seems likely that, in the long term, a wide range of applications that support human health and wellness will emerge. There will certainly be many applications where the selection and development of herb and crop plants is promoted. However, the true power of metabolite profiling may be found when plant profiling is combined with approaches to understand and enhance human health and nutrition.

References

Annesley TM (2003) Ion suppression in mass spectrometry. Clin Chem 47:1041–1044

Azprioz-Leehan R, Feldmann KA (1997) T-DNA insertion mutagenesis in Arabidopsis going back and forth. Trends Genet 13:152–156

Bino RJ, Hall RD, Fiehn O, Kopka J, Saito K, Draper J, Nikolau BJ, Mendes P, Roessner-Tunali U, Beale MH, Trethewey RN, Lange BM, Wurtele ES, Sumner LW (2004) Potential of metabolomics as a functional genomics tool. Trends Plant Sci 9:418–425

Carrari F, Urbanczyk-Wochniak E, Willmitzer L, Fernie AR (2003) Engineering central metabolism in crop species: learning the system. Metab Eng 5:191–200

Dallüge J, Beens J, Brinkman UAT (2003) Comprehensive two-dimensional gas chromatography: a powerful and versatile analytical tool. J Chromatogr A 1000:69–108

Davis CD, Milner J (2005) Frontiers in nutrigenomics, proteomics, metabolomics and cancer prevention. Mutat Res Fund Mol Mech Mutag 570:305

Fernie AR, Trethewey RN, Krotzky AJ, Willmitzer L (2004) Metabolite profiling: from diagnostics to systems biology. Nat Rev Mol Cell Biol 5:763–769

Fiehn O (2003) Metabolic networks of *Cucurbita maxima* phloem. Phytochemistry 62:875–886

Fiehn O, Kopka J, Dörmann P, Altmann T, Trethewey RN, Willmitzer L (2000) Metabolite profiling for plant functional genomics. Nat Biotechnol 18:1157–1161

German JB, Bauman DE, Burrin DG, Failla ML, Freake HC, King JC, Klein S, Milner JA, Pelto GH, Rasmussen KM, Zeisel SH (2004) Metabolomics in the opening decade of the 21st century: Building the roads to individualized health. J Nutr 134:2729–2732

Hayen H, Karst U (2003) Strategies for the liquid chromatographic-mass spectrometric analysis of non-polar compounds. J Chromatogr A 1000:549–565

Horning EC, Horning MG (1971) Metabolic profiles: gas-phase methods for analysis of metabolites. Clin Chem 17:802–809

Huhman DV, Sumner LW (2003) Metabolic profiling of saponins in *Medicago sativa* and *Medicago trunculata* using HPLC coupled to an electrospray ion-trap mass spectrometer. Phytochemistry 59:347–360

Kopka J, Schauer N, Krueger S, Birkemeyer C, Usadel B, Bergmuller E, Dormann P, Weckwerth W, Gibon Y, Stitt M, Willmitzer L, Fernie AR, Steinhauser D (2005) GMD@CSB.DB: the Golm Metabolome Database. Bioinformatics 21:1635–1638

Lange BM, Ketchum REB, Croteau RB (2001) Isoprenoid biosynthesis. Metabolite profiling of peppermint oil gland secretory cells and application to herbicide target analysis. Plant Physiol 127:305–314

Matuszewski BK, Constanzer ML, Chavez-Eng CM (2003) Strategies for the assessment of matrix effect in quantitative bioanalytical methods based on HPLC-MS/MS. Anal Chem 75:3019–3030

Niessen WM (2003) Progress in liquid chromatography-mass spectrometry instrumentation and its impact on high-throughput screening. J Chromatogr A 1000:413–436

Niwa T (1986) Metabolic profiling with gas chromatography-mass spectrometry and its application to clinical medicine. J Chromatogr 20:313–345
Ott KH, Aranibar N, Singh BJ, Stockton GW (2003) Metabonomics classifies pathways affected by bioactive compounds. Artificial neural network classification of NMR spectra of plant extracts. Phytochemistry 62:971–985
Ratcliffe RG, Shachar-Hill Y (2005) Revealing metabolic phenotypes in plants: inputs from NMR analysis. Biol Rev Cambr Philos Soc 80:27–43
Roessner U, Wagner C, Kopka J, Trethewey RN, Willmitzer L (2000) Simultaneous analysis of metabolites in potato tuber by gas chromatography-mass spectrometry. Plant J 23:131–142
Santos FJ, Galceran MT (2003) Modern developments in gas chromatography-mass spectrometry-based environmental analysis. J Chromatogr A 1000:125–151
Sato S, Soga T, Nishioka T, Tomita M (2004) Simultaneous determination of the main metabolites in rice leaves using capillary electrophoresis mass spectrometry and capillary electrophoresis diode array detection. Plant J 40:151–163
Sauter H, Lauer M, Fritsch H (1991) Metabolic profiling of plants – a new diagnostic technique. In: Baker DR, Fenyes JG, Moberg WK (eds) Synthesis and chemistry of agrochemicals II. American Chemical Society, Washington, DC, pp 288–299
Schaneberg BT, Crockett S, Bedir E, Khan IA (2003) The role of chemical fingerprinting: application to *Ephedra*. Phytochemistry 62:911–918
Schauer N, Zamir D, Fernie AR (2005) Metabolic profiling of leaves and fruit of wild species tomato: a survey of the Solanum lycopersicum complex. J Exp Bot 56:297–307
Schwartz JC, Senko MW, Syka JE (2002) A two-dimensional quadrupole ion trap mass spectrometer. J Am Soc Mass Spectrom 13:659–669
Sumner LW, Mendes P, Dixon RA (2003) Plant metabolomics: large-scale phytochemistry in the functional genomics area. Phytochemistry 62:817–836
Taylor J, King RD, Altmann T, Fiehn O (2002) Application of metabolomics to plant genotype discrimination using statistics and machine learning. Plant J 18:241–248
Tolstikov VV, Lommen A, Nakanishi K, Tanaka N, Fiehn O (2003) Monolithic silica-based capillary reversed-phase liquid chromatography/electrospray mass spectrometry for plant metabolomics. Anal Chem 75:6737–6740
Trethewey RN (2004) Metabolite profiling as an aid to metabolic engineering in plants. Curr Opin Plant Biol 7:196–201
Von Roepenack-Lahaye E v., Degenkolb T, Zerjeski M, Franz M, Roth U, Wessjohann L, Schmidt J, Scheel D, Clemens S (2004) Profiling of arabidopsis secondary metabolites by capillary liquid chromatography coupled to electrospray ionization quadrupole time-of-flight mass spectrometry. Plant Physiol 134:548–559
Wilson ID, Plumb R, Granger J, Major H, Williams R, Lenz EA (2005a) HPLC-MS-based methods for the study of metabonomics. J Chrom B 817:67–76
Wilson ID, Nicholson JK, Castro-Perez J, Granger JH, Johnson KA, Smith BW, Plumb RS (2005b) High resolution "Ultra performance" liquid chromatography coupled to oa-TOF mass spectrometry as a tool for differential metabolic pathway profiling in functional genomic studies. J Proteome Res 4:591–598
Wink M (1988) Plant breeding: importance of plant secondary metabolites for protection against pathogens and herbivores. Theor Appl Genet 75:225–233
Zhang J, McCombie G, Guenat C, Knochenmuss R (2005) FT-ICR mass spectrometry in the drug discovery process. Drug Discovery Today 10:635–642

Subject Index

Biotechnology in Agriculture and Forestry

Volumes already published

Volume 1: Trees I (1986)
Volume 2: Crops I (1986)
Volume 3: Potato (1987)
Volume 4: Medicinal and Aromatic Plants I (1988)
Volume 5: Trees II (1989)
Volume 6: Crops II (1988)
Volume 7: Medicinal and Aromatic Plants II (1989)
Volume 8: Plant Protoplasts and Genetic Engineering I (1989)
Volume 9: Plant Protoplasts and Genetic Engineering II (1989)
Volume 10: Legumes and Oilseed Crops I (1990)
Volume 11: Somaclonal Variation in Crop Improvement I (1990)
Volume 12: Haploids in Crop Improvement I (1990)
Volume 13: Wheat (1990)
Volume 14: Rice (1991)
Volume 15: Medicinal and Aromatic Plants III (1991)
Volume 16: Trees III (1991)
Volume 17: High-Tech and Micropropagation I (1991)
Volume 18: High-Tech and Micropropagation II (1992)
Volume 19: High-Tech and Micropropagation III (1992)
Volume 20: High-Tech and Micropropagation IV (1992)
Volume 21: Medicinal and Aromatic Plants IV (1993)
Volume 22: Plant Protoplasts and Genetic Engineering III (1993)
Volume 23: Plant Protoplasts and Genetic Engineering IV (1993)
Volume 24: Medicinal and Aromatic Plants V (1993)
Volume 25: Maize (1994)
Volume 26: Medicinal and Aromatic Plants VI (1994)
Volume 27: Somatic Hybridization in Crop Improvement I (1994)
Volume 28: Medicinal and Aromatic Plants VII (1994)
Volume 29: Plant Protoplasts and Genetic Engineering V (1994)
Volume 30: Somatic Embryogenesis and Synthetic Seed I (1995)
Volume 31: Somatic Embryogenesis and Synthetic Seed II (1995)
Volume 32: Cryopreservation of Plant Germplasm I (1995)
Volume 33: Medicinal and Aromatic Plants VIII (1995)
Volume 34: Plant Protoplasts and Genetic Engineering VI (1995)
Volume 35: Trees IV (1996)
Volume 36: Somaclonal Variation in Crop Improvement II (1996)
Volume 37: Medicinal and Aromatic Plants IX (1996)
Volume 38: Plant Protoplasts and Genetic Engineering VII (1996)
Volume 39: High-Tech and Microprogation V (1997)
Volume 40: High-Tech and Microprogation VI (1997)
Volume 41: Medicinal and Aromatic Plants X (1998)
Volume 42: Cotton (1998)
Volume 43: Medicinal and Aromatic Plants XI (1999)
Volume 44: Transgenic Trees (1999)
Volume 45: Transgenic Medicinal Plants (1999)
Volume 46: Transgenic Crops I1 (1999)
Volume 47: Transgenic Crops II (2001)
Volume 48: Transgenic Crops III (2001)

Volumes 1-48 were edited by Y.P.S. Bajaj†

Volume 49: Somatic Hybridization in Crop Improvement II (2001)
T. Nagata and Y.P.S. Bajaj (Eds.)
Volume 50: Cryopreservation of Plant Germplasm II (2002)
L.E. Towill and Y.P.S. Bajaj (Eds.)
Volume 51: Medicinal and Aromatic Plants XII (2002)
T. Nagata and Y. Ebizuka (Eds.)
Volume 52: Brassicas and Legumes: From Genome Structure to Breeding (2003)
T. Nagata and S. Tabata (Eds.)
Volume 53: Tobacco BY-2 Cells (2004)
T. Nagata, S. Hasezawa, and D. Inzé (Eds.)
Volume 54: *Brassica* (2004)
E.C. Pua and C.J. Douglas (Eds.)
Volume 55: Molecular Marker Systems in Plant Breeding and Crop Improvement (2005)
H. Lörz and G. Wenzel (Eds.)
Volume 56: Haploids in Crop Improvement II (2005)
C.E. Palmer, W.A. Keller, and K.J. Kasha (Eds.)
Volume 57: Plant Metabolomics (2006)
K. Saito, R.A. Dixon, and L. Willmitzer (Eds.)

Volumes in preparation

Tobacco BY-2 Cells: A New Treatise
T. Nagata, K. Matsuoka, and D. Inzé (Eds.)

Transgenic Crops IV
E.C. Pua and M.R. Davey (Eds.)

Transgenic Crops V
E.C. Pua and M.R. Davey (Eds.)

Transgenic Crops VI
E.C. Pua and M.R. Davey (Eds.)

The manufacturer's authorised representative in the EU is Springer Nature Customer Service Centre GmbH, Europaplatz 3, 69115 Heidelberg, Germany. If you have any concerns regarding our products, please contact ProductSafety@springernature.com

Printed and bound by CPI Group (UK) Ltd, Croydon, CR0 4YY
30/09/2026
02225195-0011